T0141920

Wireless Networks

Series Editor
Xuemin Sherman Shen, University of Waterloo, Waterloo, ON, Canada

The purpose of Springer's Wireless Networks book series is to establish the state of the art and set the course for future research and development in wireless communication networks. The scope of this series includes not only all aspects of wireless networks (including cellular networks, WiFi, sensor networks, and vehicular networks), but related areas such as cloud computing and big data. The series serves as a central source of references for wireless networks research and development. It aims to publish thorough and cohesive overviews on specific topics in wireless networks, as well as works that are larger in scope than survey articles and that contain more detailed background information. The series also provides coverage of advanced and timely topics worthy of monographs, contributed volumes, textbooks and handbooks.

Haipeng Yao • Mohsen Guizani

Intelligent Internet of Things Networks

 Springer

Haipeng Yao
Beijing University of Posts and
Telecommunications
Beijing, China

Mohsen Guizani
Mohamed bin Zayed University of Artificial
Intelligence (MBZUAI)
Abu Dhabi, United Arab Emirates

ISSN 2366-1186   ISSN 2366-1445   (electronic)
Wireless Networks
ISBN 978-3-031-26986-8   ISBN 978-3-031-26987-5   (eBook)
https://doi.org/10.1007/978-3-031-26987-5

This Springer imprint is published by the registered company Springer Nature Switzerland AG
The registered company address is: Gewerbestrasse 11, 6330 Cham, Switzerland

Preface

Internet of Things (IoT), as the most powerful and exciting technology, has quickly become a disruptive force, reshaping how we live and work. The Internet of Things refers to the billions of physical devices that are now connected to and transfer data through the Internet without requiring human-to-human or human-to-computer interaction. These connected IoT devices are slowly entering every aspect of our lives, ranging from healthcare to industrial manufacturing. According to Gartner's prediction, there will be more than 37 billion IoT connections in the future year of 2025. However, with large-scale IoT deployments, IoT networks are facing challenges in the aspects of scalability, privacy, and security. The ever-increasing complexity of the IoT makes effective monitoring, overall control, optimization, and auditing of the network difficult. Hence, there is a need for more powerful approaches to solve the challenges faced in IoT network design, deployment, and management.

Recently, artificial intelligence (AI) and machine learning (ML) approaches have achieved huge success in many fields of computing, including computer vision, natural language processing, and voice recognition. These successes suggest that machine learning techniques could be successfully applied to problems in the IoT network space. Besides, with the development of the emerging network technologies (e.g., Mobile Edge Computing, Blockchain, and Programmable Network), it is feasible to perform flexible processing logic and intelligent computing inside the IoT networks. Therefore, in this book, we design a new intelligent IoT network architecture. We use different machine learning approaches to investigate solutions. In this book, we focus on three scenarios of successfully applying machine learning in IoT networks.

IoT Network Awareness Network awareness is the prerequisite for network optimization control. Some machine leaning approaches have already been applied to IoT network awareness problems, including traffic classification, anomaly traffic identification, and traffic prediction. Part of these works is introduced in Chap. 3.

IoT Network Routing Control The ever-increasing network complexity makes effective routing control extremely difficult. In particular, current control strategies largely rely on manual processes, which exhibit poor scalability and robustness. Chap. 4 discusses how machine learning approaches (e.g., reinforcement learning) can deal with the routing control problem (e.g., QoS routing, load balance) in IoT networks.

Resource Optimization How to distribute network resources in order to accommodate different and possibly contrasting business requirements on a single physical network infrastructure is incredibly challenging. In Chap. 5, we introduce the network slicing technique in IoT networks and focus on how machine learning can enhance network utility.

Besides, in Chap. 6, we discuss the Mobile Edge Computing-aided intelligent IoT. Mobile Edge Computing is one of the defining technologies for envisioning the IoT. It essentially implies computational speed and intelligent processing power at the edge of the IoT network. Moreover, Chap. 7 covers blockchain-aided intelligent IoT. Blockchain has the potential to help address some of the IoT security and scalability challenges. At its core, a blockchain system consists of a distributed digital ledger, shared between participants in the system, that resides on the Internet: transactions or events are validated and recorded in the ledger and cannot subsequently be amended or removed. It provides a way for information to be recorded and shared by a community of users. We discuss how blockchain can improve the performance of intelligent IoT networks.

Beijing, China   Haipeng Yao
Abu Dhabi, United Arab Emirates   Mohsen Guizani

Contents

Acronyms

3AUS	Adaptive Action Aggregation Update Strategy
5G	5th-Generation
AI	Artificial Intelligence
AODV	Ad hoc On-demand Distance Vector
AP	Access Point
AR	Augmented Reality
AuR	Auxiliary Routing algorithm
BE	Best Effort
BFS	Breadth-First-Search
BP	Back Propagation
BSs	Base Stations
BW	Bandwidth
cIoT	consumer Internet of Things
CNN	Convolutional Neural Network
CSAT	Carrier-Sensing Adaptive Transmission
CTDE	Centralized Training and Distributed Execution
CV	Computer Vision
DA	Dynamic Adaptive
DAO	Decentralized Autonomous Organizations
DDoS	Distributed Denial of Service
DDPG	Deep Deterministic Policy Gradient
DENs	Deep Edge Networks
DFS	Depth-First Search
DNS	Domain Name System
DoS	Denial of Service
DPG	Deterministic Policy Gradient
DPI	Deep Packet Inspection
DQL	Deep Q-learning
DQN	Deep Q-Network
DR	Dynamic Random
DRL	Deep Reinforcement Learning

DRQN	Deep Recurrent Q-Learning
DSR	Dynamic Source Routing
DTN	Delay-Tolerant Network
ECN	Explicit Congestion Notification
eMBB	enhanced Mobile Broadband
EPT	Equal Power Transmission
ESS	Evolutionary Stable Strategy
FIFO	First In First Out
GA	Genetic Algorithm
GMM	Gaussian Mixture Model
GSMA	Global System for Mobile communications Association
HAN	Hierarchical Attention Network
IaaS	Infrastructure as a Service
IAGT	Isotone Action Generation Technique
IANA	Internet Assigned Numbers Authority
IDS	Intrusion Detection System
IIoT	Industrial Internet of things
IIoTD	IIoT Devices
IoE	Intelligent-of-Everything
IoT	Internet of Things
IPN	Interplanetary Internet
ISPs	Internet Service Providers
LBT	Listen-Before-Talk
LoRaWAN	Long-Range Wide Area Network
LR	Logistic Regression algorithm
LSTM	Long Short Term Memory
MADDPG	Multi-Agent Actor-Critic Policy Gradient
MARL	Multi-Agent Reinforcement Learning
MBSs	Macro Base Stations
MCC	Mobile Cloud Computing
MCP	Multiple-Constrained Path
MDP	Markov Decision Process
MDs	Mobile Devices
MEC	Mobile Edge Computing
MIT	Massachusetts Institute of Technology
mMTC	massive Machine Type Communications
MPTCP	MultiPath TCP
MSE	Mean Squared Error
NB	Naive Bayes algorithm
NDN	Named Data Network
NDVB	Naive Distributed Variational Bayes algorithm
NFV	Network Functions Virtualization
NLP	Natural Language Processing
NS	Network Service
NSR	Non-Shortest Reachable Route

NV	Network Virtualization
OFDMA	Orthogonal Frequency Division Multiple Access
ONE	Opportunistic Network Environment
OVS	OpenVSwitch
P2P	Peer-to-Peer
P4	Programming Protocol-independent Packet Processors
PaaS	Platform as a Service
PBFT	Practical Byzantine Fault Tolerance
PCA	Principal Component Analysis
PG	Policy Gradient
PISA	Protocol Independent Switch Architecture
PLR	Packet Loss Rate
PoS	Proof of Stake
PoW	Proof of Work
PPP	Poisson Point Process
QoE	Quality of Experience
QoS	Quality of Service
QU	Queue Utilization
R2L	Root to Local
RA	Reliability Aware
RCP	Rate Control Protocol
RD	Replicator Dynamics
RF	Random Forest algorithm
RHC	Rolling Horizon Control
RL	Reinforcement Learning
RNN	Recurrent Neural Network
RSSI	Received Signal Strength Indicator
RSU	Roadside Units
SBNE	Symmetric Bayesian Nash Equilibrium
SBSs	Small-cell Base Stations
SDAE	Stacked Denoising AutoEncoder
SDN	Software-Defined Network
SDWN	Software-Defined Wireless Networks
SF	Spreading Factor
SFC	Service Function Chain
SINR	Signal-to-Interference-plus-Noise Ratio
SNR	Signal-to-Noise Ratio
SR	Shortest Route
TD	Time Difference
TMDDPG	Transfer learning-based Multi-agent DDPG
TP	Transmission Power
UDNs	Ultra-Dense Networks
UEs	User Equipments
UHD	Ultra-High-Definition
UR	Unreachable Route

URLLC	Ultra-Reliable Low Latency Communications
VAE	Variational AutoEncoder
VMs	Virtual Machines
VNE	Virtual Network Embedding
WoLF-PHC	WoLF Policy Hill Climbing

Chapter 1
Introduction

Abstract We provide a detailed review of the Internet of Things network and machine learning technologies. The Internet of Things (IoT) is the interconnected Internet of Things, an extension of the Internet. Through sensor equipment and the agreed protocol, any item is connected to the Internet for information exchange and communication, so as to realize intelligent identification, monitoring, and management. The combination of machine learning and the Internet of Things enables various types of information collected by sensors in real time to be intelligently analyzed through machine learning in terminal devices, edge domains, or cloud centers. This helps the IoT reach its full potential and enables the Internet of Everything. In addition, we discuss how emerging network technologies (e.g., Mobile Edge Computing, Blockchain, and Programmable Networks) can benefit IoT network control and management.

Keywords Internet of things architecture · Machine learning · Emerging network technologies · Programmable networks

1.1 Background

The Internet of Things (IoT) as the most powerful and exciting technology has quickly become a disruptive force, reshaping how we live and work. The Internet of Things describes the network of physical objects with sensors, processing ability, software, and other technologies for the purpose of connecting and transferring data with other devices over the Internet [1]. By combining these connected devices with automated systems, it offers the potential for a "fourth industrial revolution," and experts predict that more than half of new businesses will run on the IoT by 2020. According to Gartner's prediction, there will be more than 37 billion IoT connections in the future year of 2025. With the large-scale IoT deployments, it present huge challenges for current IoT network in the aspects of scalability, privacy, and security. The ever-increasing complexity of the IoT application makes effective monitoring, overall control, optimization, and auditing of the network difficult.

© The Author(s), under exclusive license to Springer Nature Switzerland AG 2023
H. Yao, M. Guizani, *Intelligent Internet of Things Networks*, Wireless Networks,
https://doi.org/10.1007/978-3-031-26987-5_1

There is a need for more powerful approaches to solve the challenges faced in IoT network design, deployment, and management.

Recently, machine learning technology has emerged as a viable solution to address this challenge. Machine learning is the science of getting computers to act more accurate from studying data and statistics without being explicitly programmed to do so. It is a method of data analysis that automates analytical model building. Machine learning approaches have achieved huge success in many fields of computing, including computer vision, natural language processing, and voice recognition. These successes suggest that machine learning techniques could be successfully applied to problems in the IoT networks space. Machine learning for IoT networks can be used to detect anomalies, optimize traffic, and augment intelligence by ingesting network state. Therefore, in this book, we design a new intelligent IoT network architecture. We use different machine learning approaches to investigate solutions.

1.2   Overview of Internet of Things Network and Machine Learning

In this section, we provide a more detailed view of Internet of Things network and machine learning technologies. The Internet of Things (IoT) is the interconnected Internet of Things, an extension of the Internet. Through sensor equipment, according to the agreed protocol, any item is connected to the Internet for information exchange and communication, so as to realize intelligent identification, monitoring, and management. The combination of machine learning and the Internet of Things enables various types of information collected by sensors in real time to be intelligently analyzed through machine learning in terminal devices, edge domains, or cloud centers [2]. This helps the IoT reach its full potential and enables the Internet of Everything.

1.2.1   Internet of Things Architecture

The industry and academia usually divide the IoT architecture into three layers, including the perception layer, the network layer, and the application layer, as shown in Fig. 1.1. The perception layer usually monitors network nodes, collects essential data in real time through various information sensing devices, and integrates multiple information technologies. The perception layer usually monitors network nodes and collects essential data in real time by integrating various information technologies with different information sensing devices. The network layer uses Internet technologies to form an extensive network by combining the perception layer's sensing devices and the application layer's terminal devices to realize the

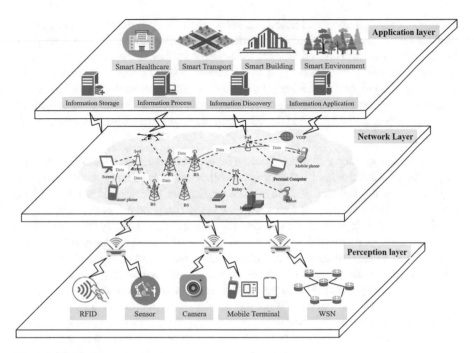

Fig. 1.1 The architecture of IoT

transmission of essential data in production and life. The application layer is located at the top of the three-layer structure of the Internet of Things. Its function is to process the data generated by the perception layer [3]. The application layer can perform calculation and knowledge mining on the data collected by the perception layer to realize real-time control, efficient management, and intelligent decision-making of terminal equipment.

(A) Perception Layer The perception layer consists of sensing nodes responsible for collecting and forwarding controlled data under control. In order to meet the needs of ubiquitous sensing, sensing nodes must have the characteristics of simplicity, economy, low energy consumption, flexible arrangement, and simple operation. Perception layer technologies include QR code tags and readers, RFID tags and readers, cameras, GPS, sensors, M2M terminals, sensor gateways, etc. Among them, the common technology has radio frequency identification communication technology. Its working principle is to identify the designated object through radio waves and obtain information about the thing [4]. In addition, it also includes sensor technology, which is a perception device used by sensors to collect and process information. As an essential means of sensing and acquiring data information in the Internet of Things, sensors play a significant role in the Internet of Things.

(B) Network Layer The network layer is usually divided into two parts: the communication technology and the communication protocol of the Internet of

Things. Communication technology is responsible for physically linking devices and enabling communication [5]. Communication protocols are responsible for establishing rules and unified formats for communication. Communication technology is divided into wired networks and wireless networks from the medium. The communication distance is divided into ultra-short-distance, short-distance, medium long-distance, and ultra-long-distance. The current Internet already uses part of the communication technology, and the other part is created according to the Internet of Things.

(C) Applications Layer   The application layer can connect with users by responding to the relevant needs of industry informatization and realizing the multifaceted application of the Internet of Things. The application layer provides rich applications based on the Internet of Things, which is the fundamental goal of developing the Internet of Things. It combines IoT technology with industry informatization needs to achieve a solution set for a wide range of intelligent applications. Due to the wide variety of electronic manufacturing and other equipment, it is very cumbersome for the perception layer to realize the monitoring process of the equipment. The application layer needs middleware to eliminate data heterogeneity generated by IoT end devices. With the development of information technology, two major IoT application technologies have been derived: cloud computing and big data [6]. Cloud computing can help store and analyze massive data in the Internet of Things. According to the service types of cloud computing, the cloud can be divided into infrastructure as a service (IaaS), platform as a service (PaaS), and software as a service (SaaS). Big data can perform calculation, processing, and knowledge mining on the data collected at the perception layer to provide better service. Machine learning will also help the further development of the Internet of Things.

1.2.2   Internet of Things Network Technologies

In recent years, with the rapid development of the Internet of Things, its application scenarios cover smart homes, smart cities, smart medical care, smart industry, and smart agriculture. Compared with the traditional Ethernet, the Internet of Things can combine various sensing devices with the network to realize the interconnection of people, computers, and objects. Meanwhile, the Internet of Things facilitates various utilities of the system, such as sensor identification, data transfer, information communication, and knowledge management. IoT network technologies include sensor technology, intelligent technology, embedded technology, etc. IoT platforms can easily integrate various IoT-enabled technologies. Various forms of IoT protocols can realize the interconnection between IoT devices. They have different protocol stacks, which makes IoT protocols often show different characteristics.

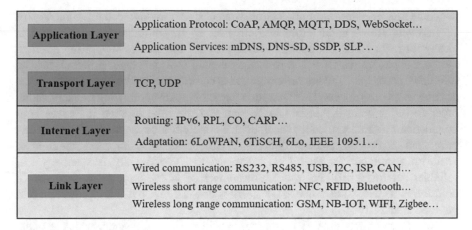

Application Layer	Application Protocol: CoAP, AMQP, MQTT, DDS, WebSocket...
	Application Services: mDNS, DNS-SD, SSDP, SLP...
Transport Layer	TCP, UDP
Internet Layer	Routing: IPv6, RPL, CO, CARP...
	Adaptation: 6LoWPAN, 6TiSCH, 6Lo, IEEE 1095.1...
Link Layer	Wired communication: RS232, RS485, USB, I2C, ISP, CAN...
	Wireless short range communication: NFC, RFID, Bluetooth...
	Wireless long range communication: GSM, NB-IOT, WIFI, Zigbee...

Fig. 1.2 IoT protocol stack

There are various IoT protocols. The most common protocol layering is shown in Fig. 1.2. It is comprised of four layers: application layer, transport layer, Internet layer, and network link layer, as described in detail below [7].

(A) Application Layer IoT application developers have a wealth of protocols available at the application layer, both from the traditional Internet and from those built specifically to support IoT applications. The IoT application layer uses the analyzed and processed sensory data to provide users with different types of specific services. Its main functions include the collection, transformation, and analysis of the data. The application layer is expanded into two layers in the IoT protocol stack: Application Protocol and Application Services. Application protocol mainly includes MQTT, CoAP, AMQP, etc. Application Services mainly include mDNS, SSDP, etc. [8]. Several key protocols are described in detail below.

MQTT (Message Queuing Telemetry Transport) is a protocol designed for a large number of sensors or controllers with limited computing power, limited working bandwidth, and unreliable network environments.

CoAP is synonymous with Constrained Application Protocol. It is based on the UDP protocol, and its original intention is to convert it to the HTTP protocol as easily as possible. CoAP allows users to use the CoAP protocol in resource-constrained IoT devices just like the HTTP protocol.

mDNS realizes mutual discovery and communication between hosts in a local area network without a traditional DNS server. mDNS can also be used in conjunction with (DNS-SD) DNS Service Discovery.

(B) Transport Layer The transport layer is the channel through which IoT devices connect and is responsible for connecting terminal devices, edges, and clouds. With the rapid increase in the number of IoT devices and the increasingly rich application scenarios, the market has put forward higher requirements for network connectivity. The transport layer contains two protocols, TCP and UDP. TCP is a connection-oriented communication protocol. It provides reliable delivery of services and allows applications on both sides of the communication to send data at any time. It can accommodate layered protocol hierarchies supporting multiple network applications. UDP is a message-oriented transport layer protocol. It is an unreliable protocol. UDP is a faster method of communication because it reduces the acknowledgment process. UDP has no congestion control, allowing the application layer to better control the data to be sent and the time to send it. It is a non-connection-oriented protocol that does not establish a connection with the other party and sends data packets directly. Therefore, there is no delay required to establish a connection. Therefore, compared to TCP/IP, UDP is relatively less reliable, but faster [9]. For quick prototyping of M2M projects, a very simple solution is to use UDP, since UDP headers contain few bytes and consume less payload than TCP.

The aforementioned MQTT protocol is built on the TCP stack, which is open, simple, lightweight, and easy to implement. CoAP is built on top of the UDP stack, which can be faster and better resource-optimized, rather than resource-intensive.

(C) Internet Layer The Internet layer protocol is responsible for data forwarding, from the source to the destination, to provide communication services for IoT devices. The IoT Internet layer is divided into two sub-layers: the routing layer and the adaptation layer. The routing layer handles the transmission of packets from source to destination, and the adaptation layer is responsible for forming packets. The following focuses on two key technologies in the Internet of Things, IPv6 and 6LoWPAN.

With the popularity of the Internet of Things, the era of the Internet of Everything has come. However, each device will use a single IP address. In terms of the volume of IPv4, IPv4 is not enough to support the market of tens of billions of devices. IPv6 has a huge address space to support the future development of the Internet of Things. IPv6 also provides a Quality of Service (QoS) mechanism to provide smooth and orderly network transmission, which can classify network resources, and maintain smooth network transmission with the help of traffic control.

Due to the existence of a large number of low-power nodes and low latency nodes in the IoT network, there are higher control requirements for the access capability and investment budget of the large-scale deployment of IoT devices, which requires flexible deployment in complex operating environments. Therefore, a new technology, 6LoWPAN, is proposed. 6LoWPAN is a low-speed wireless personal area network standard based on IPv6, namely IPv6 over IEEE 802.15.4 [10]. 6LoWPAN allows each node to connect to the Internet with an IPv6 address. This allows nodes to connect directly to the Internet using open standards. The Internet protocols can be applied even on the smallest resource-constrained devices, and low-power devices with limited processing power should be able to participate in IoT [11].

(D) Link Layer The link layer refers to the physical medium of data transmission, and its role is how bits of data are transferred from one device to another. This layer is the foundation of the IoT protocol. The physical link layer protocol of the Internet of Things can be divided into wired and wireless, and wireless can be divided into short-distance and long-distance [12]. Wired communication includes RS232, RS485, USB, I2C, ISP, CAN, M-Bus, etc. Wireless communication can be divided into short-distance and long-distance. Wireless short-range communication includes NFC, RFID, Bluetooth, etc., and long-distance includes GSM, GPRS, 2 5G, NB-IoT, Wi-Fi, LoRa, Zigbee, etc. Next, we will focus on several key technologies that are widely used in the Internet of Things network.

RFID is an acronym for Radio Frequency Identification. The principle is the non-contact data communication between the reader and the tag to achieve the purpose of identifying the target [13]. The main function of RFID technology is that it has a large data memory capacity, can be scanned quickly, and can be reused. These capabilities make it possible to automatically identify every object on the Internet of Things. NFC is short for Near Field Communication. It is a short-range high-frequency wireless communication technology that allows non-contact point-to-point data transmission (within 10 cm) to exchange data between electronic devices. The RFID introduced above is essentially an identification technology, while NFC is communication technology. Although NFC is more convenient and safe and the cost is lower than that of Bluetooth, the method of establishing a connection that requires close proximity or even contacts is a shortcoming of it.

WiFi represents the wireless local area network networking technology (the main standard is IEEE 802.11), it is widely used in the home, commercial, and industrial scenarios. It has become one of the most mainstream networking methods. Wi-Fi provides great convenience for networking in buildings, avoiding physical wiring and construction. At the same time, the range of Wi-Fi networks can be as small as the range of a room or as large as the level of towns, and the bandwidth can be from tens of Mbps to several hundred Mbps.

1.2.3 Emerging Network Technologies for IoT

Emerging technologies that bring the digital and physical worlds closer together are becoming increasingly important as IoT solutions expand into new applications and environments. Most organizations already consider emerging technologies: Artificial Intelligence (AI), Software-Defined Networking (SDN), Mobile Edge Computing (MEC), and blockchain technologies, as part of IoT solutions. These technologies are changing the nature of connected devices, how they run IoT applications, and how they communicate with each other and create benefits by connecting the digital and physical realms.

(A) Software-Defined Networking for IoT With the rapid growth of IoT devices and applications, IoT networks are becoming increasingly complex. Software-defined networking (SDN) that can be centrally managed, scalable, and flexible can address the enormous data flow brought about by IoT devices. SDN offers a cost-effective approach to managing IoT and securing networks, while also optimizing application performance and analysis. Due to the explosive growth of IoT and the

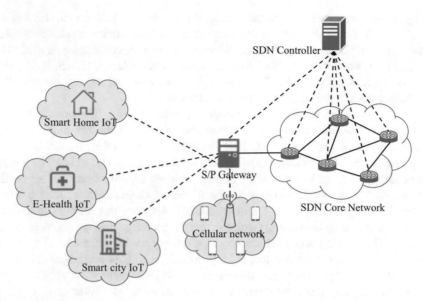

Fig. 1.3 SDN-based IoT network

chaotic nature of the public Internet, this traffic needs to be migrated to private dedicated channels; otherwise, critical communication services and applications will experience latency issues [14]. For this reason, some people propose to use SDN to solve these existing problems of the Internet of Things. The SDN combined with the IoT architecture is shown in Fig. 1.3.

SDN is short for software-defined networking. SDN is a programmable network architecture. It separates network control functions from forwarding functions. It separates network control functions from forwarding functions, so as to get rid of the hardware restrictions on the network architecture and improve the scalability and flexibility of the network [15]. The key point of the SDN network is that an SDN controller is added to the network architecture, and the centralized control of the network is realized by this centralized controller. The SDN network architecture has three basic features: separation of forwarding and control, centralized control, and open interfaces.

The number of network devices involved in the construction of the Internet of Things is very large. The use of SDN technology can separate data forwarding and network control, reducing the frequency and cost of equipment replacement. The network structure is virtualized through SDN, and an IoT network is virtualized into multiple networks with software. Network resources are dynamically allocated and managed. SDN can optimize the environmental configuration of storage and processing centers and ensure that the IoT network can operate in a unified network. It works properly in the architecture. The Internet of Things contains a huge amount of data and faces security issues in practical applications. SDN can strip data from the control plane through the controller and control the entire network. Compared with traditional networks, SDN has great advantages over traditional networks in the real-time push of security policies, fine-graininess, and traffic monitoring, ensuring the safety and reliability of network operations.

(B) Mobile Edge Computing for IoT The expansive development of the IoT devices promotes the Mobile Edge Computing architecture as a necessary architecture for enterprises. Traditional machine-to-machine communication (M2M, the originator of the Internet of things) has existed for decades. And in recent years, the rapid growth of IoT devices has caused a rapidly increasing in their data transmission quantity and transmission speed. Besides, recently many machine learning methods can be integrated with the IoT devices (called AIoT), which further increases the computing demand of the network. Traditional business computing methods rely on a centralized computing center, which will cause high transmission delays. However, IoT devices require a millisecond-level data processing ability. Their urgent need for response speed makes the cloud computing architecture more and more unrealistic due to the huge delay. Generally, cloud data centers are located several kilometers away from IoT devices, which will bring performance problems to IoT applications with high bandwidth and low latency requirements.

Mobile Edge Computing technology provides a feasible way to solve the above problems. The distributed computing capability provided by MEC provides the flexibility and possibility to process data at the data source or critical delivery points. By reasonably arranging computing resources between the cloud and the edge IoT devices, the network can provide low latency, balanced computing power supply, and sufficient bandwidth. By enabling the computing ability of data sources (Internet of Things devices), MEC enables localized processing of lightweight computing tasks at the network edge, rather than relying on centralized cloud data centers. Specifically, MEC technology embeds machine learning and artificial intelligence methods into IoT devices for agile and efficient data processing. Due to the localized processing of data, the MEC architecture reduces congestion pressure in primary links. Therefore, it also improves the service quality and elasticity of some complex IoT applications that need to be transferred to the cloud data center for processing. In addition to empowering IoT devices with lightweight computing ability, MEC technology also supports the deployment of some edge computing infrastructures on demand. Those infrastructures are located between IoT devices and cloud data centers. Therefore it can calculate and respond to some requests initiated by distributed IoT devices. Because MEC enables the computing power closer to the data source or the critical position of commercial business, it brings faster speed, higher reliability, and greater flexibility to IoT applications. Moreover, integrating MEC and cloud computing has greatly optimized the implementation speed and quality of IoT applications and further improved the scalability of the Internet of Things.

(C) Blockchain for IoT In recent years, with the increasing number of IoT devices and the emergence of IoT applications, the security of IoT has gradually attracted

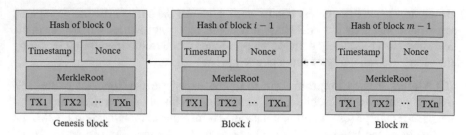

Fig. 1.4 The composition of the blockchain

the attention of academia and industry. Due to the wide variety of IoT devices and server devices, the potential security problems of the IoT are mainly concentrated in the edge devices and the cloud center.

Blockchain technology provides an effective solution to the above problems. Blockchain technology adopts a unique chain architecture to realize distributed data recording and storage, thus providing strict protection for terminal security and privacy [16]. Specifically, blockchain can be regarded as a distributed ledger. Each "block" contains transaction records of anonymous users. Blockchain technology links these transactions together to create a comprehensive history of each transaction. Since each block is encrypted to the previous block, any changes in the past affect all future blocks. Therefore, changing the previous blocks in the blockchain becomes challenging, so the security is improved. This feature makes the blockchain applicable to various IoT applications, such as currency encryption in the trading system, as shown in Fig. 1.4.

Blockchain technology's distributed trust, security, and invariance are crucial to building an effective financial trading system. In recent years, some chip-level blockchain security technologies have been widely deployed in IoT devices. By storing and signing data in the chip and encrypting data transmission links with cryptography, the blockchain fundamentally eliminates data tampering and can realize data traceability and tracking. In addition, another important application of blockchain-based IoT is to prevent service providers from tampering with data. When the assets of service providers need to be pledged or used for financing, the accuracy and reliability of this data are very important. In particular, as the data owner, the service provider can operate the data with higher authority, so it is very critical to prevent the service provider from tampering with the data. With the continuous promotion of the "Internet of Things," IoT devices have entered a geometric growth state. Efficient management of massive devices and mining the commercial value of enormous data safely and reliably are the fundamental advantages of blockchain technology. The blockchain module connects the blockchain application platform and the IoT, thus allowing IoT devices to access smart contracts. In short, if the IoT achieves low-cost access to massive data, blockchain enables us to trust these data on a large scale, with high efficiency and low cost. Undoubtedly, blockchain is of great significance to the Internet of Things.

1.2.4 Machine Learning Technologies

Machine learning can explore the potential capacity of the computer, enable the computer to self-study by utilizing the input data, and can therefore solve the problems with high time complexity and spatial complexity. Also, machine learning is an effective tool to achieve deep-level data mining and fine-grained feature extraction. Based on the above advantages, we believe that machine learning can break through the technical bottleneck of IoT technology and promote the development of the Internet of Things.

The evolution history of machine learning can be found in Fig. 1.5. Long before computers, problems closely related to machine learning have been discussed at the Dartmouth Conference in 1956 [17]. At that time, the researchers generally believed that machines could simulate human thinking to some degree. Based on this, they also proposed an evaluation criterion, the Turing test. It can be summarized as whether machines can judge the actor as a human or another machine without knowing the identity of the actor. In 1990, the 23 unsolved mathematical problems proposed by Hilbert also include the relevant contents of machine learning.

During that period, some basic research results about machine learning have been proposed, such as automatic theorem proving, checkers programs, LISP voice, and so on. Besides, the proposal of some advanced machine learning algorithms like the Multilayer Neural Network and the Back Propagation algorithm also enhanced the ability of machine learning to deal with some more complex problems.

However, the critical question of how can machines imitate human thinking to judge is still unsolved, and in 1973, the artificial intelligence report made by Lighthill points out that the development of machine learning has fallen into a bottleneck. The emergence of the expert system has improved this situation. The expert system no longer requires machines to simulate the whole situation of human

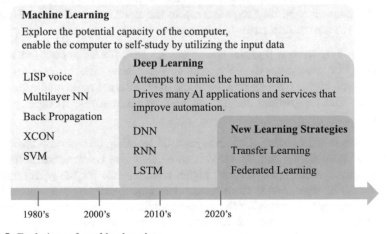

Fig. 1.5 Evolutions of machine learning

thinking but only focuses on a specified field and solves the problems in it. It greatly reduces the achievement difficulty of machine learning and promotes the development of machine learning from the theoretical stage to the implementation stage. In 1980, CMU developed the multilayer neural network expert system XCON which allows the expert system to play an efficient role in specific fields. Therefore XCON is also considered a milestone for machine learning in this period.

As mankind enters the twenty-first century, the rapid development of the Internet promotes the innovation of machine learning technology. In 2006, the Learning Multiple Layers of Representation by Jeffrey Sington establishes a new architecture of deep neural networks which is still used today. Many new achievements of artificial intelligence have sprung up, like chat robot Alice, the Deep Blue computer developed by IBM, and the AlphaGo programs. Some new technologies like cloud computing, blockchain, and big data also provide application scenarios for machine learning technology.

As a discipline with a huge architecture, machine learning can be divided into multiple small branches. As shown in Fig. 1.6, based on whether the training data is labeled, the existing machine learning techniques can be generally divided into supervised learning, unsupervised learning, and reinforcement learning [18]. The algorithms belonging to the supervised learning use labeled data to train, and the unsupervised learning algorithms train with unlabeled data. As for the reinforcement learning algorithms, they train the neural network relying on the feedback function, but the training data is unlabeled, too. We will introduce the above three types of machine learning methods below.

(A) Supervised Learning Supervised learning is the oldest machine learning method, and by using the labeled date to train, it can achieve good performance when the amount of data is small. At present, there are many kinds of supervised learning algorithms with their own advantages and disadvantages. To solve specific problems, the supervised learning algorithms can be chosen by considering the following factors: the amount of data, continuity or dispersion of data, the dimension of data, the accuracy and time requirements of the algorithm, and so on. Figure 1.7 is proposed for further reference:

From Fig. 1.7, we can find the characteristics of different supervised learning algorithms. For example, when the data is continuous, the logistic regression algorithm has a higher training efficiency and the results are more explicable, while the random forest algorithm and the support vector machine method are more accurate.

(B) Unsupervised Learning Unlike supervised learning, unsupervised learning is a method with unlabeled data, and this feature challenges the machine to make the right decisions and also puts forward higher requirements for the performance of the machine. However, robust unsupervised learning algorithms can process large-scale data, thus greatly reducing the human workload. Without loss of generality, the unsupervised learning method can be divided into the clustering algorithm and the dimension reduction algorithm.

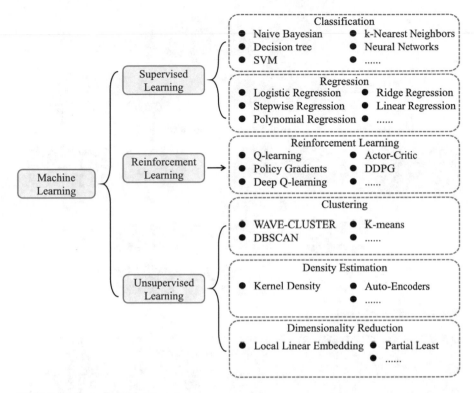

Fig. 1.6 Various of machine learning

The clustering algorithm aims to cluster the data according to their similarity. Although the classification results usually have no intuitive meaning, the detect abnormal data can be detected easily in this way. It mainly includes the K-MEANS algorithm, the DBSCAN algorithm, and the Hierarchical Clustering algorithm, and those algorithms cluster the data from different aspects. While the dimension reduction algorithm aims to reduce the dimension of data by data compression. During this process, some redundant information can be deleted to reduce the amount of calculation. The dimension reduction algorithms mainly include the PAC algorithm and the LDA algorithm.

(C) Reinforcement Learning Reinforcement learning is an important part of machine learning, different from other machine learning algorithms, and the purpose of reinforcement learning is to learn a robust strategy to maximize rewards or achieve specific goals [19]. By interacting with the environment, the strategy can be updated directionally. Adopting the method of reinforcement learning is like arranging a virtual guide for the computer, it can judge the advantages and disadvantages of computer behavior and score the computer based on this. The computer, on the other hand, will try to make the score higher and higher, so that the

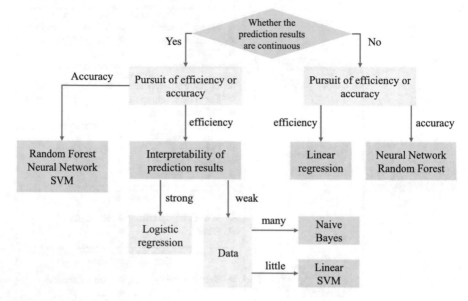

Fig. 1.7 Supervised learning

Model Free	Get feedback and learn directly from the training environment
Model-Based RL	The virtual environment is built based on the real environment and has certain predictive ability
Policy-Based RL	After analyzing the environment, output the probability of each behavior in the next step, and determine the behavior according to the probability
Value-Based RL	After analyzing the environment, output the value of all behaviors, and determine the behavior according to the value
Monte-Carlo Update	Update the code of conduct at the end of a training session
Temporal-Different Update	Update your code of conduct after each step of your training
On-Policy	The behavior strategy in training is the same as the goal strategy
Off-Policy	Behavioral strategies and target strategies can be different in training

Fig. 1.8 Reinforcement learning

behavior will be correct eventually. Without loss of generality, we classify several classical reinforcement learning algorithms as shown in Fig. 1.8:

As for the value-based RL algorithms, the Q-Learning, Sarsa, and Deep Q Network algorithms are the most commonly used methods. While for the policy-based RL algorithms, the Policy Gradient algorithms are the most widely applicable.

1.3   Related Research and Development

Recently, the combination of ML and IoT has attracted much attention in civilian applications. An intelligent IoT network is a network that improves resource utilization and decreases manual approaches between trillions of bits of data. For example, the service delivery system in our daily life achieves the above objectives. Another example is companies such as Amazon and Uber have deployed an intelligent IoT network to better manage and handle the service requests by analyzing the user's feedback and experiences.

1.3.1   IEEE CIIoT

IEEE SSCI is a flagship annual international conference on computational intelligence. And it contains two panels: Artificial Intelligence and Intelligence in IoT [20]. IEEE Symposium on Computational Intelligence in IoT and Smart Cities (IEEE CIIoT) academics discuss Intelligent IoT. IoT network has many kinds of related technologies such as wireless networks, communication modules, sensors, and smart terminals. And it is necessary to develop the above related technologies for achieving a better IoT network. However, with the increase of abundant sensors and devices, the growing volume of data is a challenging problem for intelligent IoT networks. Another challenge for the IoT network is how to make the devices more intelligent.

To solve the above problems, it is necessary to seek more intelligent methods and technologies. At the same time, Computational Intelligence (CI) plays an important role in managing and storing huge data flows in the network. Thus, as IoT gains its full potential CI will be at the forefront to facilitate the potential of IoT. IEEE CIIoT emphasizes the gathering of Computational Intelligence and IoT. And its topic mainly contains CI-based Scalability Solutions for IoT, CI Control Schemes in IoT, Resource Management Techniques Using AI for IoT, CI Applications in Industrial IoT, and so on.

1.3.2   Aeris

In 2022, Aeris, the leading global IoT solutions provider, announced the next generation of intelligent IoT networks [21]. The Aeris intelligent IoT network provides users with the most dynamic, flexible, reliable, and secure IoT network on the market. Moreover, it connects with approximately 400,000 IoT devices that enable millions of people to deliver affordable, reliable, and clean energy. It has the following properties:

(1) Intelligent IoT Connectivity. With the Aeris intelligent IoT network, it can
    realize the boundary-free connection with Seamless 5G-ready, LTE-M, 3G, and
    2G. At the same time, the technology of Aeris intelligent IoT can also increase
    uptime.
(2) Intelligent IoT security. The Aeris intelligent IoT network can protect its IoT
    devices from becoming compromised. At the same time, cyberattacks on IoT
    networks can be prevented simply and cost-effectively.
(3) Intelligent IoT service. The Aeris IoT Services platform is a comprehensive
    IoT solution that provides intelligent data management, asset management,
    and automotive services from device to application. The Aeris IoT Services
    platform is composed of event generation, IoT connectivity, data ingestion, data
    storage and management, analytics, presentation and action, and rating and IoT
    billing.
(4) Intelligent IoT applications. The Aeris's next-generation intelligent IoT net-
    work will realize Advanced Security insurance. With the increase of connected
    IoT programs, Aeris is also developing its next-generation applications in its
    intelligent IoT network.

1.3.3 Google GCP

Google also proposed an intelligent Internet of Things, which makes it easy to
connect devices to the cloud platform in a few clicks. Google Cloud Platform (GCP)
is a Google-provided suite of cloud computing services that operates on Google's
same platform internally for its end-user products such as Gmail, YouTube, Google
One, and Google search [22]. With Google's intelligent IoT platform, data can be
connected, stored, and analyzed at the edge and inside the cloud. The greatest part of
adopting the Google Cloud Platform is the unique Big Data, Artificial Intelligence,
with the Internet of Things (IoT) capabilities.

   The Google IoT solution can realize the intelligent network from the edge to
the cloud by executing real-time analysis. For example, many applications can be
achieved by Google IoT solution, such as fleet management, inventory tracking,
cargo integrity monitoring, and other key business functions. Moreover, the IoT
solution of Google can contribute to the smart city by spanning billions of sensors
and edge devices. There are also several top industries that use Google Cloud IoT
services to explore how they are deploying secure, scalable IoT solutions on the
cloud, such as information technology and services, smart cities, computer software,
smart home solutions, and manufacturing. In the future, the importance, utility, and
deployment of an IoT-enabled network intelligence system will increase along with
the increase in the amount of data.

1.3.4   3GPP RedCap

RedCap UE, the full name of reduced capability UE, is a terminal with reduced capability. It is a 5G standard technology proposed by 3GPP in the Rel-17 version standard [23]. When 3GPP first proposed this issue, it also used NR Light (NR lite). Briefly, the RedCap UE technology is to meet the requirements of specific application scenarios, by reducing the air interface capability of the terminal, reducing complexity, reducing costs, reducing power consumption, and other requirements.

The reason for the emergence of RedCap is that existing radio specifications have blind spots that cannot be covered. RedCap is positioned to address use cases that cannot be optimally serviced today with Ultra-Reliable Low Latency Communication (URLLC), massive Machine Type Communications (mMTC), or enhanced Mobile Broadband (eMBB) solutions. These use cases include wearable medical devices, smart watches, industrial wireless sensors, video monitors, etc., which are common use cases in intelligent IoT applications. These use cases are less stringent than enhanced mobile broadband (eMBB) use cases and do not require strict or deterministic latency requirements as time-critical communication use cases. Therefore, RedCap balances cost and performance, appropriately sacrifices some indicators and requirements, and finally achieves cost reduction. The Internet of Things is divided into high-speed, medium-speed, and low-speed. RedCap, in fact, corresponds to more medium-speed or medium-high-speed. At present, LTE Cat.1 and Cat.4 have already covered this part of the demand. Since the establishment of the RedCap project in 2021, it has developed rapidly. Although not yet fully commercialized, the RedCap market is expected to take shape in 2022. The IoT industry will also develop rapidly.

1.3.5   SwRI Intelligent Networks and IoT

Southwest Research Institute(SwRI) focuses on researching solutions for network applications and wireless communication technologies. Meanwhile, because of its decades-long research on embedded systems, it also provides intelligent IoT solutions for government or other commercial customers [24]. During the entire project life cycle, it can provide services for solution customization, software and hardware platform development, system verification, and so on. In addition, in order to ensure the normal operation of industrial networks, the institute has also been researching IoT security issues in manufacturing systems and critical infrastructure. The team developed an intrusion detection system (IDS) for industrial control. Specifically, they discover the cyber threats across network protocols through unique algorithms. This system can detect the transmission of all industrial control data from pipelines to robotic manufacturing.

Today, industrial networks need to utilize the Internet of Things to transmit massive data. It is the development direction of modern industry to closely integrate traditional industrial networks with AI. IoT connects various embedded sensing devices to the network. These devices continuously exchange data to realize the interconnection of people, machines, and things. Once industrial equipment is connected to the Internet, it brings cyber threats into the industrial network. Criminals may exploit vulnerabilities in equipment, protocols, or software to conduct cyberattacks. SwRI uses the Modbus/TCP protocol to discover the attacks. The protocol can monitor the controls and data acquisition (SCADA) systems equipment and provide a guarantee for the normal operation of industrial control. The detection algorithm can distinguish and identify normal Modbus/TCP traffic and network attack traffic such as data obfuscation and out-of-band timing. It can also determine the packet type based on the source of the packet. The algorithm considers the packet to be normal if it comes from a normally used industrial device; otherwise, the algorithm considers it an attack.

1.3.6 IBM Smart City

International Business Machines Corporation (IBM) is one of the world's largest information technology and business solutions companies. The company was the first to coin the term "smart city" [25]. Through the Smarter City Challenge project, the company has specified a vision based on data centralization and focusing on security. A smart city refers to taking full advantage of network interconnection and big data information to operate and optimize the city with limited resources. According to IBM's vision, becoming a smart city is a revolutionary change in urban development. Nowadays, benefiting from the fact that computing power is endowed with a wide range of devices, IBM uses the development of instrumentation, interconnection, intelligence, and other aspects to build a smart city. IBM's solution integrates the city's systems with the intelligent IoT. It improves the efficiency of urban resource utilization and optimizes urban management and services.

In general, IBM uses an intelligent and innovative approach to experiment with new business models or technologies operating in a city. The company has successfully implemented several smart city solutions. For example, they use artificial intelligence to predict public transportation and use big data to digitize operations for ports. They aim to modernize infrastructure, build digital cities for governments and people, and provide sustainable support using the latest innovations.

1.4 Organizations of This Book

This book is organized as depicted in Fig. 1.9. We first design a hybrid network control architecture of the Intelligent Internet of Things Networking, where intelligent in-network devices can automatically adapt to network dynamics and generate

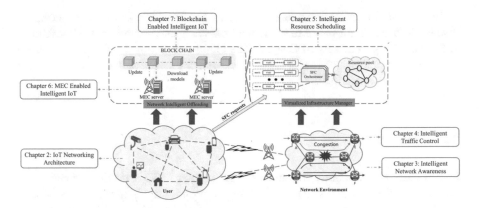

Fig. 1.9 Book organization

control strategies, and the centralized platform is used to ease the training process of distributed in-network devices for achieving network global intelligence [26]. With the advancement of programmable network hardware, it is possible to implement personalized network functions inside the network. In addition, to enhance the collaboration among distributed in-network devices, a centralized management plane is introduced to ease the training process of distributed switches.

Then, we discuss the promising machine learning methodologies for intelligent network awareness [27]. With the ability of the network awareness to migrate from end-hosts to the network core, it becomes sustainably important to strengthen the management of data traffic in-network [28]. As a critical part of massive data analysis, an efficient traffic classification mechanism plays an important role in guarding network security and defending traffic attacks. Moreover, the flexible and programmable data plane opens one way to aware the network more fine-grained.

Furthermore, we discuss how machine learning can achieve network intelligent control. Deep learning and reinforcement learning make it possible to find optimal solutions in highly complex network topologies [29]. Examples include routing decisions, QoS strategies, load balance, and so on.

However, various task scenarios in IoT bring new challenges of dynamic scheduling and on-demand allocation to network resource management. Therefore, to address this issue, in Chap. 5, we apply several intelligent approaches for efficient resource scheduling in networks.

In Chap. 6, we focus on how mobile edge computing enables the intelligent IoT, where the computing and communication resources of devices are limited [30]. We mainly discuss the resource sharing and edge computation offloading problems in mobile edge networks. In order to jointly optimize task offloading and resource allocation, we propose a second-price auction scheme for ensuring fair bidding for resources rent. With the aid of the new isotone action generation technique (IAGT) and adaptive action aggregation update strategy (3AUS) based on the proposed DRL framework, an efficient edge intelligent method is proposed to offload tasks and manage edge resources.

In order to incentive more IoT devices to participate in the network task, we discuss the blockchain-based IoT which allows fairly and securely renting resources and establishing contracts in Chap. 7. The cloud computing service is introduced into the blockchain platform for the sake of assisting to offload computational task from the IoT network itself.

1.5 Summary

In this chapter, we mainly introduce the background. We state the motivation of this book at first. Based on the motivation, we propose the hybrid network control architecture of the Intelligent Internet of Things Networking and discuss the key technologies and challenges in the architecture. Then, the related research and development are provided. Finally, we give the organizations of this book [31].

References

1. A. Tamrakar, A. Shukla, A. Kalifullah, F. Reegu, K. Shukla, Extended review on internet of things (IoT) and its characterisation. Int. J. Health Sci. **10**, 8490–8500 (2022)
2. Q. Jiang, Q. Lin, R. Zhang, H. Huang, Shared massage chair application in the context of IoT: take "lemobar" as an example, in *Journal of Physics: Conference Series*, vol. 1631, no. 1 (IOP Publishing, Bristol, 2020), p. 012173
3. H. Tao, Design and implementation of vehicle data transmission protocol based on present algorithm, in *2021 IEEE Asia-Pacific Conference on Image Processing, Electronics and Computers (IPEC)* (IEEE, Piscataway, 2021), pp. 968–971
4. H.A. Khattak, M.A. Shah, S. Khan, I. Ali, M. Imran, Perception layer security in internet of things. Future Gener. Comput. Syst. **100**, 144–164 (2019)
5. O. Bello, S. Zeadally, M. Badra, Network layer inter-operation of device-to-device communication technologies in internet of things (IoT). Ad Hoc Netw. **57**, 52–62 (2017)
6. S.N. Swamy, D. Jadhav, N. Kulkarni, Security threats in the application layer in IoT applications, in *2017 International Conference on i-SMAC (IoT in Social, Mobile, Analytics and Cloud)(i-SMAC)* (IEEE, Piscataway, 2017), pp. 477–480
7. A. Čolaković, M. Hadžialić, Internet of things (IoT): a review of enabling technologies, challenges, and open research issues. Comput. Netw. **144**, 17–39 (2018)
8. M. Asim, A survey on application layer protocols for internet of things (IoT). Int. J. Adv. Res. Comput. Sci. **8**(3), 996–1000 (2017)
9. M. Masirap, M.H. Amaran, Y.M. Yussoff, R. Ab Rahman, H. Hashim, Evaluation of reliable UDP-based transport protocols for internet of things (IoT), in *2016 IEEE Symposium on Computer Applications & Industrial Electronics (ISCAIE)* (IEEE, Piscataway, 2016), pp. 200–205
10. X. Wang, Y. Feng, J. Sun, D. Li, H. Yang, Research on fishery water quality monitoring system based on 6lowpan, in *Journal of Physics: Conference Series*, vol. 1624, no. 4 (IOP Publishing, Bristol, 2020), p. 042057
11. A. El Hajjar, Securing the internet of things devices using pre-distributed keys, in *2016 IEEE International Conference on Cloud Engineering Workshop (IC2EW)* (IEEE, Piscataway, 2016), pp. 198–200

12. R. Abdelmoumen, A review of link layer protocols for internet of things. Int. J. Comput. Appl. **182**(46), 0975–8887 (2019)
13. X. Jia, Q. Feng, T. Fan, Q. Lei, RFID technology and its applications in internet of things (IoT), in *2012 2nd International Conference on Consumer Electronics, Communications and Networks (CECNet)* (IEEE, Piscataway, 2012), pp. 1282–1285
14. M. Du, K. Wang, An SDN-enabled pseudo-honeypot strategy for distributed denial of service attacks in industrial internet of things. IEEE Trans. Ind. Inf. **16**(1), 648–657 (2019)
15. D. Kreutz, F.M. Ramos, P.E. Verissimo, C.E. Rothenberg, S. Azodolmolky, S. Uhlig, Software-defined networking: a comprehensive survey. Proc. IEEE **103**(1), 14–76 (2014)
16. H. Si, C. Sun, Y. Li, H. Qiao, L. Shi, IoT information sharing security mechanism based on blockchain technology. Future Gener. Comput. Syst. **101**, 1028–1040 (2019)
17. S. De Bruyne, M.M. Speeckaert, W. Van Biesen, J.R. Delanghe, Recent evolutions of machine learning applications in clinical laboratory medicine. Crit. Rev. Clin. Lab. Sci. **58**(2), 131–152 (2021)
18. B. Mahesh, Machine learning algorithms-a review. Int. J. Sci. Res.[Internet] **9**, 381–386 (2020)
19. N.C. Luong, D.T. Hoang, S. Gong, D. Niyato, P. Wang, Y.-C. Liang, D.I. Kim, Applications of deep reinforcement learning in communications and networking: a survey. IEEE Commun. Surv. Tutor. **21**(4), 3133–3174 (2019)
20. J.S. Raj, A. Bashar, S. Ramson, *Innovative Data Communication Technologies and Application*. Lecture Notes on Data Engineering and Communications Technologies (Springer, Berlin, 2019)
21. P.P. Ray, A survey of IoT cloud platforms. Future Comput. Inf. J. **1**(1–2), 35–46 (2016)
22. E. Bisong, An overview of google cloud platform services, in *Building Machine Learning and Deep Learning Models on Google Cloud Platform* (2019), pp. 7–10
23. S. Moloudi, M. Mozaffari, S.N.K. Veedu, K. Kittichokechai, Y.-P.E. Wang, J. Bergman, A. Höglund, Coverage evaluation for 5G reduced capability new radio (NR-RedCap). IEEE Access **9**, 45055–45067 (2021)
24. V. Bali, V. Bhatnagar, D. Aggarwal, S. Bali, M.J. Diván, *Cyber-physical, IoT, and Autonomous Systems in Industry 4.0* (CRC Press, Boca Raton, 2021)
25. V. Scuotto, A. Ferraris, S. Bresciani, Internet of things: applications and challenges in smart cities: a case study of IBM smart city projects. Business Proc. Manag. J. **22**(2), 357–367 (2016)
26. T. Mai, S. Garg, H. Yao, J. Nie, G. Kaddoum, Z. Xiong, In-network intelligence control: toward a self-driving networking architecture. IEEE Netw. **35**(2), 53–59 (2021)
27. X. Li, H. Yao, J. Wang, X. Xu, C. Jiang, L. Hanzo, A near-optimal UAV-aided radio coverage strategy for dense Urban areas. IEEE Trans. Vehic. Technol. **68**(9), 9098–9109 (2019)
28. M. Shen, Y. Liu, L. Zhu, K. Xu, X. Du, N. Guizani, Optimizing feature selection for efficient encrypted traffic classification: a systematic approach. IEEE Netw. **34**(4), 20–27 (2020)
29. C. Qiu, H. Yao, F.R. Yu, C. Jiang, S. Guo, A service-oriented permissioned blockchain for the internet of things. IEEE Trans. Serv. Comput. **13**(2), 203–215 (2019)
30. Y. Gong, H. Yao, J. Wang, D. Wu, N. Zhang, F.R. Yu, Decentralized edge intelligence-driven network resource orchestration mechanism. IEEE Netw. https://doi.org/10.1109/MNET.120.2200086
31. H. Yao, C. Jiang, Y. Qian, *Developing Networks Using Artificial Intelligence* (Springer, Berlin, 2019)

Chapter 2
Intelligent Internet of Things Networking Architecture

Abstract The Internet of Things (IoT) has many compelling applications in our daily lives. With the explosion of IoT devices and various applications, the demands on the performance, reliability, and security of IoT networks are higher than ever. Current end-host-based or centralized control frameworks generate excessive computational and communication overhead, and the dynamic response of IoT networks is sluggish and clumsy. Recently, with the advancement of programmable network hardware, it has become possible to implement IoT network functions inside the IoT network. However, current in-network schemes largely rely on manual processes, which exhibit poor robustness, flexibility, and scalability. Therefore, in this chapter, we present a new IoT network intelligent control architecture, in-network intelligence control. We design intelligent in-network devices that can automatically adapt to IoT network dynamics by leveraging powerful machine learning adaptive abilities. In addition, to enhance the collaboration among distributed in-network devices, a centralized management plane is introduced to ease the training process of distributed switches. To demonstrate the technical feasibility and performance advantage of our architecture, we present three use cases: in-network load balance, in-network congestion control, and in-network DDoS detection.

Keywords In-network intelligent control · Software-defined network · Congestion control

The Internet of Things (IoT) has many compelling applications in our daily lives. With the explosion of IoT devices and various applications, the demands on the performance, reliability, and security of IoT networks are higher than ever. Current end-host-based or centralized control frameworks generate excessive computational and communication overhead, and the dynamic response of IoT networks is sluggish and clumsy. Recently, with the advancement of programmable network hardware, it has become possible to implement IoT network functions inside the IoT network. However, current in-network schemes largely rely on manual processes, which exhibit poor robustness, flexibility, and scalability. Therefore, in this chapter, we present a new IoT network intelligent control architecture, in-network intelligence control [1]. We design intelligent in-network devices that can automatically adapt to IoT network dynamics by leveraging powerful machine learning adaptive abilities. In addition, to enhance the collaboration among distributed in-network devices, a centralized management plane is introduced to ease the training process of distributed switches. To demonstrate the technical feasibility and performance advantage of our architecture, we present three use cases: in-network load balance, in-network congestion control, and in-network DDoS detection.

2.1 In-network Intelligence Control: A Self-driving Networking Architecture

Recently, the Internet of Things (IoT) has become a disruptive force reshaping our lives and works, ranging from industrial manufacturing to healthcare. IoT refers to massive physical devices around the world that are connected to the IoT network for the sake of connecting and exchanging data. According to Gartner's prediction, it is expected that more than 25 billion IoT devices will be connected to the Internet in the future year 2025. With the exponential growth of diverse IoT devices, as well as various applications (e.g., intelligent manufacturing, smart agriculture), the expectations for the performance, reliability, and security of IoT networks are greater than ever [2]. For example, the industrial control system developed a 2 ms network delay and 1 μs jitter stringent target [3]. With users elevating their expectations for low latency, high bandwidth, ubiquitous access, and resilience to attack, IoT network control and management have become even more challenging.

Current IoT network control and management are largely built on the end-to-end mechanism and the closed-loop control framework (i.e., measurement, decision-making, and action). The IoT network control agency, which can be deployed at the end-host or the software-defined networking (SDN) controller, works tirelessly to (1) continuously collect troves of heterogeneous data from the IoT network, (2) analyze this data to infer characteristics about the IoT network, and (3) decide how to adapt the network's configuration in response to IoT network conditions (e.g., a shift in traffic demand, intrusion detection). However, this out-network control framework makes the control loop take place above the IoT network in a large timescale. The centralized control agencies have to collect and analyze a massive amount of data to respond to a single IoT network event, which incurs too much communication and computation overhead, and therefore exhibit clumsy and tardy in response to network dynamics.

Recently, with the development of the programmable IoT network hardware (e.g., SmartNIC, Barefoot's Tofino switches, Cavium XPliant switches), it is feasible to perform flexible processing logic inside the IoT network. The operator can reconfigure the IoT network hardware on the fly through high-level programming languages like P4, thus exercising control over how network packets are processed. Benefiting from this flexible data processing capacity and programmability, it is possible to consider that the control loop can be implemented in an integrated way with a set of abstractions around measuring, analyzing, and controlling the traffic at a small timescale directly as the packets are flowing through the IoT network. In other words, the programmable network hardware opens up new possibilities for in-network control, where the control loop can be directly implemented inside the IoT network (e.g., in-network load balance, in-network security).

There has been a series of works on IoT network control and management from the perspective of the in-network control paradigm, such as balancing the traffic load (e.g., CONGA [4]), adjusting the end-hosts transmission rates (e.g., FlexSwitch [5]), and DDoS detection. Compared to the end-host-based or centralized schemes, the in-network solutions are more effective at scale and more responsive to network events and dynamics. For example, in [4], the in-network load balance scheme CONGA achieves $5\times$ better flow completion times and $2 - 8\times$ better throughput than traditional solutions (e.g., MultiPath TCP). However, current in-network schemes are largely dependent on the manual process, where the operators need to meticulously analyze network behavior and design corresponding control policies (requiring the least weeks). This handcrafted method presents poor scalability and robustness, especially with the IoT network becoming extremely complicated and flexible.

Recently, machine learning has attracted a large amount of attention from both academia and industry. Machine learning can automatically learn and optimize strategy directly from experience without following predefined rules. Therefore, it is promising to apply machine learning in IoT network control and management to leverage the powerful machine learning adaptive abilities for higher IoT network performance. In this work, we present a new IoT network control paradigm, in-network intelligent control. We desire to design intelligent in-network devices that can automatically adapt to IoT network dynamics, thus improving IoT network performance and satisfying the users' requirements.

However, as an inherently distributed system, each in-network device can only perform its local control loop of "measuring-analyzing-acting" over a small portion of the whole system. It behooves us to ask the question: "how could the distributed in-network devices learn the cooperative control policy in a distributed fashion?". To address this problem, we proposed a hybrid in-network intelligent control architecture. In our architecture, we adopt the centralized training and distributed execution framework, where a centralized management platform is introduced to ease the training process of distributed in-network devices. Besides, to evaluate the feasibility and performance of our architecture, we present three relevant use cases: in-network load balance, in-network congestion control, and in-network DDoS detection.

2.1.1 In-network Functionality

In this section, we will analyze the driving factors of the development of in-network functionality from both the technique push (i.e., programmable network hardware) and the users' pulls (i.e., the higher expectations for the performance, reliability, and security).

2.1.1.1 Programmable Network Hardware

Recently, the advance of programmable network hardware leads to a more flexible packet processing architecture. Examples include the Barefoot's Tofino 2 switch, the Cavium's XPliant switch, and the Broadcom's Trident 4 switch. They enable network users to customize the processing logic on data packets, including the type, sequence, and semantics of processing operations, while maintaining line speed forwarding (e.g., Tofino 2 has 12.8 Tb/s throughputs).

Current programmable network hardware largely relies on Protocol Independent Switch Architecture (PISA). The PISA allows the users to configure the packet parsing and processing operation through domain-specific language like P4. A typical PISA switch contains a programmable parser, ingress pipeline, egress pipeline, and deparser. When a packet arrives, the parser first converts packet data into the metadata (i.e., parsed representation). Then, the ingress/egress pipeline performs user-defined processing operations on the metadata through match-action tables. Finally, the deparser converts metadata back into a serialized packet.

In the PISA, the multiple data packet processing stages (i.e., match-action units) work in a pipelined manner. That is, each stage can process a small number of data packets independently with each other, so multiple data packets can exist in different stages of the pipeline at a certain time. This programmable pipeline design guarantees line speed packet processing performance. At the same time, though, it also incurs rigid restrictions toward data plane programming, such as the limited set of operations, limited concurrent memory access, and limited stages. Therefore, we need to carefully identify what functions should be migrated to the programmable network hardware.

2.1.1.2 In-network Functionality

Current IoT network functions are largely built on the end-to-end principle [6]. This principle states that IoT network functions' operations should reside at the end-hosts rather than the IoT network core. This design reduces the IoT network core's complexity, and facilitates the generality in the IoT network, that any new functions can be added at the end-hosts without having to change the core of the IoT network. However, with the explosion of IoT devices and various applications (e.g., 4K/8K, industrial control), a lot of new requirements have emerged for the IoT network.

Firstly, as the number of end devices broadens, especially the lightweight IoT devices, it requires an easier used IoT network. In light of the end-to-end principle, the current IoT network was conceived as a dumb pipe, where the intermediary nodes are designed only to forward IoT network packets. By contrast, the end-hosts have to constantly adjust their action (e.g., balancing the traffic load, adjusting the transmission rates) to respond to IoT network events and dynamics. As a result, substantial software or protocol stacks are deployed on the end devices. These software or protocol stacks must be installed, configured, upgraded, and maintained separately by each user, which adds too much complexity to users. Meanwhile, it requires all end-hosts to cooperate to achieve optimal performance and fairness, thus leading to inefficiencies and poor performance isolation.

Secondly, the current Internet has become an untrustworthy world. There is less and less reason to believe that we can trust other end-hosts. The untrustworthy end-points lead to various IoT network security problems (e.g., Distributed Denial of Service Attack, Trojan). These security problems are becoming increasingly serious with the dramatic growth of IoT network users. As the intensity of attacks increases, current end-host-based security mechanisms are becoming ineffective and inefficient. Therefore, it urgently requires new security mechanisms to make a more trustworthy IoT network.

As discussed above, the network users elevate their expectations to network for ease of use, ubiquitous access, high performance, and resilience to attack. To meet these requirements, deploying new functions inside the network becomes a feasible solution. In recent years, there has been an increasing amount of literature on in-network function deployment. We list the representative works of in-network functions in Table 2.1. These successes show that in-network functions are more effective at scale and more responsive to quickly adapt to network events and dynamics.

Table 2.1 Representative works of in-network functionality

Ref.	Function	Performance
[4]	Load balancing	Achieving more than $5\times$ better flow completion time for a realistic datacenter workload and $2-8\times$ better than MPTCP in incast scenarios
[7]	Load balancing	Achieving $10\times$ better performance than ECMP at 60% network load
[8]	Congestion control	Achieving $4\times$ better average flow completion time and $10\times$ better tail latency compared to TCP
[9]	Fairness	Improving the latency observed by web traffic by almost 50% compared to end-host solution
[10]	QoS scheduling	Keeping the packet delay closer to the target value across the full range of bandwidths and the measured link utilizations are consistently near 100% of link bandwidth
[11]	Heavy-Hitter detection	Detecting 95% of the heaviest flows with less than 80 kb memory

2.1.2   In-network Intelligent Control

While in-network functionality has many advantages, current in-network solutions are largely dependent on the manual process. Inspired by the recent success of machine learning in the control field (e.g., robotic control, autonomous vehicles, and Go), in this section, we propose a new IoT network control paradigm, in-network intelligent control. We introduce machine learning to in-network devices to leverage the powerful self-adaptive learning abilities for responding to IoT network events and dynamics. In addition, a centralized management platform is used to enhance global IoT network intelligence.

2.1.2.1   Hybrid In-network Intelligence Architecture

As shown in Fig. 2.1, it shows an overview of the in-network intelligent control architecture and its functional planes. In our architecture, we design an intelli-

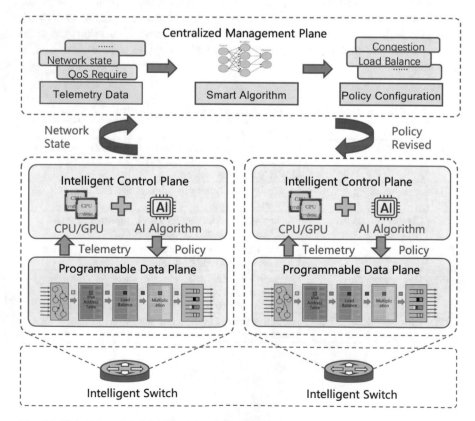

Fig. 2.1 Hybrid in-network intelligent control architecture

gent switch for hosting intelligent in-network control functions. In the switch, the programmable data plane is responsible for measuring IoT network events and dynamics and performing corresponding processing operations on packets. Concretely, the data plane uses telemetry probes (e.g., in-band network telemetry) to monitor and report flows and IoT network information (e.g., link utilization, microburst flow). These probes travel periodically overall desired adjacent links (e.g., adjacent equal-cost multipath). Note that this telemetry data only contains a small portion of the IoT network state (i.e., local observation information). Subsequently, the data plane performs specific processing operations according to its local observation and current strategy. The processing logics rely on the configuration instructions issued by the control plane.

The intelligent control plane is responsible for learning the behavior of the IoT network and automatically generating control strategies accordingly. The control plane consists of the high-performance CPU and GPU to provide computing power for the training process. Machine learning algorithms can constantly learn and optimize control strategies through the operating data reported from the data plane. Then, the updated control strategies (programmed by P4) will be feedback to the data plane.

As discussed above, these two planes constitute a closed-loop control of the "measurement-learning-decision-action" process, which contributes to control the local IoT network autonomously. While this control loop enables quick to adapt to IoT network dynamics in local, the IoT network as a distributed system requires the cooperation of all nodes to achieve optimal performance. Learning from nodes with only a local observation and control is a complex task, especially with the goal of global optimization. Therefore, for the sake of achieving network global intelligence, in our architecture, we introduce a centralized management plane to coordinate these switches. The management plane constantly collects IoT network information and learns global knowledge and then shares the knowledge to each switch to revise the learning process of control strategies, thus improving IoT network performance at the global level. Note that compared to the centralized control paradigm, our centralized plane is only used to revise the strategies.

In this section, we will present three specific use cases to demonstrate the feasibility and performance advantage of the in-network intelligent control paradigm.

2.1.2.2   In-network Load Balance

The first use-case is the in-network load balance. Recently, with the dramatic increase of connected IoT devices, the traffic volume of IoT applications is growing exponentially. To improve the throughput, the use of multiple wireless technologies (e.g., Bluetooth, Zigbee, 802.11a/b/g/n) to realize parallel forwarding of data. Therefore, efficiently distributing traffic among multiple available paths (i.e., load balance) is the key point for maximizing IoT networks' throughput.

However, achieving optimal load balancing is not trivial due to the complexity and dynamics of IoT networks. In recent years, there has been an increasing amount

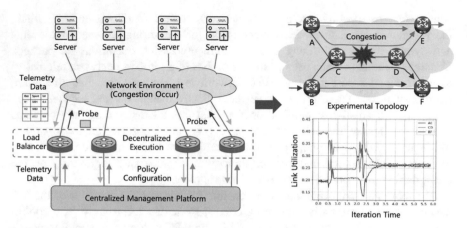

Fig. 2.2 Hybrid in-network load balance scheme

of literature focused on designing better load balance schemes. The prior works can be classified into centralized scheduling, host-based transport protocols, and in-network load balance. Combining with the SDN architecture, some schemes, such as Hedera [12] and SWAN [13], use a centralized SDN controller to constantly collect network events and dynamics and calculate the optimal decision. While the advantages of the centralized mechanism are clear, it brings computation and communication overhead for flow-level traffic control. They are too slow to respond to rapidly varying traffic. Besides, host-based schemes, such as MultiPath TCP (MPTCP) [14], make the end-host protocol stack too complex and therefore add processing burdened to high-performance applications.

With the development of programmable network hardware, in-network load balance schemes (e.g., CONGA [4], HULA [7]) have received an amount of attention from both academia and industry. For example, in [7], Katta et al. use specific probes periodically to collect the link utilization and automatically shift traffic to less-congested links directly in switches. The experiment results show that the in-network scheme is more responsive in a few microseconds to adapt to the volatility of traffic.

In this work, compared to the current in-network schemes, we present a reinforcement learning aided in-network load balance scheme. As shown in Fig. 2.2, we adopt a centralized learning and distributed execution framework and propose a multi-agent actor-critic policy gradient (MADDPG)-based load balance algorithm. In our architecture, a centralized plane is reinforced with the global network state (i.e., global link utilization), which is uploaded from each switch, to ease the distributed agent training process and thus help distributed switches to act in a globally coordinated way. Note that each switch makes decisions only according to the local observation, which is collected by the telemetry probes. The algorithm details can be found in [15].

To assess the validity of this architecture, we carried out the following simple experiment. Our simulation environment is based on a discrete-time simulator Omnet++ and TensorFlow 1.8.0. As shown in Fig. 2.2, we simulated a network topology with 6 nodes, which consists of 2 ingress switches, 2 intermediate switches, and 2 egress switches. The in-network load balance agencies are implemented in the ingress switches. We set all link capacity as 50 Mbps and the traffic volume of each source–destination pair as 20 $\pm$ 2 Mbps. The experiment result shows that our algorithm exhibits a good convergence performance and can be stable in traffic fluctuation.

2.1.3   In-network Congestion Control

The second use case is in-network congestion control. In IoT networks, affecting by the lossy communication medium, dense deployment, and frequent topology change, congestion control has become a challenging problem. Nowadays, in IoT transport layer protocol (e.g., TCP, TIMELY), congestion control is primarily achieved by end-to-end-based schemes, where end-hosts react to congestion signals from the network. For example, the well-known additive-increase/multiplicative-decrease (AIMD) algorithm in TCP is a feedback-based congestion control algorithm. When congestion is detected (i.e., receiving a congestion signal from the network, such as Explicit Congestion Notification (ECN)), the sender will reduce the congestion window size and therefore ease congestion. While such a scheme simplifies network design, massive end-hosts must cooperate to achieve fair bandwidth allocation and isolation among competing flows, thus leading to inefficiencies and poor convergence.

Compared to the end-host, enforcing fair allocation and isolation among competing flows directly in switches seems more efficient. Congestion can be controlled at in-network devices through scheduling and queueing algorithms. Recently, several success cases proved that in-network congestion control presents perfect performance. Examples include FlexSwitch and Approximating Fair Queueing. In [5], Sharma et al. implement Rate Control Protocol (RCP) on programmable network hardware to control congestion, where RCP relies on explicit network feedback to control sending rate. The switch computes sending rate $R(t)$ periodically and sends it back to the end-host. In an ideal scenario, the rate can be simply calculated by $R(t) = C/N(t)$, where $N(t)$ is the number of ongoing flows and C is the link capacity. The experiment results show that the flow completion time of RCP is 10 times faster compared to traditional solutions.

Compared to prior works, we consider a more complex congestion control scenario. We focus on a kind of microburst traffic that occurs on the timescales 10 to 100 ms, which are too short for traffic engineering to react but still large enough to cause congestion to occur. This microburst will lead to the condition that the arrival rate momentarily exceeds the maximum process rate of a particular switch (e.g., $s3$ in Topology 2 of Fig. 2.3). To handle this problem, the switch can use a queue to

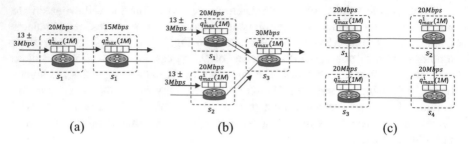

Fig. 2.3 Multiple switch congestion control. (**a**) Topology 1. (**b**) Topology 2. (**c**) Experimental topology

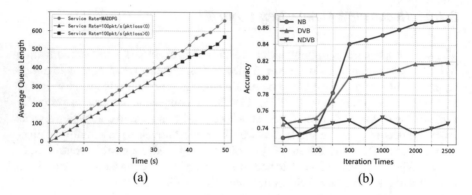

Fig. 2.4 The performance analysis. (**a**) In-network congestion control. (**b**) In-network DDoS detection

smooth the fluctuation and relieve downstream switch processing pressure, where the packet can be cached in peak periods and be released at other times.

For example, as shown in Topology 1 of Fig. 2.3, we assume that the arrival rate of *switch1* is 13 Mbps with a standard deviation 3 Mbps. The maximum transmission rate of upstream *switch1* and downstream *switch2* is 20 Mbps and 15 Mbps, where the queue capacity of both is 1 M. At peak flow, the arrival rate in *switch1* is 16 Mbps, which is smaller than the 20 Mbps but larger than 15 Mbps. If *switch1* sets the service rate as 13 Mbps, *switch2*'s queue will be quickly full. Therefore, *switch1* needs to reduce its sending rate to relieve the processing pressure of *switch2*. Meanwhile, the excessive decrease in the process rate will bring too much transmission delay and even packet loss of locality switch. The multiple switches need to work collaboratively to achieve global optimal.

Adopting the hybrid intelligent architecture and MADDPG algorithm we described above, we present a multi-agent reinforcement learning aided in-network congestion control scheme. Our simulation environment is built on a discrete-time simulator Omnet++. As shown in Fig. 2.3, the experiment topology contains four switches and a centralized management platform. As shown in Fig. 2.4a, in the baseline, the packet loss occurs in about 37 s. In contrast, our algorithm did not experience packet loss until the 50 s.

2.1.4   In-network DDoS Detection

The third use case is in-network distributed denial of service (DDoS) detection. The DDoS attack is a malicious attempt to disrupt the legitimate requests of a targeted network or server, through flooding the targeted machine or resource with a large number of superfluous requests. The DDoS attacks achieve effectiveness by utilizing massive, hijacked internet-connected devices as sources of the attack traffic. Especially, with the development of the Internet of Things, vulnerable IoT devices have been the primary force behind the DDoS botnet attacks. For example, in October 2016, a series of IoT DDoS attacks caused widespread disruption of legitimate Internet activity in the United States. As the intensity of DDoS attacks increases, the current DDoS detection engine, which relies on proprietary hardware appliances deployed at network edge, is becoming ineffective and inefficient.

Recently, emerging programmable switches provide an opportunity to address these limitations by using in-network security functionality to mitigate DDoS attacks. In-network security schemes are able to detect anomaly traffic immediately in the switch as packets are forwarded without needing to go through proprietary hardware, thus presenting more effective at scale [16]. Therefore, in this work, based on our architecture, we design a variational Bayesian-based in-network DDoS detection schemes to identify malicious traffic. The intelligent switches are able to constantly learn defense tactics (i.e., Bayesian classifier) and detect DDoS attacks in a distributed and automated fashion. Besides, the centralized management plane is used to synchronize a global set of observations and therefore realize multiple switches cooperation effectively. The algorithm details can be found in [17].

To validate our algorithm, we carried out a simple experiment. We use the KDD 99 DATA set to evaluate our algorithm, which is a standard data set for evaluating the performance of the DDoS detection algorithm. In addition, we construct a network with 20 network nodes and set two different baseline algorithms, the centralized naive Bayes algorithm (NB) and distributed variational Bayes algorithm (NDVB, without centralized sharing platform). As shown in Fig. 2.4b, the average accuracy of the centralized naive Bayes and our algorithm is increasing with the process of learning. This result shows that our hybrid learning architecture can enhance cooperation effectively.

2.1.4.1   What Functions Should Be Implemented Inside the Network

While programmable network hardware provides a powerful primitive for developing specific in-network functions, they are not a panacea for improving performance. In this section, we will discuss the question of what functions should be implemented inside the network. Firstly, considering the limited computational and storage capabilities and rigid restrictions of programmability, functions that require a great deal of memory or computing on per-packet processing are more suitable to implement at the end-host rather than the switch, such as virtual network mapping.

This is because that the end-host has more memory and computing power and its relative ease of programmability.

Secondly, functions should be implemented as close as possible to the data where they need to access. For example, it will be more accurate and efficient to directly measure the queuing delay and link utilization of the particular nodes. By contrast, end-host approaches require collecting a massive amount of data distributed across the IoT network, thus incurring inaccurate and overheads. Another example is congestion control. As discussed above, Rate Control Protocol, which can directly access queue size data in the switch, presents a better performance compared to the end-host approaches (e.g., TCP congestion control).

Thirdly, functions that need to respond to network dynamics at very short timescales should be implemented inside the network. For example, in the data center network, the load balancing agency needs to respond to microbursts in a millisecond. End-host and centralized solutions are too slow to respond to rapidly varying traffic. These guidelines may help the network operators to govern where (end-host vs. network core) a network function should be implemented.

2.2 Summary

Recently, machine learning has attracted a large amount of attention from both academia and industry. Machine learning can automatically learn and optimize strategy directly from experience without following predefined rules. Therefore, it is promising to apply machine learning in network control and management to leverage the powerful machine learning adaptive abilities for higher network performance. The Internet of Things can make the network self-adjusting through the hybrid in-network intelligent control architecture of centralized learning and distributed learning. The centralized learning engine adapts to unstable network conditions and dynamically controls the network, making it break through the fixed network configuration. Self-driving networking is expected to boost the intelligent connectivity capabilities of the Internet of Things with explosive growth of devices and various applications, thereby improving service quality.

References

1. T. Mai, S. Garg, H. Yao, J. Nie, G. Kaddoum, Z. Xiong, In-network intelligence control: toward a self-driving networking architecture. IEEE Netw. 35(2), 53–59 (2021)
2. S. Guan, J. Wang, H. Yao, C. Jiang, Z. Han, Y. Ren, Colonel blotto games in network systems: models, strategies, and applications. IEEE Trans. Netw. Sci. Eng. 7(2), 637–649 (2019)
3. T. Mai, H. Yao, S. Guo, Y. Liu, In-network computing powered mobile edge: toward high performance industrial IoT. IEEE Netw. 35 (2021)
4. M. Alizadeh, T. Edsall, S. Dharmapurikar, R. Vaidyanathan, K. Chu, A. Fingerhut, V.T. Lam, F. Matus, R. Pan, N. Yadav, G. Varghese, CONGA: distributed congestion-aware load balancing for datacenters. ACM SIGCOMM Comput. Commun. Rev. 44, 503–514 (2014)

5. N.K. Sharma, A. Kaufmann, T. Anderson, C. Kim, A. Krishnamurthy, J. Nelson, S. Peter, Evaluating the power of flexible packet processing for network resource allocation, in *Proceedings of the 14th USENIX Conference on Networked Systems Design and Implementation*, ser. NSDI17 (2017), p. 6782

6. J.H. Saltzer, D.P. Reed, D.D. Clark, End-to-end arguments in system design. ACM Trans. Comput. Syst. **4**, 277288 (1984)

7. N. Katta, M. Hira, C. Kim, A. Sivaraman, J. Rexford, Hula: scalable load balancing using programmable data planes, in *Proceedings of the Symposium on SDN Research*, ser. SOSR16 (2016)

8. N.K. Sharma, M. Liu, K. Atreya, A. Krishnamurthy, Approximating fair queueing on reconfigurable switches, in *15th {USENIX} Symposium on Networked Systems Design and Implementation ({NSDI} 18)* (2018), pp. 1–16

9. A. Shieh, S. Kandula, A.G. Greenberg, C. Kim, B. Saha, Sharing the data center network. NSDI **11**, 23–23 (2011)

10. K. Nichols, V. Jacobson, Controlling queue delay. Commun. ACM **55**(7), 42–50 (2012)

11. V. Sivaraman, S. Narayana, O. Rottenstreich, S. Muthukrishnan, J. Rexford, Heavy hitter detection entirely in the data plane, in *Proceedings of the Symposium on SDN Research* (2017), pp. 164–176

12. M. Al-Fares, S. Radhakrishnan, B. Raghavan, N. Huang, A. Vahdat, Hedera: dynamic flow scheduling for data center networks, in *Proceedings of the 7th USENIX Symposium on Networked Systems Design and Implementation* (2010), p. 19

13. C.-Y. Hong, S. Kandula, R. Mahajan, M. Zhang, V. Gill, M. Nanduri, R. Wattenhofer, Achieving high utilization with software-driven wan, in *SIGCOMM '13: Proceedings of the ACM SIGCOMM 2013*, ser. SIGCOMM13 (2013), pp. 15–26

14. D. Wischik, C. Raiciu, A. Greenhalgh, M. Handley, Design, implementation and evaluation of congestion control for multipath TCP, in *8th USENIX Symposium on Networked Systems Design and Implementation* (2011), pp. 99–112

15. T. Mai, H. Yao, Z. Xiong, S. Guo, D.T. Niyato, Multi-agent actor-critic reinforcement learning based in network load balance, in *GLOBECOM 2020 2020 IEEE Global Communications Conference* (2020), pp. 1–6

16. R. Harrison, Q. Cai, A. Gupta, J. Rexford, Network-wide heavy hitter detection with commodity switches, in *Proceedings of the Symposium on SDN Research* (2018), pp. 1–7

17. C.W. Fox, S.J. Roberts, A tutorial on variational Bayesian inference. Artif. Intell. Rev. **38**(2), 85–95 (2012)

Chapter 3
Intelligent IoT Network Awareness

Abstract IoT devices are everywhere sensing, collecting, storing, and computing massive amounts of data. In the Internet of Things scenario, diversified services will generate traffic with different characteristics and put forward different business requirements. The application based on network intelligent awareness plays a key role in effectively managing network and deepening the control of network. In this chapter, we propose an end-to-end IoT traffic classification method relying on a deep learning aided capsule network for the sake of forming an efficient classification mechanism that integrates feature extraction, feature selection, and classification model. Then, we propose a hybrid IDS architecture and introduce a machine learning aided detection method. In addition, we model the time-series network traffic by the recurrent neural network (RNN). The attention mechanism is introduced for assisting network traffic classification in the form of the following two models: the attention aids long short term memory (LSTM) and the hierarchical attention network (HAN). Finally, we propose to design a machine learning-based in-network Distributed Denial of Service (DDoS) detection framework. Benefit from switch processing performance, the in-network mechanism could achieve high scalability and line speed performance.

Keywords Network awareness · Traffic classification · Encrypted traffic · Recurrent neural network · DDoS detection

IoT devices are everywhere sensing, collecting, storing and computing massive amounts of data. In the Internet of Things scenario, diversified services will generate traffic with different characteristics and put forward different business requirements. The application based on network intelligent awareness plays a key role in effectively managing network and deepening the control of network. In this chapter, we propose an end-to-end IoT traffic classification method relying on a deep learning aided capsule network for the sake of forming an efficient classification mechanism that integrates feature extraction, feature selection and classification model [1]. Then, we propose a hybrid IDS architecture and introduce a machine learning aided detection method [2]. In addition, we model the time-series network traffic by the recurrent neural network (RNN). The attention mechanism is introduced for assisting network traffic classification in the form of the following two models, the attention aids long short term memory (LSTM) as well as the hierarchical attention network (HAN) [3]. Finally, we propose we design a machine learning-based in-network Distributed Denial of Service (DDoS) detection framework [4]. Benefit from switch processing performance, the in-network mechanism could achieve high scalability and line speed performance.

3.1 Capsule Network Assisted IoT Traffic Classification Mechanism

In recent years, the Internet of Things (IoT) has witnessed its success in a range of compelling applications [5–9]. It is estimated that by 2020, the global market value of IoT may reach 7.1 trillion dollars [10]. Particularly, smart cities will be the future direction of cities, where IoT techniques and artificial intelligence (AI) algorithms play critical roles in prospering the construction of the smart cities [11]. With the gradual development of smart cities, more and more IoT devices are included [12–14], and the interactive data traffic also grows rapidly [15]. Hence, the analysis of this massive data is of great significance and has a large workload [16–19]. Fortunately, the method of using AI technique can be a beneficial solution for massive data analysis.

As a critical part of massive data analysis, traffic classification plays an important role in ensuring network security and defending traffic attacks. Moreover, the classification of different traffic can help IoT devices improve their work efficiency and quality of service (QoS). It is noted that the data traffic in smart cities is more complex in comparison to other IoT application scenarios. First, the scale of data traffic is much larger because there are more IoT devices and users. Second, more diverse data flows are transferred between devices since a variety of services are offered. Third, there are more sources of traffic attacks, and some uses the camouflage technique, which leads to more complex malware traffic [20]. All of these reasons account for the complexity of data traffic characteristics and the difference in fine details in smart cities, which impose a huge challenge on traffic classification in this scenario. Based on the characteristics of data traffic in smart cities, in this section, we propose an end-to-end traffic classification mechanism based on the capsule network. Our model is illustrated in Fig. 3.1.

Fig. 3.1 Two methods of data flow classification. (**a**) The conventional method. (**b**) The proposed method

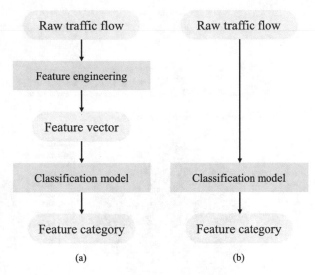

(a)   (b)

3.1.1   Methodology

3.1.1.1   Brief Introduction of Dataset

As mentioned before, the data traffic in smart cities has the following three characteristics: huge amount, various data types, and diverse aggressive data sources. Therefore, the dataset used in this work should have related characteristics to represent the traffic data in smart cities. Hence, we select the UTSC-2016 dataset for our experiments [21], which contains both the malware data and the benign data. More specifically, it includes total 10 kinds of malware data traffic, such as Cridex, Virut, etc., and 10 kinds of benign data traffic, i.e., BitTorrent, FTP, etc., which beneficially match the fine-grained characteristics of the traffic in smart cities. There are total 242,211 benign raw data flows and 179,252 malware raw data flows, respectively. The pie charts in Fig. 3.3 portray the proportion of the data flows of both benign dataset and the malware dataset.

In this work, the end-to-end traffic classification method is proposed. In order to verify the feasibility of our proposed model, we omit the step of data feature extraction and directly put the raw data into the model for further classification. However, data flows in smart cities are usually of different sizes, which may be in conflict with the requirement of our classification model having the uniform size of input data. Hence, the overall flow of the traffic classification is designed as shown in Fig. 3.2, where we first of all perform a simple unified preprocessing including both padding and segmentation operations based on raw data flows for the sake of making input data have a uniform size. Then, the processed input data flows are represented in the form of the two-dimensional matrix.

Suppose the raw data flows are represented as F_i, $i = 1, 2, 3, \ldots, I$, where I is the number of categories of flows. Each flow F_i contains M_i packets, denoted as

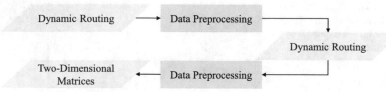

Fig. 3.2 The diagram of data processing

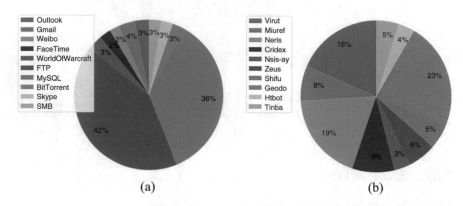

Fig. 3.3 Pie chart of the proportion of various flows in both benign (**a**) and malware (**b**) dataset, where benign dataset contains 242,211 raw data flows, while malware dataset contains 179,252 raw data flows

P_j, $j = 1, 2, 3, \ldots, J$. Each packet P_j consists of N_j bytes. Moreover, each byte is represented as a decimal number from 0 to 255 and is normalized into [0, 1]. For the purpose of unifying data size, we assume that each data flow F_i contains the same number of packets, i.e., M, and each packet P_j contains the same number of bytes, i.e., N. For the case where M_i is smaller than M or N_j is smaller than N, we pad the data with 0. For the case where M_i is greater than M or N_j is greater than N, we segment the data.

After the padding and segmentation operations, we transform the data flows into two-dimension matrices [24], which have the uniform size of $N \times M$. Suppose we choose the hyperparameters $M = 10$ and $N = 1000$, and then a flow can be visually represented as shown in Fig. 3.4.

3.1.1.2 Data Classification

Once these data flows are represented as two-dimensional matrixes, they act as the input of our proposed capsule network classification model, which will be elaborated in detail in Sect. 3. The output of the model is the predicted category of each input data flow. As is shown in Fig. 3.5, the entire traffic classification process can be executed automatically without any manual participation.

Fig. 3.4 The traffic flows and their classification in smart city scenario

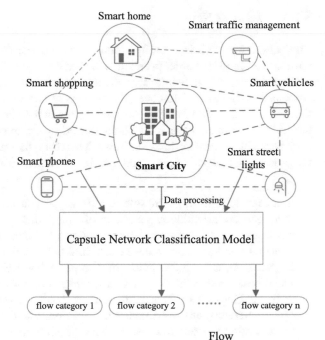

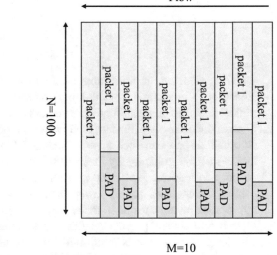

Fig. 3.5 The representation of a data flow in the form of two-dimensional matrix, where "PAD" means that the packet is pad with 0. The uniform size of processed flow is 1000×10

3.1.2   The Capsule Network Architecture

In this section, we will introduce the concept of the capsule network, which is the basic part of our classification model in Sect. 3. Furthermore, we will analyze the benefits of the capsule network in traffic classification in smart cities in Sect. 3.

3.1.2.1 Capsule Network

The capsule network is a family member of deep learning networks proposed by Hinton et al. in [22] in order to enhance the performance of CNN. As a deep learning model, the capsule network is applicable to process big data, which is exactly what smart cities need. Moreover, the concept of capsules and the dynamic routing algorithm are designed for efficiently processing of data flows. As mentioned before, the traffic data in smart cities is characterized by complexity and diversity. There may be slight differences between flows. Therefore, the capsule network is suitable for processing massive fine-grained data in the context of smart cities.

1. Capsule
   In a capsule network, a capsule is a group of neurons whose activation vector represents an instantiation parameter of a particular type of an entity. Moreover, the length of it represents the probability of occurrence of an entity, and its direction position denotes the generalized pose of the entity [22]. In this way, the vector-based representation of a capsule has richer information than the scalar-based representation of CNN, which provides a prerequisite for better distinguishing the difference of data in detail. In order to achieve this, the capsule network utilizes a nonlinear function, namely the squash function, as the activation function of capsules. The squash function can be expressed as

$$v_j = \frac{\|s_j\|^2}{1 + \|s_j\|^2} \times \frac{s_j}{\|s_j\|}, \tag{3.1}$$

   where s_j denotes the total input of a capsule in layer j, and v_j is its output. Equation (3.1) ensures that the length of short vectors can be compressed into almost zero, and the length of long vectors is slightly below 1.
2. Dynamic Routing Mechanism CNN relies on pooling algorithm to transfer data information between each layer, which is a bottom-up and passive mechanism lacking of the guidance by tasks [23]. By contrast, in a capsule network, the output of a capsule is a vector, which makes it possible to invoke a dynamic routing mechanism for ensuring that the output vector can be sent to an appropriate parent capsule in the higher layer. Initially, the output vector in a low layer is routed to all parent capsules and is scaled down by coupling coefficients. The low-layer capsule is multiplied by a weight matrix to obtain a prediction vector, if the prediction vector has a large scalar product with the output of the high-layer capsule. Then, the coupling coefficient of the low-layer capsule to the high-layer capsule is increased relying on a feedback mechanism, and the coupling coefficient of the other high-layer capsules is correspondingly reduced. Thus, the contribution of the low-layer capsule to the high-layer capsule can be increased by iterations. Hence, valuable information can be transmitted more effectively, while shotten information, by contrast, can be reduced accordingly. This mechanism is far more effective than traditional pooling algorithm, such as

max-pooling, which abandons all but the most active feature detector in a local pool in the low layer.

In the following, we focus on the dynamic routing mechanism. Let u_i represent the output of a capsule of layer i and W_{ij} be a weight matrix. The prediction vector $\hat{u}_{ij}$ from capsule i to capsule j can be calculated by

$$\hat{u}_{ij} = W_{ij} u_i. \tag{3.2}$$

The input s_j of capsule j is a weighted sum over all prediction vector $\hat{u}_{ij}$, and c_{ij} denotes the coupling coefficient between s_j and $\hat{u}_{ij}$, which is updated during the dynamic routing process. In our model, s_j can be given by

$$s_j = \sum_i c_{ij} \hat{u}_{ij}, \tag{3.3}$$

where c_{ij} is determined by a softmax function, i.e.,

$$c_{ij} = \frac{\exp(b_{ij})}{\sum_k \exp(b_{ik})}, \tag{3.4}$$

where b_{ij} represents the log prior probabilities coupling capsule i to capsule j. We use a_{ij} to indicate the agreement between prediction vector $\hat{u}_{ij}$ and current output v_j, which can be calculated by

$$a_{ij} - v_j \hat{u}_{ij}. \tag{3.5}$$

Hence, b_{ij} can be updated by

$$b_{ij} = b_{ij} + a_{ij}. \tag{3.6}$$

The pseudo-code of the dynamic routing algorithm is shown in Algorithm 3.1. The iteration of the dynamic routing algorithm refers to Fig. 3.6b.

3. Capsule Network Based Classification Model

Our capsule network based classification model includes five layers from bottom to top, i.e., one-dimensional CNN layer, convolutional capsule network layer, fully connected capsule network layer, long short term memory (LSTM) layer [25], and output layer. The data is calculated through a five-layer network and the result is finally obtained [26]. In the following, we will introduce them separately. The overall model flow is shown in Fig. 3.6a.

4. One-Dimensional CNN Layer

CNN has been widely used in the field of computer vision and has achieved superior results in image classification. In recent years, CNN has been applied to the field of natural language processing [27] with satisfactory results. The success of CNN results from the use of convolutional kernels for extracting local features of the input data. As the increase of the number of convolutional layers,

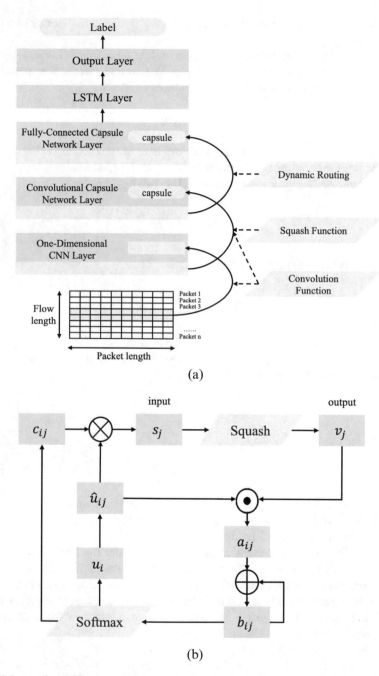

Fig. 3.6 (a) The architecture of the capsule network based classification model. (b) A detailed iterative process of dynamic routing, where ⊕ is an add operation, ⊙ is an inner product operation, and ⊗ is an element-wise product operation. Moreover, W is the weight matrix

---

Algorithm 3.1 Dynamic routing algorithm

Input: The output u_i of capsule i in layer l,
        Number of iteration t
Output: The output v_j of capsule j in layer $(l+1)$
initialize the logit of coupling coefficient $b_{ij} = 0$,
        $\hat{u}_{ij} = W_{ij} u_i$
for the t-th iteration **do**:
    for all capsule i of layer l **do**:
        $c_{ij} = \text{softmax}(b_{ij})$, where 'softmax' refers to (3.4);
    for all capsule j in layer $(l+1)$ **do**:
        $s_j = \sum_i c_{ij} \hat{u}_{ij}$;
        $v_j = \text{squash}(s_j)$, where 'squash' refers to (3.1);
    for capsule i of layer l and capsule j in layer $(l+1)$ **do**:
        $a_{ij} = v_j \hat{u}_{ij}$;
        $b_{ij} = b_{ij} + a_{ij}$;
return v_j

---

the extracted features are more advanced. Therefore, CNN is considered to be applicable to process the data with the following characteristics, i.e., there is a certain spatial relationship between the data, and the characteristics of the data do not change with its spatial rotation or distortion [28].

In the data processing, we represent the one-dimensional data flow by a two-dimensional matrix. The vertical direction denotes the number of packets in a data flow, while the length of each packet is represented by the horizontal direction. The sequence characteristics of the data packets in a data flow are reflected in the longitudinal direction of the two-dimensional matrix, and the spatial features in the data packets are reflected in the horizontal direction of the two-dimensional matrix. In this way, we are capable of comparing the data flow and of extracting features relying on CNN. We use a one-dimensional CNN as the first layer of our model for the sake of extracting features of input data.

Let x_i represent the input flow. One-dimensional CNN layer employs convolution operation on the input to extract features. The output of the one-dimensional CNN $Conv_i$ can be determined by

$$Conv_i = \sigma(W_{conv} \cdot x_i + b_{conv}), \qquad (3.7)$$

where W_{conv} is the filter of the CNN, while b_{conv} denotes its bias. Moreover, $\sigma()$ is a nonlinear activation function, such as the Relu.[1]

---

[1] https://en.wikipedia.org/wiki/Rectifier_(neural_networks).

Convolutional Capsule Network Layer

The input of the convolutional capsule network layer is the feature vector $Conv_i$, say the output from the CNN layer. In order to obtain high-order features, we perform a convolution operation to further extract features. Then, a squash function is used to compress them for constructing capsules. The capsule Cap_{conv} of the convolutional capsule network can be calculated by

$$Cap_{conv} = \rho(W_{cap} \cdot Conv_i + b_{cap}), \tag{3.8}$$

where W_{cap} is the filter of the convolutional capsule network, while b_{cap} is its bias. Moreover, $\rho()$ represents the combination of convolution function and the squash function mentioned in (3.1).

5. Fully Connected Capsule Network Layer

Each capsule in the fully connected capsule network layer is connected to the capsules in the convolutional capsule network layer via a weight matrix. The weight matrix is utilized to generate the prediction vector $\hat{u}_{ij}$ of an input. The input-to-output relationship is adjusted relying on the dynamic routing mechanism. The fully connected capsule network layer uses a dynamic routing mechanism instead of the traditional pooling operation, which enables an efficient layer-to-layer information delivery among capsules. The output of the fully connected capsule network Cap_{full} can be given by

$$Cap_{full} = \Phi(W_{full} \cdot Cap_{conv} + b_{full}), \tag{3.9}$$

where W_{full} is the weight matrix, while b_{full} is the bias. Furthermore, $\Phi()$ represents the dynamic routing algorithm mentioned in Sect. 2.

6. LSTM Layer

LSTM is an improved network structure for RNN. RNN is originally designed to process time-series sequence data, which performs well for short sequences. However, for long sequences, RNN may cause a poor performance due to the short memory. LSTM adds a gate structure based on RNN, which effectively solves the problem and achieves a beneficial effect on long sequences. Here, we employ the LSTM layer to connect the underlying fully connected capsule network layer, which helps increase the correlation between capsules. We put a list of capsules of the fully connected capsule network layer into the LSTM network layer. And the output of LSTM directly is the input of the next layer.

The LSTM layer is beneficial of accurately differentiating the traffic with nuances in smart cities. Suppose there are two flows belonging to different categories. Their differences are so small that the output of the fully connected capsule network layer has only one capsule with subtle differences. Even so, the LSTM layer can distinguish and widen the subtle differences and finally influence the output.

7. Output Layer

The last layer of the model in this article relies on a fully connected layer followed by a softmax function to obtain a prediction vector $\hat{y}_i$. The result of the LSTM output is input of the fully connected layer, and the predicted result is the final output. We then use the cross-entropy loss function for calculating the loss L of the result, which can be expressed as

$$L = -\sum_{i=1}^{n} y_i \log \hat{y}_i, \tag{3.10}$$

where y_i is the true label vector, and $\hat{y}_i$ is the prediction vector.

We employ TensorFlow[2] to build and train our five-layer network structure model. In the training step, processed traffic data is directly put into the model for training model parameters.

3.1.3 Experiments and Result Analysis

3.1.3.1 Experimental Environment

The environment of our experiments is as follows: Ubuntu 16.04OS, Python 2.7, TensorFlow 1.8.0, 4-core CPU, and 64G memory. The experimental dataset has been described in detail in Section 2.1.1.1. In terms of the dataset partition, we randomly extract 80% of each type of data as the training set, 10% as the validation set, and 10% as the test set. The batch size is 64, and the number of training epoch is 30.

3.1.3.2 Evaluation Metrics

In order to measure the performance of our proposed method, we use four evaluation indicators commonly used in classification problems, i.e., accuracy, precision, recall rate, and $F1$ value. We use "TP" to represent the result of viewing the positive case as a positive case, "FP" to represent the result of viewing a negative case as a positive case, "FN" to denote the result of identifying a positive case as a negative case, and "TN" to denote the result of identifying a negative case as a negative case. Hence, aforementioned four evaluation indicators can be calculated by

$$\text{accuracy} = \frac{TP + TN}{TP + FN + FP + TN}, \tag{3.11}$$

---

[2] https://www.tensorflow.org.

Fig. 3.7 The outline of two experiments

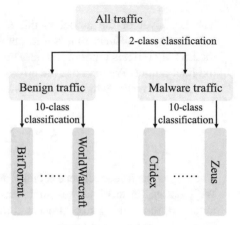

$$P = \frac{TP}{TP + FN}, R = \frac{TP}{TP + FP}, F1 = \frac{2PR}{P + R}. \qquad (3.12)$$

Here the accuracy represents how many samples are correctly predicted, while the precision P shows how many of the samples with positive predictions are correct. Moreover, the recall rate R indicates how many positive examples in the sample are predicted correctly, and $F1$ value is the harmonic mean of the precision and the recall.

3.1.3.3   Experimental Result Analysis

The main purpose of this work is to design an appropriate and reasonable model for classifying IoT data traffic. Specifically, we expect that our proposed model can achieve two goals, i.e., to ensure that the aggressive traffic and normal traffic are efficiently distinguished and to classify the data traffic with a high accuracy rate. Hence, we conduct our experiments relying on two scenarios. Firstly, we do a two-class experiment for classifying aggressive traffic and normal traffic based on the dataset containing malware traffic and benign traffic. Secondly, we conduct a multi-classification experiment on identifying each sub-category of 10-class aggressive traffic and 10-class normal traffic. Figure 3.7 outlines above-mentioned two experiments. In the following, we focus our attention on two parts of the experiment. The preliminary experiment is used to determine the best representation of the data flow and to determine two hyperparameters. By contrast, the main experiment is used to evaluate the effects of the proposed method. Our proposed model is then applied to classify the flow data based on aforementioned two scenarios in comparison to a pair of traditional deep learning models as well as a machine learning model.

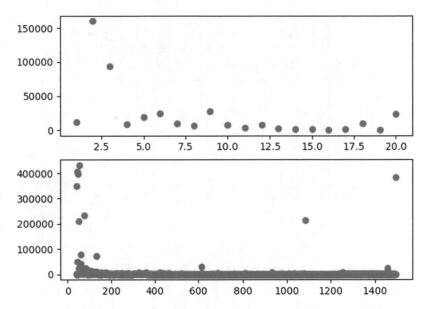

Fig. 3.8 The distribution of both the length of the traffic flows and the length of the packets. The figure above describes the relationship between the length of the data flow (the number of packets contained in a flow) and the number of flows. The figure below describes the relationship between the length of the packet (the number of bytes in a packet) and the number of packets

3.1.3.4 Preliminary Experiment

In this section, in order to unify the representation, raw flows are segmented or padded. It is necessary to determine the appropriate number of packets M of each flow and the number of bytes N in each data packet. In order to determine the two hyperparameters, we first perform a preliminary experiment and use the grid search method to determine the optimal hyperparameter value. In the first place, we perform statistical analysis of the raw traffic distribution in the dataset, including the distribution of the number of packets in each data flow and the distribution of the number of bytes in each packet, which are shown in Fig. 3.8. From the distribution of the traffic in Fig. 3.8, we can find that the number of packets contained in each flow is less than 20, and the majority of them are concentrated within 10, while most of them are concentrated within 5. Hence, as for hyperparameter M, we take $M = 5$, $M = 10$ and $M = 20$ for the grid search. As to the length of the packets, i.e., N, it can be found that all of them are within 1500, and the majority of them are within 1100, while most of them are less than 700. Hence, we select $N = 700$, $N = 1100$, and $N = 1500$ for the grid search. After executing the grid search, our experimental results are shown in Table 3.1. We demonstrate the average accuracy rate of the grid search for each type of the traffic in Fig. 3.9a. It can be seen that when the hyperparameters $M = 20$ and $N = 1100$, we achieve the highest accuracy rate both in the malware dataset and in the whole dataset and obtain a relatively high

Table 3.1 The length of different flows and packets

Class	20,1500	20,1100	20,700	10,1500	10,1100	10,700	5,1500	5,1100	5,700
BitTorrent	99.68	100	100	100	100	99.21	99.68	99.57	99.68
FaceTime	100	100	100	100	100	100	100	100	100
FTP	100	100	99.98	99.98	99.98	99.98	100	99.98	100
Gmail	99.53	99.55	99.81	96.98	99.81	99.81	99.67	99.75	99.81
MySQL	100	100	100	100	100	100	99.97	99.97	99.99
Outlook	98.08	99.2	100	97.26	100	98.68	98.06	97.86	99.03
Skype	100	100	100	100	100	100	100	100	100
SMB	98.38	98.37	98.38	98.81	98.23	99.34	97.64	97.57	97.98
Weibo	97.44	97.41	97.41	98.17	97.39	99.15	96.06	96.38	97.54
WorldWarcraft	99.81	100	99.81	99.8	99.8	99.81	99.81	99.81	99.86
Cridex	99.97	100	99.89	100	100	99.93	99.59	99.89	99.97
Geodo	99.69	98.91	99.76	99.75	99.84	99.64	99.61	99.74	99.72
Htbot	98.11	90.91	97.04	98.05	98.55	95.9	97.94	96.77	98.76
Miuref	99.82	100	99.73	99.77	99.68	99.82	99.96	99.61	99.41
Neris	86.93	91.49	88.55	86.15	88.18	87.79	87.94	87.12	87.83
Nsis-ay	97.38	100	97.29	96.52	95.44	97.41	96.36	96.6	96.84
Shifu	99.45	97.98	99.32	99.61	99.77	99.62	99.18	99.67	99.59
Tinba	99.76	100	99.76	99.76	99.7	99.76	99.76	99.76	99.59
Virut	88.17	93.27	89.53	88.03	89.51	89.19	88.89	87.95	88.7
Zeus	99.43	100	99.39	99.29	100	99.94	99.96	99.43	99.94

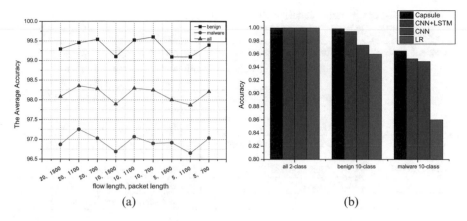

Fig. 3.9 The accuracy analysis. (**a**) The average accuracy. (**b**) The accuracy

accuracy rate in the benign dataset. Therefore, in the main experiment, we select the hyperparameters with $M = 20$ and $N = 1100$.

3.1.3.5 Main Experiment

In the experiment, we use the capsule network model proposed in Sect. 3 to classify the traffic of data flows. In addition, we use a pair of traditional deep learning models, i.e., the CNN and the CNN combined with LSTM (CNN-LSTM), as well as the logistic regression model for comparative experiments. For the data representation, we simply segment and pad original data flows and then represent each data flow as a two-dimensional matrix. The length of the data flow and the length of the packet are selected as $M = 20$ and $N = 1100$.

In order to better evaluate the performance of the proposed method, we conduct experiments in two different scenarios, i.e., a two-class experiment on the whole dataset and two 10-classification experiments on malware traffic dataset and benign traffic dataset, respectively. Furthermore, we use the accuracy, precision, recall, and $F1$ values as evaluation indicators of our experiments.

In the following, we provide the hyperparameters commonly used in our proposed model, which is shown in Table 3.2. Specifically, the number of filters used by the one-dimensional CNN layer of the capsule network is 64. Moreover, the size of the kernel is 3, and the size of stride is 1. By contrast, in the convolutional capsule network layer, the size of the above three parameters is 128, 2, 1. And the dimension of a capsule vector is 8. The number of output capsules is 16. In the fully connected capsule network layer, the dimension of the capsule vector is also 8, and the number of output capsules is determined by the final number of classifications (2 or 10). In LSTM layer, the number of hidden layer units is 32 and the size of forget bias is 1.0. For the CNN model, we used 64 filters, and kernel sizes are 3, 4, and 5, respectively. Moreover, the size of stride is 1. A dropout with a ratio of 0.5 is utilized to optimize

Table 3.2 The configuration details of classification models

Model	Layer	Configuration details
Capsule network	One-dimensional CNN layer	Filters=64, kernel=3, stride=1
	Convolutional capsule network layer	Filters=128, kernel=2, stride=1, capsule vector=8, capsule=16
	Fully connected capsule network layer	Capsule vector=8, capsule=number of classes (2 or 10)
	LSTM layer	Hidden units=32, forget bias=1.0
CNN	One-dimensional CNN layer	Filters=64, kernel=3,4,5, stride=1
CNN LSTM	One-dimensional CNN layer	Filters=64, kernel=3,4,5, stride=1
	LSTM layer	Hidden units=32, forget bias=1.0
LR	—	l2 regularization coefficient=1.0

the model. Learning rate is set as 0.1. For the CNN+LSTM model, a layer of LSTM was added to the CNN model, where the parameters are consistent with those used in the capsule network. For the LR model, we use a l2 penalty with a regularization coefficient of 1.0. Other non-numeric parameters employ the default values of the LogisticRegression interface provided in the sklearn[3] library.

The average accuracy of each type of flow is shown in Fig. 3.9b. We can conclude that in terms of identifying the malware traffic and benign traffic, all of four models have achieved 100% accuracy, that is, for the two-class classification task, our proposed model, two deep learning models, and the machine learning model can accurately complete the task. Moreover, as for the multi-classification experiment, the accuracy rate of classifying benign traffic is higher than that of classifying malware data because the benign data is the normal flow dataset, where data characteristics of each flow are generally similar and follow standard protocols. By contrast, the malware data flow varies substantially and often does not have the consistency, and hence it is difficult for identifying. Furthermore, we can see that the capsule network based model proposed in this work has the highest accuracy on all three datasets. This is because that the capsule network uses the dynamic routing mechanism, which is more suitable for classifying traffic with subtle difference thereby getting higher accuracy. The results indicate that the capsule network based model has a beneficial performance on traffic classification and the end-to-end method of classification architecture is feasible.

In order to better evaluate the model and demonstrate the experiment results, the precision, recall, and $F1$ value of the two 10-class experiments are shown in Figs. 3.10 and 3.11. In Fig. 3.10, we analyze the experiment results in classifying benign traffic. We can find that the capsule network has a high score for all 10 kinds of traffic in terms of three evaluation indicators, especially in some traffics, such as Weibo, SMB, and Outlook. The score of the other three models is around 0.8 in

---

[3] https://scikit-learn.org/stable/.

Fig. 3.10 The $F1$ score, recall, and precision of each class in benign traffic. (**a**) The $F1$ score of benign traffic. (**b**) The recall score of benign traffic. (**c**) The precision score of benign traffic

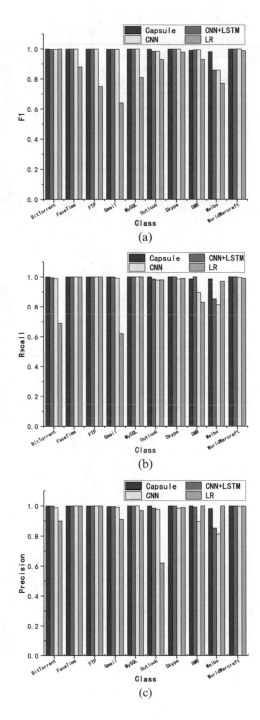

Fig. 3.11 The $F1$ score, recall, and precision of each class in malware traffic. (**a**) The $F1$ score of malware traffic. (**b**) The recall score of malware traffic. (**c**) The precision score of malware traffic

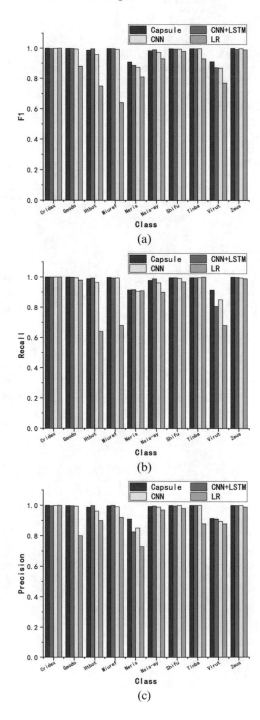

some contexts, but the capsule network always has a score close to 1. Hence, the capsule network not only has a high classification accuracy rate but also has a very stable classification effect.

We conclude that some classification effect is relatively poor because some traffics, such as Weibo and SMB, tend to vary greatly in the size of packets and flows. Traditional models do not handle this situation well, but the capsule network using the dynamic routing mechanism can give higher weight to the more important parts of the packets and flows, and then it is less affected by size changes, so it can classify traffic more accurately.

As shown in Fig. 3.11, we can conclude that all of the scores of four models in malware traffic are lower than those in benign traffic. The reason is that the malware traffic data flow is more complicated compared with other traffics. However, the capsule network still scores higher than the other three models. We can see that the capsule network can reach the score of 0.9 on three evaluation indicators for the data flow of Virut, Neris, etc., which is significantly higher than the other three models. The capsule network relies on the dynamic routing mechanism for extracting the detailed features, which can accurately and efficiently process the nuanced data.

In addition, we can see that the linear classifier LR does not perform well on both datasets, which indicates that most of the traffics in the dataset are nonlinearly separated. While the capsule network has high scores in terms of three evaluation indicators, indicating that the capsule network has strong classification ability for data.

In a nutshell, our proposed capsule network based model as well as the end-to-end classification method can achieve a superior classification accuracy, the recall rate, and the $F1$ value in comparison to the CNN model, the CNN-LSTM model, and the LR model.

3.2 Hybrid Intrusion Detection System Relying on Machine Learning

Recently, with the concept of Industry 4.0, the industrial Internet of Things (IIoT) has been widely developed. It is estimated that the value created by the IIoT will exceed 12 trillion dollars by 2030. As a family member of the Internet of Things (IoT), IIoT has a range of compelling applications, namely smart factories, smart grids, etc. Edge-based IIoT, by definition, relies on a large amount of edge devices for the sake of sensing, computing, and storing, of which the scenario is shown in Fig. 3.12. However, decentralized data interaction makes it easy to result in data leakage, manipulation, and other network attacks. Hence the security issue of edge-based IIoT has become the focus of our attention.

For the sake of ensuring the security of IIoT, some researches have paid their attention to providing data consistency, identity authentication, etc. However, these methods were not able to deal with attacks resulting from malicious traffic, such as

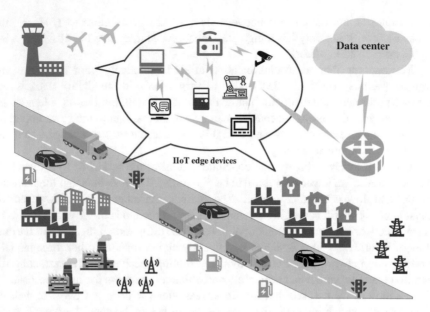

Fig. 3.12 Edge-based IIoT scenario

the deny of service (DoS), root to local (R2L), etc. Therefore, intrusion detection becomes a fundamental task for ensuring network security. An intrusion detection system (IDS) is designed for detecting detrimental intrusions. Generally, an IDS is used for identifying the traffic in the network, especially for distinguishing normal and malicious traffic, and hence is beneficial to eliminating malicious traffic. As for the IDS at edge-based IIoT, considering its limited storage and computational power of each small edge device, resource utilization efficiency should be considered when designing detection methods as well as the system architecture.

In edge-based IIoT, the design of IDS mainly considers the following two aspects: namely, the detection method and the system architecture. To elaborate a little further, detection methods can be classified into knowledge-based methods and anomaly-based methods, where knowledge-based methods are based on off-the-shelf databases for detection which cannot identify unknown attacks, while anomaly-based methods typically invoke machine learning algorithms for identifying traffic. As for the system architecture, the architecture of IDSs can be classified into three categories, i.e., the distributed architecture, the centralized architecture, and the hybrid architecture. Specifically, the distributed IDS operates on each physical device in the network and conducts the intrusion detection independently. By contrast, the centralized IDS runs on several central devices, such as central servers [29]. Only these so-called central devices are capable of detecting malicious traffic in the network. The hybrid IDS combines the advantage of the above-mentioned two architectures, where different detection methods are operated among the lower layer network devices and the upper layer network devices, respectively.

However, there are still some challenges imposed for the edge-based IIoT IDS. In terms of the detection method, for the sake of reducing the computational complexity, some traditional machine learning algorithms have to decrease the detection accuracy. Moreover, the deep learning based intrusion detection algorithms can improve the accuracy, while they also result in high power consumption, which is a key limitation of the edge-based IIoT devices. As for the system architecture, distributed architectures require a large amount of network resources and also reduce the overall monitoring capability of the whole network. Centralized architectures often generate amounts of request-type information between edge devices and central devices, increasing bandwidth resource consumption. Centralized architectures cannot detect intrusion or malicious traffic between edge devices. Therefore, in this work, we propose a new hybrid IDS architecture for edge-based IIoT, where a new machine learning algorithm and a deep learning algorithm are employed in the lower layer network and the upper layer network, respectively. In comparison to existing IDS architectures and intrusion detection methods, the main contributions of our proposed architecture and detection algorithm are summarized as follows. A new machine learning based intrusion detection algorithm is applied to the lower layer network, where the detection accuracy can be substantially improved without increasing the training time. Deep learning algorithm applied to the resource-rich upper layer network, yielding a higher detection accuracy with the aid of its powerful learning capability. The hybrid architecture combines the advantage of both machine learning algorithms and deep learning algorithms for improving the security of the network. Moreover, the hierarchical information interaction and resource allocation lower the current limit of the network's bandwidth and energy.

3.2.1 Traditional Machine Learning Aided Detection Methods

In this section, we will introduce anomaly-based detection methods relying on traditional machine learning algorithms and analyze their pros and cons. On basis of that, we will propose our method for detecting intrusion. Table 3.3 shows some typical machine learning based intrusion detection methods and their contributions.

In the IIoT, the methods of intrusion detection are mainly divided into knowledge-based and anomaly-based methods. Knowledge-based methods cannot detect unknown traffic because of utilization of existed databases for detection, which does not meet the needs of IIoT to detect multiple types of intrusion. The anomaly-based method can solve the problems faced by the knowledge-based method. It detects intrusion by learning the general features of intrusive traffic and does not stick to exited traffic databases [30]. Once the learning of intrusive traffic features is completed, the known and unknown intrusion traffic can be effectively detected based on anomaly-based method.

The employment of machine learning algorithms for anomaly-based methods is an important direction. Machine learning algorithms learn the general features of the training traffic data. Based on the learned features, the input traffic data will

Table 3.3 Machine learning and deep learning based intrusion detection methods

	Reference	Algorithms	Contributions
Machine learning	[1]	Naive Bayes	Introduce the feature reduction to intrusion detection
	[2]	Logistic regression Naive Bayes	Develop an anomaly-based IDS relying on feature engineering
	[3]	Random forest	Propose a hybrid intrusion detection method
Deep learning	[4]	CNN, neural network	Propose a novel android malware detection system using CNN
	[5]	CNN, LSTM	Develop two models employing CNN and LSTM for intrusion detection
	[6]	CNN, RNN LSTM, GRU	Apply CNN and RNN and its variants for intrusion detection

be detected correctly. This generalized learning mode allows the machine learning algorithms to process data that has never appeared before. At the same time, the traditional machine learning algorithms do not require high computational power of the hardware, and the training time is relatively short, which is more suitable for the requirements of the IIoT edge devices.

A range of researches have introduced machine learning algorithms for intrusion detection. Specifically, the authors of [31] employed Naive Bayes (NB) algorithm to detect intrusion. First, four different feature reduction methods were utilized to reduce the feature dimension of the original data feature. Then the dimensionality-reduction data feature, as the representative of the flow, was input to the NB for intrusion detection. Experimental results showed that this method had relatively high recognition accuracy. The Logistic Regression (LR) algorithm was adopted by Subba et al. [32]. They first utilized an evaluation tool to map data feature in the original dataset and selected the top 23 features to generate a new dataset. Then several machine learning algorithms were performed on the dataset for a two-class experiment. Experimental results showed that LR had a higher accuracy. Zhang et al. [33] designed an intrusion detection system that integrated online phases and offline phases. The offline phase employed Random Forest (RF) algorithm. The input data was first preprocessed, and the most important data features were selected as feature vectors, which were used as input to the RF to perform intrusion detection task.

Although these studies combined with machine learning algorithms had achieved good experimental results, improvement still exits for intrusion detection tasks in the edge-based IIoT. First, the high experimental accuracy of these studies is based on a combination of feature engineering including feature reduction, feature mapping, etc. and classification algorithms. However, feature engineering is a time-consuming task, which is not a good choice for time-critical edge-based IIoT and edge devices with limited computing resources. In addition, the detection accuracy of machine learning algorithms employed in the studies still has room for improvement.

In view of the above two problems, this work utilizes a new machine learning algorithm LightGBM[4] for intrusion detection of original data features and improves the detection accuracy on the basis of ensuring no increase in time consumption, which can be viewed as a fast, distributed, high performance decision-tree-based gradient lifting framework. Based on the decision tree algorithm, LightGBM employs the optimal leaf-wise strategy to split the leaf nodes, while other lifting algorithms split trees generally relying on depth or horizontal-wise rather than leaf-wise. The leaf-wise algorithms reduce more losses than the horizontal-wise algorithm. This helps LightGBM get higher accuracy in varieties of tasks, which no other existing lifting algorithms can reach. As far as we know, this is the first time that LightGBM has been applied to detect intrusion, which implements tree boosting in parallel computing, which can be seen as a final classification model formed by a combination of multiple tree classification models. In this work, LightGBM is used for intrusion detection, which can reduce both the training time of model and the cost of communication, which is beneficial of improving parallel computing efficiency. Compared to traditional machine learning algorithms, LightGBM has higher computational efficiency, supports distributed computing, and encapsulates feature selection algorithms in it, which means it does not require additional feature engineering. To demonstrate the performance of the LightGBM, we conduct comparison experiments relying on a dataset using several traditional machine algorithms. The experimental content is to perform a two-class classification experiment for intrusion detection of network traffic data. Four evaluation indicators, i.e., accuracy, precision, recall, and F1, are utilized for evaluating the performance. The experimental results are shown in Fig. 3.13. As we can see, LightGBM has a higher detection accuracy than other traditional machine learning algorithms. LightGBM's low-time-consumption and high-detection-accuracy features meet the requirements for fast-traffic-detection and high-quality-security in edge-based IIoT. In addition, LightGBM can generate the sorting results of the importance of features without increasing computation consumption, which contributes to further work mentioned later.

3.2.2 Deep Learning Aided Detection Methods

In this section, we will introduce the researches of intrusion detection using deep learning algorithms and analyze the feasibility of deep learning algorithms applied in edge-based IIoT. Table 3.3 summarizes the related typical researches.

The application of deep learning algorithms is also an anomaly-based intrusion detection method. As like machine learning algorithms, deep learning algorithms actively learn the general features of input data, which help to learn model parameters. Based on this, a task model can be built. Once the model parameters

---

[4] https://github.com/Microsoft/LightGBM.

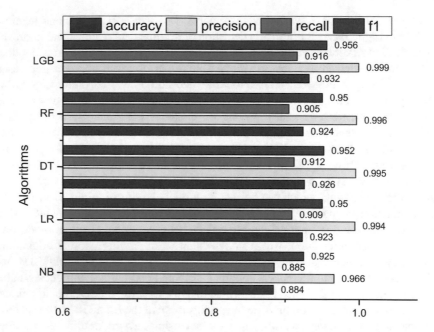

Fig. 3.13 The performance comparison of five different machine learning based algorithms (NB: Naive Bayes, LR: Logistic Regression, DT: Decision Tree, RF: Random Forest, and LGB: LightGBM)

are learned, the model is not dependent on the input data, so unknown data can be processed. Compared with traditional machine learning algorithms, the deep learning model has better effects in scenarios with larger data volume. However, deep learning models are generally more complicated, resulting in longer training time and higher requirement of computing power of hardware.

In existing studies, different deep learning algorithms are employed, such as convolutional neural network (CNN), recurrent neural network (RNN), etc. Mclaughlin et al. [34] utilized a combination of CNN and fully connected neural networks to identify malicious traffic. The fully connected neural network is actually a traditional neural network, and the CNN plays the role of feature extraction in the overall architecture. The scenario targeted by the paper was malicious traffic detection in the Android system. As a lightweight operating system, the Android system is often employed on mobile devices and IIoT devices. Malicious traffic identification in the Android system is also part of the intrusion detection task in the IIoT scenario. During the experiment, Mclaughlin et al. processed the data traffic as text and then trained the CNN model on the GPU, which should be noted that the GPU is applied to train the model in order to reduce training time. Wang et al. [35] designed two models for intrusion detection. One used one-dimensional CNN to process the input data traffic to obtain feature vectors, which were utilized as the basis for classification. The other employed CNN combined with long short term memory (LSTM) network for detection. Two datasets were used to experimentally verify the proposed models. Experimental results were analyzed combined the characteristics of the datasets. Final experiment used three indicators to evaluate the test results, which showed that the proposed models had a good detection effect. Chawla et al. [36] employed the different combinations of CNN, RNN, and its variants as detection models. Experimental results showed that these combinations can achieve satisfactory detection results.

Beneficial experimental results were achieved in the above-mentioned researches, which is based on the ability of deep learning algorithms for high-accuracy classification in big data scenarios. However, as far as we mentioned earlier, deep learning algorithms also have some problems, such as higher requirements for hardware devices and long training time for models. These issues cannot be overlooked in edge based IIoT. Mclaughlin et al. [34] applied GPU to train the detection model to reduce training time. However, in the actual edge-based IIoT, most of the edge devices are equipped with CPUs rather than GPUs. Training a deep learning model on CPUs is much slower than on GPUs. Therefore, if the deep learning model is placed on the IIoT edge device equipped only with CPUs, the time spent on the overall intrusion detection task in the network will be greatly increased, which is not acceptable for edge-based IIoT. Wang et al. [35] emphasized that the proposed models have high detection accuracy but did not mention the time problem required for model training and testing. Chawla et al. [36] demonstrated the training and testing time of the model in the experimental analysis section. On the training set used by the authors, the model had a minimum training time of more than 300 s. Moreover, the training of these models was performed on the GPU, and if the same process is performed on the CPU, it will cost more time.

In order to demonstrate the training time and detecting accuracy of the deep learning algorithm more clearly, we conduct a comparative experiment using the LightGBM and CNN, which is performed on the CPU, and two datasets are used. Experimental results are showed in Fig. 3.14. In the pair of experiments, we can draw three consistent and obvious experimental conclusions. First, the training time of LightGBM is significantly less than that of CNN. Specifically, the time for training the CNN is several times longer than that of the training LightGBM. Second, it takes longer for the CNN to achieve the same classification accuracy against LightGBM. Third, with sufficient training time, CNN can achieve higher classification accuracy than that of LightGBM. By analyzing these conclusions, we can see that both CNN and LightGBM have their own strengths and weaknesses. To elaborate a little further, CNN has a higher detection accuracy but needs longer training time. LightGBM's training time is smaller than that of CNN, while its detection accuracy is lower. These facts have inspired us to consider how to combine the advantages of the two algorithms in edge-based IIoT so that it can get better security, which we discuss later.

In addition, Xiao et al. [37] proposed to use Reinforcement Learning (RL) to solve the DoS attack problem in edge networks. Moreover, Min et al. [38] employed RL to solve related problems in the IoT scenario. This suggests that our future work may consider applying RL to intrusion detection problems.

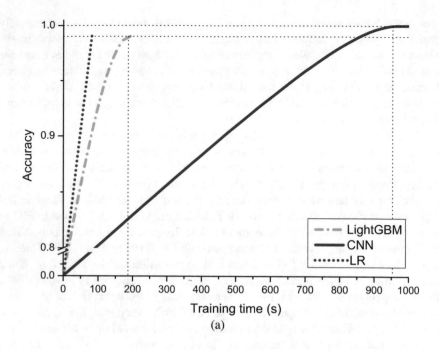

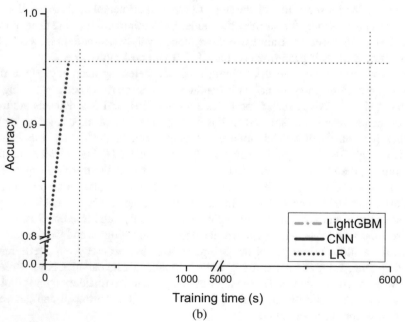

Fig. 3.14 (**a**) The accuracy vs. training time of LightGBM, CNN, and LR relying on dataset 1. (**b**) The accuracy vs. training time of LightGBM, CNN, and LR relying on dataset 2

3.2.3 The Hybrid IDS Architecture

In this section, we will review some of the existing researches on IDS architecture. On this basis, for the edge-IIoT scenario, we propose an IDS architecture and analyze its advantages over other architectures. Table 3.4 summaries the investigated effort to design IDS architectures.

The existing IDS architectures are mainly divided into distributed architecture, centralized architecture, and hybrid architecture. The distributed architecture places the IDS on each edge device in the network, and each edge device independently detects inbound and outbound traffic. For the problem of limited computing power and memory in the IoT scenario, Oh et al. [39] designed a lightweight distributed detection system using a novel malicious pattern-matching engine, which matched the signature and packet payload of malicious traffic. In order to make the proposed system run on devices with limited computing power, the memory footprint was limited. By comparing with one of the fastest pattern-matching algorithms, better performance was revealed in the proposed method. Wallgren et al. [40] chose to place IDSs in the IoT edge routers for unified management and control. A heartbeat protocol is proposed, in which the routers regularly sent detection requests to all edge devices, and the edge devices sent the information to the router for detection after receiving the request. Although the authors stated that the protocol increases no memory burden of the edge devices, this approach increased the additional traffic in the network. In a more recent study, the IDS proposed in [41] was a kind hybrid architecture. The authors divided the network into small clusters. Each cluster consisted of a small number of common nodes and a cluster head. The cluster head was responsible for collecting information about the self and neighbor nodes detected by the common nodes. On this basis, the intrusion detection task was completed by the cluster head utilizing a lightweight detection method.

In summary, the three architectures of IDS have pros and cons. The distributed architecture can quickly detect traffic passing through the node itself and does not generate many interactive information. However, distributed architectures need to balance detection accuracy and node resources, such as processing power, memory, etc. Especially for the problem of limited edge device resources in edge-based IIoT, it is more difficult to design an appropriate one. The centralized architecture

Table 3.4 The characteristics of three kinds of IDS architectures

Architecture	Reference	Pros and Cons
Distributed architecture	[10]	Decrease the communication overhead, while requiring more resources of edge devices
Centralized architecture	[11]	Be able to apply algorithms with higher detection accuracy, while generating more additional information
Hybrid architecture	[12]	Adapt the resources of different devices, while it is difficult to design in the context of IIoT

places the IDS on a central device, such as border routers, servers, etc. with strong computing power and sufficient memory, which enables IDS to perform intrusion detection tasks with a more complex algorithm. However, the centralized architecture requires additional communication between central devices and edge devices, generating amounts of additional information in the network that consumes bandwidth resources. Moreover, centralized architecture cannot detect intrusion between edge devices, which is a security risk. The hybrid architecture combines the advantages of the two architectures. But in edge-based IIoT, the design of the IDS of the hybrid architecture also faces many problems, such as how to select the central devices and edge devices to place IDS using different detection methods, how to improve the communication efficiency between these devices to reduce the additional information in the network, and so on.

Moreover, [42–45] mentioned hybrid architectures in different network application scenarios. Inspired by these works and combined with the above analysis, we propose a hybrid architecture for IDS in edge-based IIoT, as shown in Fig. 3.15. The overall IIoT scenario can be divided into central network part and edge part. The IDS architecture we will propose of which the target object is the edge-based IIoT. First, we further divide the edge-based IIoT. Devices with strong computing power and sufficient resources such as edge routers are regarded as the master nodes, while the industrial equipment of the edge part is regarded as edge nodes. Due to the limited computing power and resources of edge nodes, we apply the lightweight LightGBM algorithm on them and perform the first intrusion detection task at the edge nodes to ensure the security. At the same time, we utilize the LightGBM algorithm to extract the more advanced features as the representative of the traffic without increasing the resource consumption, so as to perform further detection on the master nodes. On master nodes, we employ deep learning algorithm with higher accuracy to perform the second intrusion detection task, further improving the detection accuracy of the overall network. Master nodes receive the advanced traffic features delivered by the edge nodes instead of the original data traffic for detection, in order to reduce the additional information in the network. We believe that master nodes have sufficient resources to meet the needs of the deep learning model to efficiently complete the intrusion detection task.

To illustrate the advantages of the proposed architecture, we compare it to distributed and centralized architecture. The distributed architecture places IDS on each device in the network. Due to the constrain of resource-constrained devices, the intrusion detection method used by IDS is usually traditional machine learning algorithms with less resource requirements. In Fig. 3.13, we compare the performance of traditional machine learning methods with LightGBM. Therefore, the architecture proposed in this work utilized LightGBM algorithm to have higher detection accuracy than distributed architecture combined with machine learning algorithms. The centralized architecture places the IDSs on master nodes, usually employing more complex deep learning algorithms. But the centralized architecture cannot detect intrusion between edge nodes, which can be naturally solved by hybrid architecture. The architecture proposed in this work utilizes LightGBM and deep learning algorithm to detect the lower level and upper level network respectively,

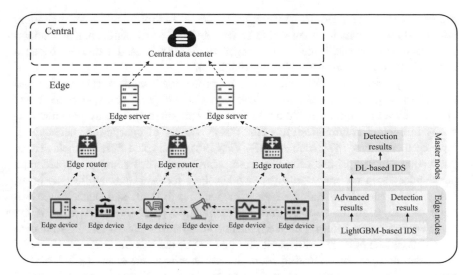

Fig. 3.15 The proposed hybrid IDS architecture

making full use of the advantages of the two algorithms, as shown in Fig. 3.14. The overall detection accuracy of the network can be guaranteed without increasing the training time. As a hybrid architecture, it solves the problem that centralized architectures cannot detect intrusion between edge nodes. In addition, the proposed architecture utilized LightGBM algorithm to select the advanced traffic features to be transmitted in the network, which effectively reduces the bandwidth burden. The amount of bandwidth reduction depends on both the proportion of actual network intrusive traffic and the proportion of the advanced features selected. For example, suppose there are 100 units of traffic, of which normal and intrusive traffic have 80 units and 20 units, respectively. The edge node correctly detects 10 units of intrusion traffic. We select advanced features that account for 10% of the original data, and then only 9 units of advanced features need to be sent to the master nodes for further detection. Compared to directly sending raw data to the upper layer, our proposed method reduces bandwidth consumption by 90%.

3.3   Identification of Encrypted Traffic Through Attention Mechanism Based LSTM

Network traffic classification plays a critical role in next-generation communication networks, which aims at classifying network traffic based on both the type of protocols, such as HTTP, FTP, and the type of applications, such as Facebook, Skype, etc. Meanwhile, network traffic classification is beneficial in terms of understanding the distribution of traffic flows, improving the utilization efficiency of network resources, enhancing quality of service (QoS), and guaranteeing network security [46]. In addition, as network traffic becomes larger and larger, operators tend to adopt big data tools for stable storage and fast processing, such as Hadoop and Spark. Besides, distributed computing methods are also used in this task because of the high requirements for real-time calculation.

Recently, with the extensive demands for protecting both data transmission and user privacy, protocols and applications prefer to adopting encryption methods. Under such a circumstance, the amount of the encrypted traffic has grown extensively in current communication networks. A variety of encryption mechanisms have been employed [47], such as SSH, VPN, SSL, encrypted P2P, VoIP etc. These encryption algorithms are different from each other, since some encrypted data packets locate in the transport layer, while others locate in the application layer, which make the classification of encrypted traffic difficult and hence impose a huge challenge on the network traffic classification [48]. Moreover, even if the same encryption algorithm is utilized, the encrypted traffic can exhibit different data distributions because of the different distributions of the original traffic.

Traditional port-based traffic identification methods have unsatisfied performance on the classification of encrypted traffic since they utilized the official standards defined by the Internet assigned numbers authority (IANA) to identify the type of applications. However, some protocols did not follow those standards, such as P2P protocol utilized the random port, and HTTP protocol used the 80 port for disguising. Moreover, some deep packet inspection (DPI) based methods conducted the traffic classification by regular expression for matching the payload data, while the payload of the encrypted packet was changed relying on the encryption algorithm. Consequently, the DPI based methods could only identify those coarse-grained protocols such as SSL but completely failed to identify the encrypted traffic. In addition, most of the machine learning aided traffic classification methods [49–51] were based on manual extraction of data packets or their statistical features at the data flow level to train the classifier, such as the duration of a flow, the total number of packets, the length of a packet, the number of the bytes contained in a flow, as well as the interval of the packet arrival. Since these characteristics require prior knowledge and experience, the extraction of them would also be time-consuming. More importantly, it cannot be guaranteed that these features are really helpful in improving the performance of classification. Therefore, to the best of our knowledge, all of these aforementioned methods cannot achieve preferable results on the encrypted traffic classification problem.

Recently, deep learning has rapidly developed and has witnessed its great success in a variety of areas, such as computer vision (CV), speech recognition, natural language processing (NLP), etc. Meanwhile, deep learning methods have been widely used in the scenario of communication networks. Network traffic classification can be regarded as a common classification problem in the field of machine learning, and some papers [52–56] have proposed the application of deep learning methods in solving such a problem. Nevertheless, most of them utilized CNN to extract the features from the traffic flow without considering the timing features among different packets. In this work, we propose to utilize LSTM for network traffic classification tasks, which cannot only omit the complex feature engineering but also automatically learn the temporal relationship between traffic flows. The specific contributions of this work can be summarized as follows:

- We treat the network traffic flow as time series and analyze it as text data with the aid of LSTM model. Experimental results yield the best representation of the network traffic, where each flow contains 10 packets and each packet contains 1500 bytes.
- Two models are proposed for encrypted traffic classification, i.e., attention based LSTM and hierarchical attention network (HAN). The attention based LSTM focuses more on important data packets in the traffic flow, while HAN is capable of distinguishing the role of different bytes in each packet during the process of classification.
- Simulation results demonstrate that the classification accuracy rate of our proposed model can achieve 91.2%, which outperforms traditional machine learning based methods.

3.3.1 Methodology

3.3.1.1 Dataset

As aforementioned, the dataset and evaluation criteria are not consistent in the network traffic classification. In such a case, the models and algorithms proposed in existing works cannot be compared with each other in terms of a common benchmark. In this work, we focus on the ISCX VPN-NonVPN dataset [57], which contains two levels of traffic classification tasks. The first level is identification of protocol types (chat, email, etc.), while the second level is the identification of application types (Facebook, Skype, etc.). We classify them from the perspective of the protocol type in this work which contains 6 kinds of non-VPN data and 6 kinds of VPN data. The dataset is saved in the form of pcap file, where the name of each file is specified by protocol. In addition, those two kinds of data, i.e., VPN and NonVPN, allow us to train the classifier for the encrypted packets. The original dataset is about 35G, so we execute the data preprocessing process on Hadoop platform, and the data after processing is about 1G. In addition, since we need to perform multiple sets of comparison experiments to select the best hyperparameters, our algorithms are distributed based on TensorFlow to reduce training time.

3.3.1.2 Data Preprocessing

In order to facilitate the training model, we need to store the original packets in the form of pcap format according to three structures, i.e., category, flow, and packet. Data preprocessing consists of the following four stages: traffic flow segmentation, unwanted field removal and data normalization, time-series data representation, and the segmentation of training set and test set. The process of data preprocessing is shown in Fig. 3.16.

Fig. 3.16 The flow diagram
of data preprocessing

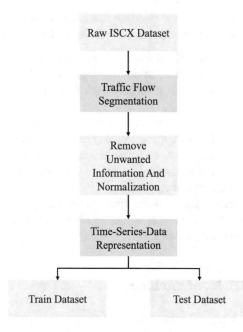

1. Traffic Flow Segmentation

   Let us take each network traffic flow F_i, $i = 1, 2, 3, \ldots, N$ for example, which consists of multiple data packets P_j, $j = 1, 2, 3, \ldots, M$. We utilize a five-tuple including the source IP, destination IP, source port, destination port, and transport layer protocol such as TCP and UDP, to identify a traffic flow. Packets having the same five-tuple belong to the same traffic flow. In particular, we put the source-to-target and target-to-source packets together to form bidirectional flows. In such a case, we can use the SplitCap tool to split the original pcap files into many bi-flows. Then, we saved each bi-flow as a small file in the form of pcap file and label the data with the corresponding file name. The number of flows that each class contains is shown in Fig. 3.17. In such a case, we can use the SplitCap tool to split the original pcap files with the aid of the above-mentioned five-tuple. Then, we save the data of each flow as a small file in the form of pcap file and label the data with the corresponding file name. The number of flows that each class contains is shown in Fig. 3.17.

   As we can see from Fig. 3.17, there is a distinct difference between the data with different category, which is extremely unbalanced. Therefore, we use cost-sensitive learning method to reduce the impact of unbalanced dataset. Furthermore, some files may generate a large amount of flows after segmentation, and thus we are able to choose a portion of the flows and abandon the rest. However, because of the difference of preprocessing methods used, the final datasets may be very different.

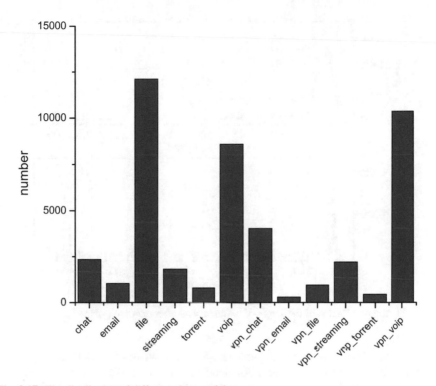

Fig. 3.17 The distribution of different classes of data

2. Unwanted Field Removal and Data Normalization

   Each packet consists of multiple protocol layers, so we can use all data or just the application layer (L7) data for classification. We will verify which is better in Sect. 3.4. In attention, because the dataset was collected by several fixed IP addresses where each IP address is responsible for collecting data of a certain protocol type, it can mingle external information so that the experimental results are greatly biased when using all data. Therefore, we chose to delete the data link layer information and IP address [53]. As for data normalization, considering the packets consisting of binary strings, each byte consists of 8 bits and can be represented as a decimal number in the range of [0–255]. The input of neural network needs to be normalized, so we normalize each byte in the range of [0–1].

3. Time-Series-Data Representation

   In order to use the traffic flow as an input to the model, we need to represent it as an $N*M$-dimensional matrix, where N means the number of packets in a traffic flow and M means the number of bytes in a packet. If a packet contains less than M bytes, we need to pad 0 after its data until its length reaches M. While if the packet length is larger than M, it needs to be truncated to retain only the first M bytes. The same method will also be used to N. This method can solve the

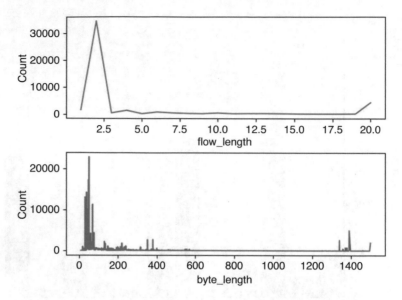

Fig. 3.18 The distribution of the length of traffic flow and the length of packet

problem of variable length sequences. Moreover, we use 0 as the padding value, which does not bring any additional information and bias to the classification result because if the input of neural network is 0, then the output is also 0. To determine the best values for N and M, we first visualize the data distribution as shown in Fig. 3.18.

From Fig. 3.18, we can conclude that most of the data consists of only two data packets because the dataset contains many domain name system (DNS) protocol packets. We will construct a comparison experiment in section 5 to verify the impact of the DNS flows. In order to determine the optimal length of traffic flow N, we take experiments with the length of $N = 5$, $N = 10$, and $N = 20$ to determine the hyperparameter, respectively.

Moreover, in Fig. 3.18, the number of bytes contained in the packet P_j is mainly distributed at both ends. In order to determine the optimal length of package M, contrast experiments with $M = 500$, $M = 1000$, and $M = 1500$ are conducted. Once the best choices for N and M are determined, a traffic flow can be represented as a matrix of $N \times M$, and the value of each element in the matrix is between 0 and 1, i.e., each traffic flow is represented as a matrix of $N \times M$ as an input sample of LSTM. Figure 3.19 portrays the representation of each traffic flow.

4. Dataset Segmentation

In the stage of dataset segmentation, in order to ensure the generalization ability of the model and the credibility of the experimental results, we adopt the tenfold cross-validation method. The dataset is randomly divided into 10 parts. Next, the

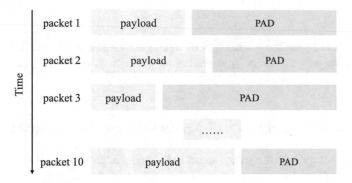

Fig. 3.19 The matrix representation of traffic flow

Table 3.5 The statistical information of dataset

	VPN	Non-VPN
Training set	15,545	22,706
Validation set	1943	2838
Test set	1943	2838

training–validation–test sets are randomly selected in the manner of 8-1-1 for 10 times, and the final result is averaged. Specifically, the validation set is used to determine the hyper parameters in the experiment, while the test set is conceived for representing the final model effect. Finally, the statistical information of dataset is shown in Table 3.5.

3.3.2 Attention Based LSTM and HAN Architecture

Considering the network traffic as time-series data, we used the improved RNN for modeling it in this section. Because the sequence of traffic data is very long, for example, there are 1500 bytes per packet in our dataset, so we use LSTM [60] instead of the original RNN. The reason is that LSTM can remove or add information to the hidden state vector with the aid of the gate function. This means that LSTM can retain important information in hidden layer vectors.

There are three gate functions, i.e., the forget gate, the input gate, and the output gate. The **forget gate** is used to control how much information in C_{t-1} is retained in the process of calculating C_t. The forget vector f_t can be given by

$$f_t = \sigma(W_f \cdot [h_{t-1}, x_t] + b_f),\tag{3.13}$$

where W_f and b_f are the parameters of forget gate, x_t is the input vector in step t, and h_{t-1} is the hidden state vector in step $t - 1$.

Moreover, the **input gate** decides how much information of x_t is added to C_t, which can be express as

$$i_t = \sigma(W_i \cdot [h_{t-1}, x_t] + b_i), \tag{3.14}$$

where W_i and b_i are the parameters of input gate, and hence C_t can be calculated relying on forget gate vector f_t as well as on the input gate vector i_t, i.e.,

$$C_t = f_t \cdot C_{t-1} + i_t \cdot \widetilde{C}_t, \tag{3.15}$$

where $\widetilde{C}_t = \tanh(W_C \cdot [h_{t-1}, x_t] + b_C)$ denotes the information represented in the hidden layer vector.

The **output gate** controls the output in C_t, and we have

$$\begin{aligned} o_t &= \sigma(W_o \cdot [h_{t-1}, x_t] + b_o), \\ h_t &= o_t \cdot \tanh(C_t), \end{aligned} \tag{3.16}$$

where W_C, W_o, b_C, and b_o are the parameters of output gate, and C_t is the internal state in step t. However, the length of the packet is large so that the LSTM model cannot memorize all of the information. Besides, the long sequence may also produce gradient explosions and gradient vanishing during the training process.

To address the long-term dependence of time-series data, in [61], Bahdanau et al. utilized the attention mechanism to the seq2seq model, which was used to calculate the weight of all hidden vectors, as shown in the following.

$$u_i = \tanh(W_p h_i + b_p), \tag{3.17}$$

$$\alpha_i = \frac{\exp(u_i^T u_s)}{\Sigma_j \exp(u_j^T u_s)}, \tag{3.18}$$

$$c = \sum_i \alpha_i h_i, \tag{3.19}$$

where W_p, b_p, and u_s are all parameters that need to be trained, while u_i is the importance score of each packet, and α_i is the normalized weight. We have $\Sigma_i \alpha_i = 1$. In fact, the above calculations are equivalent to using a fully connected neural network to calculate the weight of each vector and then weighting each hidden layer vector with the weighted value to obtain the intermediate vector c. Based on this attention mechanism, in the following, we propose the attention based LSTM and HAN [62] for network traffic classification.

3.3.2.1 Attention Based LSTM

The attention based LSTM neural network structure diagram is shown in Fig. 3.20, where each packet P_i is encoded into an input vector by a Bi-LSTM model. The hidden layer vectors $\overrightarrow{h_i}$ and $\overleftarrow{h_i}$ are connected to form the context vector $h_i = [\overrightarrow{h_i},$

Fig. 3.20 The architecture
diagram of the attention
based LSTM model

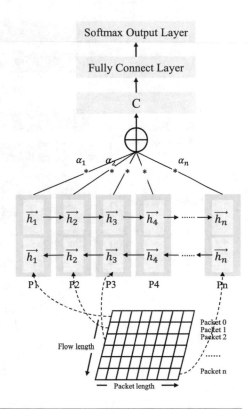

Algorithm 3.2 Attention based LSTM

1: Input: Network flow data $F = \{P_i | i = 0, 1, 2, \ldots, N\}$
2: Output: The predict label $predict$ of F
3: for P_i in F:
4:      $\overrightarrow{h_i} = LSTM_1(P_i)$
5:      $\overleftarrow{h_i} = LSTM_2(P_i)$
6:      $h_i = [\overrightarrow{h_i}, \overleftarrow{h_i}]$
7: end for
8: $c = \sum_i \alpha_i h_i$ (Eqs. (3.17)–(3.19))
9: $predict = softmax(fullyConnect(c))$

$\overleftarrow{h_i}$], where h_i represents the encoding vector of each packet and it contains the
information of the preceding and following packets. Then the attention mechanism
is used to calculate the weight of h_i. As shown in Eq. (3.19), we multiply each
vector by their weight and add them up. Finally, we obtain the encoder vector c of
the traffic flow.

The pseudo-code of Algorithm 1 is shown in Algorithm 3.2.

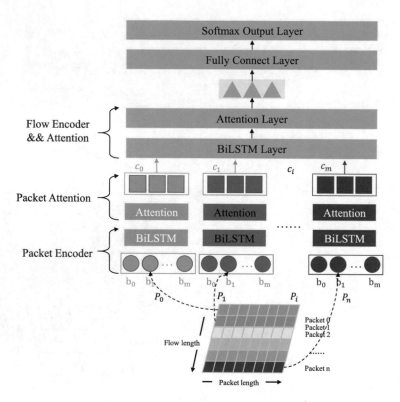

Fig. 3.21 The architecture diagram of HAN model

3.3.2.2 HAN Architecture

The HAN architecture is a kind of neural network structure proposed by Yang et al. [62] in the scenario of text classification. It is similar to attention based LSTM except that two layers of LSTM networks are used to encode each packet and flow, separately. The architecture diagram is shown in Fig. 3.21.

The first layer of the LSTM network uses each byte b_j in the packet as input and processes only one byte and encodes it at each time. Therefore, the rolled step of the LSTM is as large as the length of the packet. Each b_j is encoded as a hidden vector $h_j = [\overrightarrow{h_j}, \overleftarrow{h_j}]$. Then, it uses the attention mechanism to calculate the weight of each byte and performs a weighted summation for getting the vector p_i, which represents the information in each packet. The second layer of the LSTM network is used to exactor the encoder vector of the whole flow F, where p_i is encoded as the input of the second LSTM layer at each time, and the attention mechanism is also used to calculate the importance score of each data packet. Moreover, weighted summation is performed to get the representation vector F of the traffic flow. The pseudo-code of Algorithm 2 is shown in Algorithm 3.3.

Algorithm 3.3 HAN architecture

Input: Network flow data $F = \{P_i | i = 0, 1, 2, \ldots, N\}$
2: Output: The predict label $predict$ of F
   for P_i in F:
4:     for b_j in P_i:
             $\overrightarrow{h(b_j)} = LSTM_1(b_j)$
6:           $\overleftarrow{h(b_j)} = LSTM_2(b_j)$
             $h(b_j) = [\overrightarrow{h(b_j)}, \overleftarrow{h(b_j)}]$
8:     end for
       $c_i = \sum_j \alpha_j h(b_j)$ (Eqs. (3.17)–(3.19))
10:    $\overrightarrow{h_i} = LSTM_1(c_i)$
       $\overleftarrow{h_i} = LSTM_2(c_i)$
12:    $h_i = [\overrightarrow{h_i}, \overleftarrow{h_i}]$
   end for
14: $c = \sum_i \alpha_i h_i$ (Eqs. (3.17)–(3.19))
    $predict = softmax(fullyConnect(c))$

3.3.2.3 Output Layer and Objective Function

We can choose one of the above-mentioned two methods to encode the traffic flow into the vector c, which is the input of the full-connection layer. The processed result is then sent to the softmax layer for obtaining the probability $\hat{y}_i$. We utilize a dropout in the full-connection layer for the sake of increasing the generalization ability of the model and keep the probability of the dropout as 0.8.

Cost-sensitive learning is used because of the serious class imbalance problem in the training data. Cost-sensitive learning means that the cost of mis-classification in different class is varying, so we introduce a cost vector $c_n \in [0, \infty)^K$ for each sample where each dimension means the cost of classifying the sample as k-th class. The loss function proposed in [63] is used to calculate the loss of each sample as follows:

$$\delta_{n,k} = \ln(1 + \exp(z_{n,k} \cdot (r_k(x_n) - c_n[k]))), \qquad (3.20)$$

where $\delta_{n,k}$ represents the loss of classifying the n-th sample as k-th class, $r_k(x_n)$ represents the k-th dimension of the neural network output vector, $c_n[k]$ represents the k-th dimension of the cost vector, $z_{n,k}$ indicates whether the n-th sample is k-th class, and its calculation formula is $z_{n,k} = 2[c_n[k] = c_n[y_n]] - 1$. The advantage of this loss function is smooth and differentiable so that it can be used as the objective function of the neural network to perform the back propagation directly. Therefore, the objective function is as follows:

$$L(\theta) = \sum_{n=1}^{N} \sum_{k=1}^{K} \delta_{n,k}. \tag{3.21}$$

In order to get prediction label of the model, the following formula can be used:

$$\hat{y}_n = \underset{1 \le k \le K}{\arg\min} \, r_k(x). \tag{3.22}$$

3.3.3   Experiments and Result Analysis

3.3.3.1   Experimental Environment

The experimental environment is listed as follows: Ubuntu 14.04 OS, TensorFlow 1.4.0, Python 2.7, NVIDIA 1080Ti graphics, and 16G memory. In order to prevent the over-fitting phenomenon, the dropout technique is utilized and its probability is set as 0.8 during the training process. As for the optimization method, Adam optimization is employed and the initial learning rate is set as 0.001. Meanwhile, we have used Relu as the activation function. Moreover, the batch size is 64 and the program is trained for 30 epochs. The major parameters in both models include the $lstm\_size$ of LSTM cell, the $hidden\_size$ of the fully connected layer, and the out_size of the output layer which is selected according to the number of output classes. The $lstm\_size$ and $hidden\_size$ are both the dimension of vector which encoded the information of the flow. Because the shape of a flow is similar to the shape of an article, we can refer to the hyperparameter setting in the text classification task [62]. A vector of approximately 100 dimensions is sufficient to solve this problem. Specially, in the attention based LSTM model, we set $lstm\_size$ as 100 and $hidden\_size$ as 128, while in HAN, the first layer is used to encode the packet which is a long sequence, and the second layer is used to encode the entire flow, so we set $lstm\_size$ in both LSTM layers as 128 and $hidden\_size$ as 128. As for the input shape of the model, it is selected according to the following experiment.

3.3.3.2   Evaluation Metrics

In order to evaluate the experimental result, we use four evaluation criteria, i.e., accuracy (acc), precision (P), recall (R), and F1 score $(F1)$. Because encrypted traffic classification is a multi-category task, we need to separately calculate the above indicators for each category. Specially, we use N as the total number of training samples; TP_c is to indicate the quantity that originally belongs to category c and is predicted by the model as c; FP_c indicates the quantity that originally does not belong to category c but is predicted by the model as c; TN_c indicates the quantity that does not belong to category c and is not predicted to be class c; and FN_c indicates the quantity that it belongs to class c but is misclassified to other class. Hence, the definition of aforementioned four evaluation metrics can be given by

$$P_c = \frac{TP_c}{TP_c + FP_c}, \quad R_c = \frac{TP_c}{TP_c + FN_c}, \quad F1_c = \frac{2P_c R_c}{P_c + R_c}, \tag{3.23}$$

$$acc = \frac{\sum_{c=1}^{C} TP_c}{N}. \tag{3.24}$$

Traffic Representation Evaluation

In order to test the performance of the proposed model, four experimental scenarios in [54, 57] are used in this work. The first task is the protocol encapsulated traffic identification, which is a two-category problem. Moreover, the second task is the regular encrypted traffic classification, which is a six-category problem. The third one is the protocol encapsulated traffic classification, which is also a six-category problem. The difference between second and third tasks is the datasets used. To elaborate, the second task uses the VPN dataset, while the third task relies on NonVPN dataset. Finally, the last task is the encrypted traffic classification, which is the most difficult task, because it is a twelve-category problem. The details of aforementioned experiments are given in Table 3.6.

Table 3.7 shows the results with attention based LSTM for different dataset processing methods. There are two main variables: whether to remove the DNS flows and whether to use all data or only L7 data. All data means the data after removing data link layer and IP address. The results show that it is better to retain the DNS flows. The reason is that DNS is used for hostname resolution, which is closely related to the host. They are easier to identify than other encrypted flows and bring some bias in the results. So we should remove the DNS flows from the dataset. In addition, the accuracy of using all data is 2–3% higher than only using L7 data which is similar with [54]. This is because all data contains the transport layer and part of the network layer information, such as the port number and packet length,

Table 3.6 The description of contrastive experiments

	Description	Class num
1	Protocol encapsulated traffic identification	2-classes
2	Regular encrypted traffic classification	6-classes
3	Protocol encapsulated traffic classification	6-classes
4	Encrypted traffic classification	12-classes

Table 3.7 The accuracy of different processing methods

With DNS	Yes		No	
Packet data	All data	L7 data	All data	L7 data
Exp 1	1.000	0.964	0.997	0.951
Exp 2	0.898	0.859	0.893	0.844
Exp 3	0.970	0.934	0.948	0.920
Exp 4	0.936	0.905	0.912	0.886

Table 3.8 The accuracy result of different length of flow and packet

Parameter	5,500	5,1000	5,1500	10,500	10,1000	10,1500	20,500	20,1000	20,1500
Acc	0.891	0.897	0.891	0.904	0.908	**0.912**	0.899	0.906	0.902

We highlight the highest accuracy and lowest latency of experimental results, which correspond to the method with better performance.

which are helpful for classification. Therefore, we will use all data to represent the traffic flow and remove the DNS flows.

In order to find the best network traffic flow representation method, we perform a grid search on both the length of the traffic flow N and the length of the data packet M. The value range of N is [5, 10, 20] and that of M is [500, 1000, 1500]. The grid search performs 9 sets of experiments based on a combination of different values of N and M, while the other parameters remain unchanged. The experimental results are shown in Table 3.8.

In the Table 3.8, when $N = 10$ and $M = 1500$, the classification accuracy is highest so that we chose it as the hyperparameters. In order to show the effect of the model in different categories in detail, we draw Fig. 3.22a based on the F1 value. After analyzing the F1 score for each category considered, we can conclude that the F1 scores of NonVPN's chat and NonVPN's email protocol are very small because some relatively short traffic flows may involve more frequent interactions. The distribution difference between traffic flows is relatively large, which leads to a poor performance of the classification. Therefore, we can conclude that when $N = 10$ and $M = 1500$, the classification of the chat class is better. We need to use a larger amount of data to learn this type of protocol with a large difference in data distribution.

Performance Comparison with the State of the Art

We choose $N = 10$ and $M = 1500$ as hyperparameters and use the attention based LSTM and HAN models to perform traffic classification tasks in the above-mentioned four experimental scenarios. The experimental results are shown in Table 3.9, followed by more details in Appendix B.

From Table 3.9, we can conclude that the models can handle the two-category classification problem correctly. As described in the introduction, the traffic is encrypted to exhibit a completely different distribution, such that the model can easily distinguish them. In addition to the C4.5 decision tree proposed in [57],

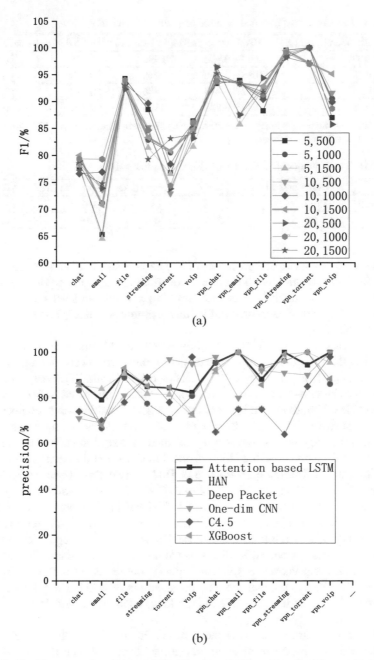

Fig. 3.22 The F1 score and precision. (**a**) The F1 score of different length of flow and packet. (**b**) The precision of all four experiments

Table 3.9 The accuracy of five models in four experimental scenarios

	Attention based LSTM	HAN	Deep packet	One-dim CNN	C4.5 decision tree	XGBoost
Exp 1	**0.997**	0.995	0.992	0.990	0.900	0.991
Exp 2	**0.893**	0.851	0.868	0.818	0.890	0.841
Exp 3	0.948	0.929	0.923	**0.986**	0.870	0.918
Exp 4	**0.912**	0.895	0.898	0.866	0.800	0.864

We highlight the highest accuracy and lowest latency of experimental results, which correspond to the method with better performance.

Table 3.10 The training runtime of five models in experimental 4

	Attention based LSTM	HAN	Deep packet	One-dim CNN	XGBoost
One batch	0.05 s	1.48 s	0.1 s	0.02 s	–
Total	1783.6 s	53,149 s	3600 s	720 s	300 s

all other methods work better on the VPN dataset than the Non-VPN dataset. As shown in [54], each protocol has different distributions after encryption, which can contribute to the distinction. The deep learning method can learn and extract the distribution features of encrypted traffic, thereby more accurately distinguishing the encrypted traffic.

In scenarios 1, 3, and 4, the attention based LSTM model achieves the best experimental results and has a significant performance improvement compared with Deep Packet [53], one-dim CNN [54], decision tree [57], and XGBoost models. The accuracy of the two-category classification is directly improved up to 0.997. The most significant improvement is the twelve-category classification in comparison to one-dim CNN, which increases by 5%, to NonVPN, which increases by nearly 7%. The overall performance of HAN is not as good as that of attention based LSTM, while it is still much better than that of one-dim CNN and decision tree models in scenarios 1 and 4. But the performance of HAN in two six-category classification problems is poorer than that of the one-dim CNN. The results of the Deep Packet model are the closest to attention based LSTM but better than the HAN model. The Deep Packet model classifies traffic flow using two-layer CNN and seven fully connected layers. It is a little complicated so that the speed of Deep Packet is slower. In terms of processing speed, XGBoost is the fastest, then one-dim CNN, attention based LSTM, Deep Packet, HAN. The training runtime of these models is shown in Table 3.10. We can see that the time consumption of HAN is very large, mainly because the length of its first LSTM is too long, while other models are relatively faster.

In order to analyze the experiment results in detail, we show in Fig. 3.22b the precision of a model in different categories. It can be seen that the attention based LSTM performs the best. However, its precision in the streaming and VoIP categories is lower than one-dim CNN. This is because the traffic in these two categories is too huge for LSTM to learn the long-term relationship.

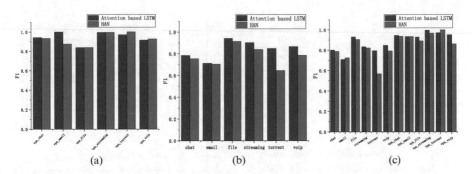

Fig. 3.23 The F1 score of each class, where the red bar represents F1 score of attention based LSTM model, while the blue bar represents that of HAN model. (**a**) The F1 score of each class in scenario 2. (**b**) The F1 score of each class in scenario 3. (**c**) The F1 score of each class in scenario 4

In addition, the overall performance of the proposed two models has a relatively high improvement compared with the previous models with the following two reasons:

- Firstly, we treat the network traffic as time-series data and use the LSTM for processing. In comparison to the CNN model, RNN is capable of learning the relationship between adjacent packets, where each hidden layer state records relevant information of all previous packets in order to learn its long-term dependency.
- Secondly, we use the attention mechanism to improve the accuracy of the LSTM classification. The weighted summation of all historical state information is used as the final coding vector. Compared with the use of the latest hidden layer vector, more historical information is included.

In order to clearly show the classification effect of attention based LSTM and HAN, we transform the F1 scores of each class in scenarios 2, 3, and 4 into bar graphs, which are shown in Fig. 3.23.

Packet Importance Distribution

For the sake of exploring the importance of each packet in the final classification, we need to visualize the weight value α_i of each packet. The result is shown in Fig. 3.24. The test set is divided according to the category, and the attention vector c is calculated to obtain the average attention vector of the entire dataset. Each category has a different attention vector, which represents the contribution of each packet imposed on different categories. The basic focus is on the first four data packets, because the packets at the end of the traffic flow hardly contribute features

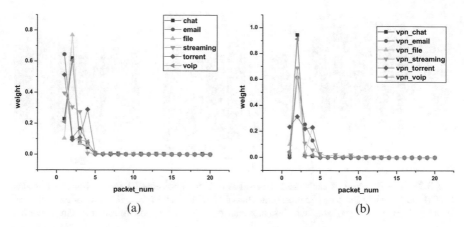

Fig. 3.24 The weight distribution of different packets in (**a**) NonVPN dataset and (**b**) VPN dataset

except those belong to protocols, such as streaming, torrent, and file. As we all know, the first few packets carry more protocol-related information in the transmission of the traffic, such as the three-way handshake of the TCP protocol. In other words, our model does learn this important feature and increase the weight of the first few packets. In addition, we also analyze the importance of different bytes of data. Similar to the distribution of packets, important bytes are concentrated in the header of the data. This is because the bytes that contain category information are often located at the header of each packet in traffic.

3.4   Distributed Variational Bayes-Based In-network Security

Recently, the Internet of Things (IoT) has become a disruptive force reshaping our lives and works. According to Gartner's perspective, it is estimated that there may be a total of 20 billion connected IoT devices by 2025 [64, 65]. However, with the numbers of IoT devices increasing, so does the number of insecure IoT devices in the network. Network security, especially the DDoS attack, has presented a great challenge [66]. DDoS attacks are designed to consume the limited resources on a target service host through a large number of legitimate and useless requests [67–69]. Vulnerable IoT devices as a DDoS vector can be easily hijacked to spread malware, recruited to form botnets to attack other Internet users. The GitHub DDoS attack of 2018 as an example. This was the first multi-terabyte/s DDoS attacks (1.3 TB/s attack) against the cloud-based code hosting platform, which have been bringing large organizations to their knees.

Traditional DDoS detection and mitigation solutions often adopted out-of-band architecture. As shown in Fig. 3.25, it is a typical out-of-band DDoS detection and

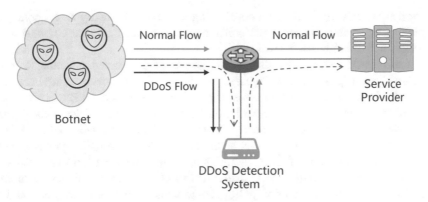

Fig. 3.25 Typical DDoS attacks and detection in IoT

mitigation solution in current networks. This design is accomplished by processes that receive monitoring data from NetFlow, sFlow, or IPFIX in each switch and then analyzes that flow data to detect attacks. In this design, due to the scale and cost, DDoS detection cannot monitor all network traffic [70]. The network operator must set some static mirror rules on the edge router to mirror part of the traffic. Therefore, for highly distributed and complex attacks, devices can either scale to detect megabits per second and millions of connections or can monitor small-scale traffic with low accuracy. Especially, facing such a huge number of IoT devices, this approach is confronted with limited processing capacity, bandwidth resources, and service assurance, especially with the number of IoT devices increasing [71–74].

Recently, with the development of programming switch, it allows the packet processing behavior, including the type, sequence, and semantics of processing operations, to be reconfigured on the fly in a systematic fashion [75]. As such, programmable switches open up new possibilities for aggregating traffic statistics and identifying abnormal traffic flows directly in the data plane [76]. The DDoS detection mechanism could be migrated from the edge to the core of networks. In other words, the deployment of security mechanism for in-network will set up programmable data plane defenses into the network paths and synchronize the whole-network defense.

The advantages of the in-network DDoS detection solution are as follows: under any type of attack, it can use the least memory and resource consumption to achieve high scalability and line speed performance; it has high accuracy and almost negligible false alarm probability; programmable data plane allows customers to flexibly customize DDoS detection methods and mitigation schemes; and fine-grained statistical data allows customers to quickly identify the attacked applications and services. The in-network solution will undoubtedly enhance the DDoS detection performance. However, with the DDoS detection responsibility shifted to each switch, how the geo-distributed switches can cooperatively learn the distributed DDoS detection strategies is a critical problem. Meanwhile, how to design the effective algorithm considering the time complexity and accuracy in matching requirements desires our concern on account of the lower computational performance of hardware devices.

In this work, inspired by the recent success of distributed machine learning in classification problems, we present a lightweight machine learning-based in-network DDoS detection scheme. We propose a hybrid variational Bayes algorithm for DDoS detection, where the variational Bayes method provides a local optimization with an approximate posterior method. With the massive data in DDoS detection, it turns the problem from statistical inference into optimization, greatly reducing the computational cost. In addition, a centralized platform is introduced to synchronize parameters among distributed switches to realize the collaborative learning of the DDoS detection policy. Some simulation results are presented to evaluate the correctness of our architecture and algorithm.

The main contributions of our work are summarized as follows:

- We construct an in-network security architecture for DDoS detection, where the distributed switches can collaboratively learn the independent DDoS detection policy through a centralized parameter synchronous platform.
- We propose a hybrid variational Bayes algorithm in our architecture for detecting the DDoS attacks in the IoT. The lightweight distributed Bayes algorithm can dynamically learn DDoS detection strategies.
- Extensive simulations are conducted to demonstrate the convergence and efficiency of our mechanism.

3.4.1   System Model and Problem Formulation

As shown in Fig. 3.26, consider a network with a set of switches denoted as $N = \{1, \ldots, n\}$. Within a time interval ΔT, the flow data collected by node i can be fitted as Gaussian mixture model (GMM) by K characteristic components. We regard flow K's characteristics of each traffic information as indicators to evaluate the flow data.

Taking node i as an example, it transmits packets. In our work, a network composed of N nodes consists of a series of nodes $V = \{1, \ldots, N\}$ and a series of edges γ. Each pair of edge $(i, j) \in \gamma$ connects a different pair of an unordered node. For each node $i \in V$, we use $N_i = \{j \mid (i, j) \in \gamma\}$ to represent the neighbor nodes of i. The distance between neighbor nodes is called the communication distance. There is no interaction when the distance between two nodes is larger than the preset value. In the node i, it will receive the flow data $x_{ij}(i = 1, 2, \ldots, n, j = 1, 2, \ldots, n_i)$. Furthermore, the μ_k is average and Λ_k^{-1} is the covariance of the Gaussian distribution for each component of data N_i.

Considering the correlation among data by order and the high dimension of the parameter matrix, we introduce the discrete hidden variable: y_i. Local distribution of each node can be expressed as

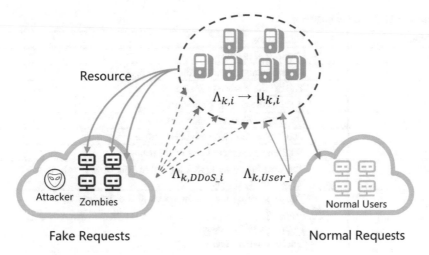

Fig. 3.26 Traffic collection in DDoS attacks

$$P(\{x_i\}_N \mid y_i, \mu, \Lambda) = \prod_{j=1}^{N_i} \prod_{k=1}^{K} N(x_{ij} \mid \mu_k, \Lambda_k^{-1})^{N_{y_{ijk}}}, \tag{3.25}$$

where $y_i = \{y_{i1}, y_{i2}, \ldots, y_{iN_i}\}_i$, $y_{ij} = \{y_{ij1}, y_{ij2}, \ldots, y_{ijk}\}$. The prior distribution of y_i is the product of multinomial distributions, conditioned on the mixing coefficient. The prior distribution of the mixing coefficient is a Dirichlet distribution, and the prior distribution of the mean is specified as multivariate Gaussian distribution, conditioned on the precision matrix. The prior distribution of the precision matrix is a Wishart distribution. Using these priors distributions and the conditional independence relationships, the generate model of each node i can be expanded as

$$
\begin{aligned}
P(\{x_i\}_N, y_i, \pi, \mu, \Lambda) = &P(\{x_i\}_N \mid y_i, \mu, \Lambda)P(y_i \mid \pi) \\
&P(\pi)P(\mu \mid \Lambda)P(\Lambda).
\end{aligned} \tag{3.26}
$$

Under the mean field theory hypothesis, the joint variational distribution of unobserved variables can be decomposed into

$$q(y_i, \pi, \mu, \Lambda) = q(y_i)q(\pi) \prod_{k=1}^{K} q(\mu_k \mid \Lambda_k)q(\Lambda_k). \tag{3.27}$$

In this way, we construct the system model according to a sets of characteristics of traffic and observe the independent parameters separately so that we can simplify the classification problem as optimization problem. For a clear understanding, we list the notations of this work in Table 3.11.

Table 3.11 Majority notations

Notation	Description
N	Number of nodes
i	The sequence of each node, $i = 1, 2, \ldots, N$
j	The neighbor node of node i
N_i	A set of the neighbor nodes of node i
x_{ij}	Number of observed data of node i
y_i	The discrete hidden variables of node i
K	Number of characteristics of the data
Λ_k	The covariance matrix of the k component of the dataset $x_{ij}, k = 1, 2, \ldots, K$
μ_k	The average of the k component of the dataset x_{ij}, $k = 1, \ldots, K$
θ	Parameter of the model
z	Unobserved variable
π_i	Mixing coefficient of node i
$\phi_{\theta,i}$	Global natural parameter vector of node i
$P(z\|x)$	Posterior probability of unobserved variables
$Q(z)$	Approximate of $P(z\|x)$
$\ell(Q)$	Lower bound of evidence $\log P(x)$
$repeat$	Number of iterations
η	Time-varying step size

3.4.2   Hybrid Variational Bayes Algorithm

In this section, we will describe an in-network security architecture and propose a hybrid variational Bayes algorithm for in-network DDoS detection.

3.4.2.1   In-network Security Architecture

With the DDoS detection responsibility shifted to each switch, it acts as an independent intelligent agent and constitutes a Multi-Agent System. How the geo-distributed switches can cooperatively learn DDoS detection strategies is a critical problem. Inspired by the recent success of distributed machine learning in classification problems, in this section, we construct a distributed machine learning-based in-network security architecture. As shown in Fig. 3.27, the attack flows mixed with normal flows are transmitted through distributed switches randomly. During a processing period, neighbor switches exchange local information and iterate to update local parameters. The centralized platform collects the current processing results, flags the abnormal traffic, and calculates the global parameter for the next processing phase. The co-work of centralized platforms and switches can effectively reduce the communication overhead among switches.

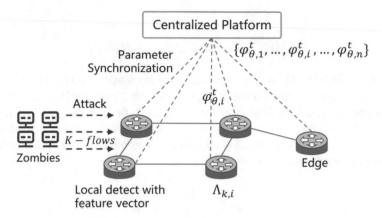

Fig. 3.27 A machine learning-based in-network security architecture

3.4.2.2   Hybrid Variational Bayes Algorithm

In this section, we propose a hybrid variational Bayes algorithm for effectively distributed DDoS detection policy generation. The hybrid variational Bayes is a kind of Bayesian inference method. While the traditional Bayesian inference can avoid the over-fitting of the maximum likelihood function to some degree, the posterior probability is difficult to calculate in the mixed model. Variational Bayesian method provides an approximate analytic solution for the posterior probability. We set θ represent model parameters, including unknown parameters and global hidden variables. Therefore, θ and $\{y\}$ can be collectively referred to as "unobserved variables," denoted as $z = \{\theta, y\}$. Variational Bayes uses the distribution $Q(z)$ to approximate the posterior probability $P(z \mid x)$ of the unobserved variables. This simple distribution $Q(z)$ is obtained by minimizing the KL divergence between these two distributions.

$$
\begin{aligned}
KL(Q(z) \mid\mid P(z \mid x)) &= \int Q(z) \log \frac{Q(z)}{P(z \mid x)} dz \\
&= -E_Q \left[\log \frac{Q(z)}{P(z \mid x)} \right] + \log P(x) \\
&= -\ell(Q(z)) + \log P(x),
\end{aligned}
\tag{3.28}
$$

where $\ell(Q)$ is the lower bound of logarithmic evidence $\log P(x)$. To minimize the divergence of KL, in which logarithmic evidence $\log P(x)$ is fixed by the corresponding Q, we just need to maximize $\ell(Q)$. By selecting the appropriate distribution of Q, $\ell(Q)$ is easy to calculate and obtain the extreme value, which gives an approximate analytic expression of the posterior $P(z \mid x)$ and the lower bound of the evidence $\ell(Q)$, also known as the variational free energy. The variational free energy can be expressed as the sum of an energy term and an entropy term:

$$\ell(Q(z)) = E_Q[\log P(z \mid x)] + H[Q(z)]. \tag{3.29}$$

Then, minimizing KL divergence is equivalent to maximizing the variational free energy, in which case the inference problem becomes an optimization problem of the distribution function. Based on the naive mean-field theory, the optimization space of the optimization problem can be limited to a distribution subset that is easy to describe. Assuming that the variational posterior distribution of unobserved variables can be decomposed into

$$Q(z) = \prod_{m=1}^{M} q_m(z_m), \tag{3.30}$$

where $z = \{z_1, z_2, \ldots, z_m\}$. Under the family distribution of conjugate exponentials, $Q(z)$ can be re-parameterized as

$$q_m^*(z_m) = h(z_m)exp\{\phi_m^{*T} u(z_m) - A(\phi_m^*)\}, \tag{3.31}$$

where q_m^* is the optimal distribution of q_m. Natural parameters can be seen as functions of super parameters, without considering local variables and hidden variables. The natural parameter vector of Dirichlet distribution is

$$\phi_{\pi_i} = [\alpha_{i1} - 1, \alpha_{i2} - 1, \alpha_{iK} - 1]^T. \tag{3.32}$$

Then we introduce the global natural parameter vector $\phi_{\theta,i}$ that belongs to the joint distribution of the exponential family $q^*(\pi_i) \prod_{k=1}^{K} q^*(\mu_{ik}, \Lambda_{ik})$.

$$\phi_{\theta,i} = \left[\phi_{\pi_i}^T, \phi_{\mu_{i,1}, \Lambda_{i,1}}^T, \ldots, \phi_{\mu_{i,k}, \Lambda_{i,K}}^T\right]^T \tag{3.33}$$

is the information interacted between two nodes. The distribution of the variational form is known and remains unchanged throughout the iterative process, and therefore, the optimal distribution given by (3.33) can be determined and represented by its natural parameter vector ϕ_m^*.

The optimization of variational free energy can be directly optimized by natural parameter vectors in parameter space instead of variational distribution optimization in probability space. The variable distribution of the hidden variable y and the model parameter θ in VB algorithm alternate to maximize the lower bound of the variable, which is similar to the EM algorithm.

$$\phi_y^* = argmax_{\phi_y} \ell(\phi_y, \phi_\theta^*). \tag{3.34}$$

$$\phi_\theta^* = argmax_{\phi_\theta} \ell(\phi_y^*, \phi_\theta). \tag{3.35}$$

The joint distribution can be decomposed by conditional independence assumption as

$$P(z, x) = P(\{x_i \mid y_i, \theta\}) P(y_i \mid \theta)$$

$$= P(\theta) \prod_i^N P(y_i \mid \theta) \prod_i^N P(x_i \mid y_i \theta) \tag{3.36}$$

$$= P(z) \prod_i^N P(x_i \mid z).$$

The objective function, also the global lower bound, is replaced by the average of a series of local lower bounds.

$$\ell(Q(z)) = E_Q[\log P(z, x)] + H[Q(z)]$$

$$= \frac{1}{N} \sum_{i=1}^N E_Q[\log P(z, x)] + E_Q\left[\frac{P(z)}{Q(z)}\right] \tag{3.37}$$

$$= \frac{1}{N} \sum_{i=1}^N \ell_i(Q(z)).$$

In the formula (3.37),

$$\ell_i(Q(z)) = E_Q\left[\log \frac{P(x_i \mid z)^N P(z)}{Q(z)}\right], \tag{3.38}$$

which is obtained from the inequality of the piano and the concavity of the logarithmic function,

$$\ell_i(Q(z)) \leq \log E_{P(z)}[P(x_i \mid z)]^N = \log P(\{x_i\}_N). \tag{3.39}$$

Only if $Q(z)$ is exactly the given repeated observation data $\{x_i\}_N$, also the posterior distribution of the unobserved variable z, for (3.33),

$$max_Q \ell(Q) = \frac{1}{N}\ell_i(Q^*) \leq \frac{1}{N}\ell_i(Q_i^*) = \frac{1}{N}\ell_i(Q), \tag{3.40}$$

where the Q^* is the optimal variational distribution that maximizes the global lower bound $\ell(Q)$, and the Q_i^* is the optimal variational distribution that maximizes the local lower bound ℓ_i. In other words, the global lower bound of logarithmic evidence of complete observation data is strictly less than or equal to the average of the lower bound of logarithmic evidence of repeated observation data on all nodes. In this case the local lower bound on each node cannot be maximized separately to achieve a global optimal.

$$\phi_{y_i}^* = argmax_{\phi_{y_i}} \ell_i(\phi_{y_i}, \phi_\theta^*). \tag{3.41}$$

$$\phi_\theta^* = argmax_{\phi_\theta} \sum_{i=1}^{N} \ell_i(\phi_{y_i}^*, \phi_\theta). \tag{3.42}$$

Given the global natural parameter ϕ_θ^*, the optimization of (3.41) can be addressed individually on each node. In the case of fixed hidden variable variational distribution, the optimal variational distribution of global model parameters for each node is $q_{\theta,i}^*$, whose corresponding natural parameter vector is $\phi_{\theta,i}^*$. That means the local lower bound ℓ_i gets the maximization at $\phi_{\theta,i}^*$ while given the fixed $\phi_{y_i}^*$.

$$\phi_{\theta_i}^* = argmax_{\phi_{y_i}} \ell_i(\phi_{y_i}, \phi_\theta). \tag{3.43}$$

Taking the derivative of ℓ_i with respect to ϕ_θ and set it to zero, we can get the general solution to (3.38):

$$\phi_\theta^* = \frac{1}{N} \sum_{i=1}^{N} \phi_{\theta,i}^*. \tag{3.44}$$

Equation (3.40) is the average value of locally optimal natural parameters calculated at each node. Usually, a centralized fusion center is the optimal choice for this type of problem, in which case we can design a centralized VB algorithm using the local optimal natural parameters of all nodes. However, considering the large amount of network resources in DDoS attack and the low cost network system, we propose a distributed estimation method to deal with this problem.

3.4.2.3 Knowledge Sharing

We explore vital features and analyze their weight of differentiating normal and attack IoT traffic. When single switch calculates local parameter, we present the process of knowledge sharing as follows:

As shown in Fig. 3.28, the gradient-based distributed estimation method does not directly use first-order conditions on each node to obtain the locally optimal $\phi_{\theta,i}^*$ but use stochastic gradient and a diffusion process to get the global natural parameter approximately. We use $\varphi_{\theta,i}$ to represent the intermediate quantity obtained by a gradient ascent on node i. For each iteration, the update equation on node i can be described as

$$\varphi_{\theta,i}^t = \varphi_{\theta,i}^{t-1} + \eta_t \nabla_{\phi_\theta} \ell_i(\phi_{y_i}^*, \phi_{\theta,i}^{t-1}), \tag{3.45}$$

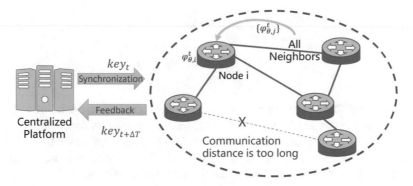

Fig. 3.28 The process of knowledge sharing at a single switch

more plainly,

$$\phi_{\theta,i}^t = \sum_{j \in N_i \cup \{i\}} \omega_{ij} \phi_{\theta,j}^t, \tag{3.46}$$

where the $\nabla_{\phi_\theta} \ell_i$ represents the natural gradient over Riemannian space, the η_t represents the step length, and $\{\omega_{ij}\}$ represents the nonnegative weights.

The iteration process (3.41)–(3.42) makes up a distributed implementation of random variational inference, with each node performing once gradient ascent step using local data. Process (3.42) diffuses all local estimates to the whole network, seen as the process of collecting global sufficient statistics gradually through the VB iteration process. The gradient-based method takes the estimator $\phi_{\theta,i}^{t-1}$ from the last step into the consideration of (3.41), which is the result of diffusion of all nodes' estimators across the whole network. Therefore, the convergence value of gradient rise (3.41) is the solution of the global target function (3.37), not the solution of the local target function of ℓ_i. Literature clarifies that a distributed parameter space has a Riemann metric structure, in which case the natural gradient follows the steepest direction. According to this conclusion, the natural gradient of local lower bound $\ell_i(\phi_{y_i}^*, \phi_\theta)$ with respect to ϕ_θ can be simplified as

$$\nabla_{\phi_\theta} \ell_i(\phi_{y_i}^*, \phi_\theta) = \nabla_{\phi_\theta}^{-2} A(\phi_\theta) \nabla_{\phi_\theta} \ell_i(\phi_{y_i}^*, \phi_{\theta,i}^{t-1})$$
$$= \phi_{\theta,i}^* - \phi_{\theta,i}^{t-1}. \tag{3.47}$$

An iterative formula for calculating global natural parameters for each node i is integrated as

$$\phi_{\theta,i}^t = \sum_{j \in N_i \cup \{i\}} \omega_{ij} \phi_{\theta,j}^t$$

$$+ \eta \sum_{j \in N_i \cup \{i\}} \omega_{ij} (\phi_{\theta,i}^* - \phi_{\theta,i}^{t-1}). \tag{3.48}$$

As our core formula of the algorithm, the optimization of the local lower bound is transmitted to iteratively calculate global natural parameters for each node i. As shown in (3.48), this process can be divided into two parts, of which the first term spreads the information throughout the network, and the second term uses local data and information from neighbors to gradually update the estimator.

In the natural gradient descent step, time-varying step size is selected as

$$\eta_t = \frac{1}{d_0 + \tau t}. \tag{3.49}$$

On the global network, each switch calculates the intermediate $\varphi_{\theta,i}^t$ using local data and then sends it to its neighbor, from which it receives the message $\varphi_{\theta,j}^t$. Since the communication dimensions of the intermediate route and the computing amount after distributed processing are much lower than that after centralized processing, the method in our work saves communication resources and energy to a great extent. The server collects and memorizes the result of each processing period and then centrally regulates the processing of the next period.

Algorithm 3.4 Distributed variational Bayes algorithm

1: Initialization: Node i collects data x_i. Initialize natural parameters with no information prior.
2: Set appropriate parameter τ based on the result of last period from superior network.
3: **for** each $t = 1, 2, \ldots$ **do**
4:     **for all** $i = 1, \ldots, N$ **do**
5:         $\phi_{y_i}^{*,t} = argmax_{\phi_{y,i}} \ell_i(\phi_{y,i}, \phi_{\theta,i}^{t-1})$
6:         $\phi_{\theta,i}^{*,t} = argmax_{\phi_\theta} \ell_i(\phi_{y,i}^{*,t}, \phi_\theta)$
7:         Calculate natural gratitude $\varphi_{\theta,i}^t$.
8:         Broadcast the $\varphi_{\theta,i}^t$ to all the neighbor nodes of node i.
9:     **end for**
10:     **for all** $i = 1, \ldots, N$ **do**
11:         Calculate $\phi_{\theta,i}^t$
12:     **end for**
13: **end for**

3.4.2.4   Complexity Analysis

As for the complexity of our proposed dVB algorithm, we use the EM algorithm with the same logic to make the analogy. As well known, EM is almost identical to the Lloyd variant of K-means, so each iteration requires $O(n*k)$ distance calculation and the worst case complexity of the EM algorithm is $O(n*k*i)$, a theoretically infinite value. However, in our GMM model, the result of classification is 0 or 1, and we selected the number of clusters used for classification as 3. Considering the interaction between switches, the complexity is approximately $T*O(3n)$, where T represents the iteration times.

3.4.3   Experiment Results and Analysis

In this section, we present experiment results to evaluate our proposed framework and algorithm. We firstly use the KDD 99 DATA to evaluate our algorithm.

3.4.3.1   Experimental Setup and Data Preprocessing

The KDD 99 DATA provides a standard dataset for evaluating the performance of different algorithms, and it contains four gigabytes raw training data of the TCP dump data from seven weeks of network traffic. In the original dataset, the DoS flows are labeled as "back," "land," "Neptune," "smurf," "teardrop," and "pod," and normal flows are labeled as "normal." Considering the limited computing power in switches, we use the random forest to rank the importance of different features and leave the top one-thousandth important features, which can reduce runtime computing complexity. Table 3.12 is listed in order of importance.

In learning process, it is common practice to divide data into training sets and test sets. We use 10% of the dataset as the training data for our algorithms training. The data label proportions are 44,118 label normal and 59,108 label attack. The test sets are data independent of the training process. It is used to evaluate the performance of the learned model. Besides, to avoid the overfitting problem in the training process, we separate part of the training data into validation data to evaluate the training effect of the model. We adopt fivefold cross-validation in our training process. The result of fivefold cross-validation is [0.98324131   0.99287998   0.98721302   0.95141445   0.99346057]. It shows that a characteristic vector consisting of the above six components can represent a single flow.

In each time period ΔT, the switch receives a set of flows, which can be represented as flow feature vectors. Then all flow data received by this node can be fitted with a Gaussian distribution on each feature, seen as the component of feature vectors. And all components of feature vector constitute a Gaussian mixing model GMM. As shown in Fig. 3.29a, we establish a network with 20 switch nodes,

Table 3.12 The characteristics and their occupation

Feature	Description	Occupation
SAME_SRV_RATE	The percentage of connections that have the same destination host as the current connection in the last two seconds that have the same service as the current connection	0.9644
DST_HOST_ SRV_COUNT	Of the first 100 connections, the number of connections that have the same destination host service as the current connection	0.0151
COUNT	Has the same number of connections to the target host as the current connection in the last two seconds	0.0085
DST_HOST_SERROR_RATE	Percentage of the first 100 connections that have the same target host as the current connection that have SYN errors	0.0060
SRV_COUNT	The number of connections that have the same service as the current connection in the last two seconds	0.0023
DST_HOST_COUNT	Of the first 100 connections, the number of connections that have the same target host as the current connection	0.0022

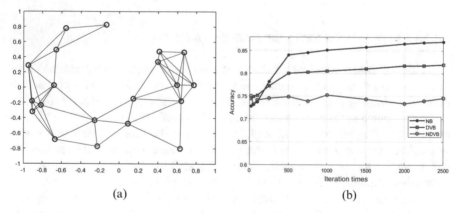

(a)   (b)

Fig. 3.29 Experiment results and analysis. (**a**) Network topology. (**b**) The accuracy vs. the iteration times

placed randomly in a $5 * 5$ square area, in which the communication limitation distance is set as 0.8. There are very few switches adjacent to the top and bottom nodes in the topology in our experiment, and in actual network deployments it is even directly connected to the data center due to communication distance. We deploy edge switches to handle unexpected but necessary traffic requests, as well as to safeguard remote communications and backup services. In our algorithms, we focus more on the interactions between switches, in which case the edge switch in real network environment is represented by sparse connection and long distance communication. There will be no information exchange between two nodes over than the communication distance.

The algebraic connectivity of the network was 0.2238 and the average degree was 5.2000. In order to judge whether the traffic is normal or abnormal globally, we use this network to collect and process traffic samples distributively. The data flow is randomly divided among the 20 nodes, with each dimension being ("dst_host_count," "srv_count," "dst_host_serror_rate," "count," "dst_host_srv_count," "same_srv_rate").

3.4.3.2 Baseline Algorithm

In this work, we adopt several baseline algorithms to evaluate our framework.

1. Naive bayes
   The first baseline algorithm is Naive bayes. One of the difficulties in statistical inference under Bayesian framework is the complexity in calculating posterior probability. The traditional Naive Bayes approach is equivalent to one fusion center and multiple nodes, whose mathematical model is expressed as (3.50).

$$P(c|K) = \frac{P(c)P(K|c)}{P(K)} = \frac{P(c)}{P(K)} \Pi_{i=1}^{d} P(k_i|c), \tag{3.50}$$

where we consider the factors of traffic characteristics to be independent of each other. Here, we use c to represent the label of flow, as 0 or 1. Obviously, the incoherence between nodes greatly save the time-consuming of the algorithm, but it is not a good result for the correlation of various features in the case of large traffic attacks.

Distributed Variational Bayes Without Cooperation

The second baseline algorithm is distributed variational bayes without cooperation between neighbor nodes to further show the superiority of our proposed algorithm. The fusion coefficient ω_{ij} is determined using the simple nearest neighbor rule as (3.51)

$$\omega_{ij} = \begin{cases} \text{loss function} = \dfrac{1}{|N_i + 1|}, & j \in N_i \cup i \\[2mm] 0, & \text{else.} \end{cases} \tag{3.51}$$

2. Variational AutoEncoder (VAE)
   The third baseline is the VAE. Moreover, VAE is derived from the variational Bayes theory as well as DVB, in which case VAE is an excellent comparison algorithm to test the merits of the algorithm we proposed. Similar to the formula (3.28), the goal of VAE is to minimize the loss function as follows:

$$KL[Q(z) \parallel P(z \mid x)] - \log[P(\hat{x} \mid z)], \qquad (3.52)$$

where the $\hat{x}$ represents the output of the network. We use the backpropagation for training VAE. VAE calculates the probability of reconstruction by deriving the random latent variables of the parameters. What is being reconstructed is the parameter of the input variable distribution, not the input variables itself. Using these parameters, VAE can calculate the probability of the original data being generated from the distribution. The average probability is named as the anomaly score or the reconstruction probability mentioned above. The flow data with a high probability of reconstruction will be classified as an attack. To deploy the experiment, the continuous data feature distribution is still the Gaussian mixture distribution.

3. Stacked Denoising Autoencoder (SDAE)

   The forth baseline we use is the SDAE with dropout [77]. In recent years, many researches have focused on the advanced autoencoder to construct an intrusion detection system [78–80]. In [81], the authors have proved that SDAE had higher accuracy and shorter running time than other deep learning methods such as DBN. In general AE, the hidden layer activation is

$$h^{(k)}(x) = g(a^{(k)}(x)), \qquad (3.53)$$

where $a^{(k)}$ represents the pre-activation value in layer k and $g(\cdot)$ is the activation function. It is easy to be overfitting when using a complex model such as deep neural network. Compute gradient of hidden layer as equation (3.53).

$$
\begin{aligned}
\nabla_{W^{(k)}} L(W, \ b; x, y) \\
= (\nabla_{a^{(k)}(x)} L(W, \ b; \ x, \ y)) h^{(k-1)}(x)^T \nabla_{b^{(k)}} L(W, \ b; x, \ y) \qquad (3.54) \\
= \nabla_{a^{(k)}(x)} L(W, \ b; \ x, y),
\end{aligned}
$$

where $W^{(k)}$ is the weight matrix in layer k and $b^{(k)}$ is the offset in layer k.

Generally, an auto-encoder uses the long short term memory (LSTM) to predict. To speed up the process, SDAE fine-tunes the parameters using stochastic gradient descent. The average probability is named as the anomaly score or the reconstruction probability mentioned above. The flow data with a high probability of reconstruction will be classified as an attack. To deploy the experiment, the continuous data feature distribution is still the Gaussian mixture distribution.

Finally, we compare the latest deep learning algorithms using the same dataset and gave a conclusion.

3.4.3.3   Performance Analysis

1. DVB vs. NB and NDVB

   According to the analysis of our algorithm, we need to set the iteration time as an appropriate number so that our model can achieve high accuracy without much resources consumption. In our experiment, we first evaluate the performance of our algorithm. We set two different baseline algorithms: the centralized naive Bayes algorithm and distributed variational Bayes (no cooperation between neighbor nodes). As shown in Fig. 3.29b, with the increasing of iteration times, the average classification accuracy is improving both our algorithm and naive Bayes algorithm. After 2000 iterations, these two algorithms can almost converge to the optimal point. The accuracy of the NDVB method does not improve significantly over time due to the lack of cooperation between neighboring nodes. This experiment results show that our hybrid variational Bayes algorithms can effectively enhance the cooperative learning ability.

   Then, we compare the performance of the three algorithms by observing the curve of average error (KL divergence) changing with iteration. As shown in Fig. 3.30a, NDVB algorithm falls into local optimality and is highly biased on account for no steps for information interaction and fusion, and only local information is used for each iteration. However, distributed variational Bayes algorithm gradually improves the estimator as the information spreads throughout the network, which can finally achieve as good a result as the centralized algorithm.

   More specifically, we notice that the centralized Bayes classifier achieves higher performance in our algorithm consuming less time. This is because the centralized solution can use more data to train the model and need not the local interaction. NDVB shows the advantage of time consumption, which is even better than centralized algorithm due to the distributed processes. In the view of resources consuming and accuracy, our algorithm combines the advantages of centralized Bayes classifier and distributed processes. However, as discussed above, the centralized solution presents poor scalability and robustness in a large-scale network environment. Thus, our algorithms trade off the performance and robustness of the whole system, where the performance of our algorithm reaches the centralized solution and can still be implemented in a distributed fashion. As shown in Table 3.13, we list the statistical average inference accuracy and average misjudgment rate of three algorithms.

2. DVB vs. VAE

   Above we compare and evaluate the DVB with the traditional naive Bayes classification algorithm. Furthermore, we analyze the difference between the DVB algorithm based on stochastic gradient descent and the VAE algorithm of deep learning in DDoS detection. We set up a single hidden layer with 400 dimensions for both the encoder and decoder. The potential size is set 200 dimen-

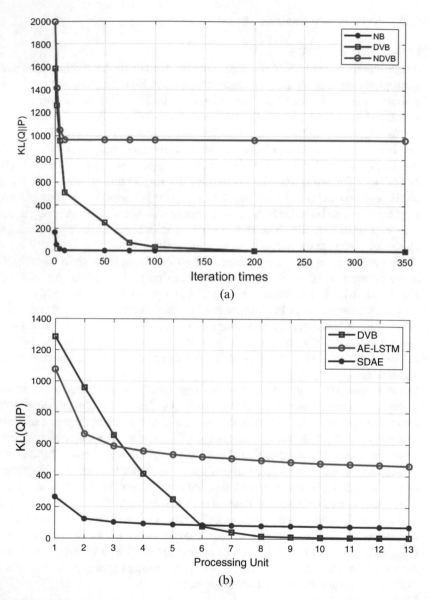

Fig. 3.30 KL divergence. (**a**) KL divergence changing with iteration. (**b**) KL divergence changing with epochs

sions. To balance the running time of the DVB algorithm, the iteration number of the DVB is set as 1000. Both VAE and DVB are trained using complete feature data as ("dst_host_count," "srv_count," "dst_host_serror_rate," "count," "dst_host_srv_count," "same_srv_rate"). To visualize the result explicitly, we

Table 3.13 Statistical average inference accuracy and average misjudgment rate

Methods	Accuracy	The average of misclassified samples
DVB	0.8171	182.9
Naive Bayes	0.8652	159.3
Nocoop-DVB	0.7342	265.8

just select the three features components with the largest proportion of weight, as shown in Fig. 3.31a and b.

We select the running time and AUC-ROC as evaluation indicators and obtain the comparison of DVB and VAE as shown in Table 3.14.

From the above experimental results, we can conclude that the operating efficiency of VAE is better than DVB because the interaction among switches can be considered more complex. That is why DVB is regarded as more robust.

3. DVB vs. SDAE

It is obvious that the time-consuming focuses on the process of stochastic gradient descent, which leads to slow convergence near local optima. In this part, the input vector uses complete flow characteristics in three models. As shown in Fig. 3.30b, we compare the KL divergence of DVB, AE-LSTM and SDAE. The KL divergence changes with epochs in AE-LSTM and changes with iteration times in DVB. It can be observed that using more characteristics as input, our algorithm can convergent faster. We select the running time and mean accuracy as evaluation indicators and obtain the comparison of DVB and SDAE as shown in Table 3.15. Since this experiment uses all the data characteristics, the DVB algorithm will reach convergence with a smaller number of iterations. We test different iteration times (repeat) to observe the performance.

From the above experimental results, we can conclude that the operating efficiency of DVB is better than SDAE because the interaction among switches can shorten the time of the stochastic gradient descent. The accuracy of general autoencoder used in the DDoS detection with LSTM only achieves 49%.

SDAE uses a very high time cost in exchange for accuracy, which is what we expected, that is, the deep learning method will have a better effect on the accuracy of the two classification of data. However, our proposed algorithm is more worthy of being deployed in the actual network environment. For 10% of the KDD99 dataset, the accuracy of deep migration learning based on rough set theory is 87.19% [82]. It can be proved that centralized data analysis is more likely to confuse attacks that conform to normal behavior patterns in big datasets, while they can be identified simply through a single switch. Additionally, our algorithm has more advantages than most deep learning algorithms in terms of time complexity, which is more in line with the concept of in-network security.

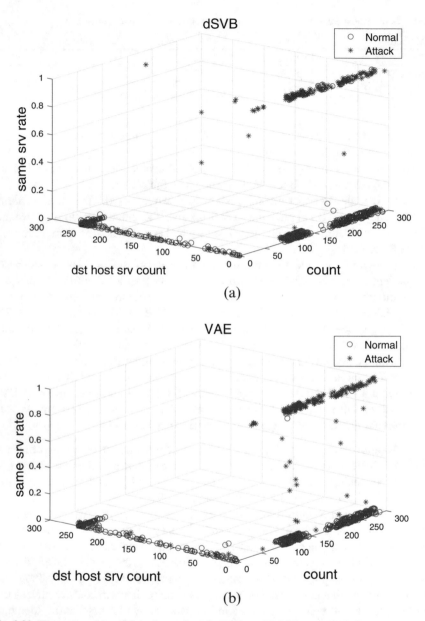

Fig. 3.31 The comparison of experimental results between (**a**) DVB and (**b**) VAE

3.4.3.4 Characteristic Analysis

To analyze different traffic features, we will visualize the three features components with the largest proportion of weight in this section. As shown in Fig. 3.32, there is a

Table 3.14 Contrast experiment of DVB and VAE

Methods	Running time	AUC-ROC
DVB	13 min 2 s	**0.907**
VAE	**7 min 14 s**	0.795

We highlight the highest accuracy and lowest latency of experimental results, which correspond to the method with better performance.

Table 3.15 Contrast experiment of DVB and SDAE

Methods	Running time (s)	Mean accuracy
DVB(repeat = 20)	**42**	0.898
DVB(repeat = 100)	104	0.868
DVB(repeat = 1000)	784	0.890
SDAE	2281	**0.933**

We highlight the highest accuracy and lowest latency of experimental results, which correspond to the method with better performance.

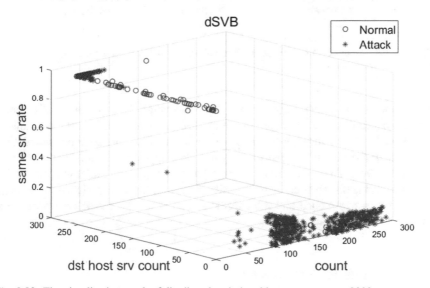

Fig. 3.32 The visualization result of distributed variational bayes as repeat = 2000

clear identification to distinguish traffic conditions in the view of the most weighted component *same_srv_rate*. In the last two seconds, the percentage of connections that have the same destination host more than 30% of total traffic will almost be considered as an abnormal traffic. The smaller the weight of the characteristic component, the more likely to appear in the detection of fuzzy discrimination. Some subsequent experiments showed that the detection performance is not increasing significantly as all characteristic considered.

As shown in Fig. 3.33, it shows that DVB and NB can achieve similar results. But after training, NB can better identify anomalies that are not easily detected. In the absence of interaction between nodes, the accuracy of NDVB in detection decreased

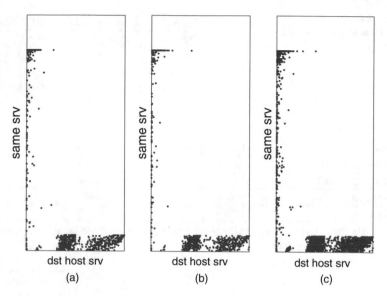

Fig. 3.33 The results of 3 models (Blue: Normal; Red: Attack). (**a**) Distributed VB. (**b**) Naive Bayes. (**c**) Noncoop-VB

significantly. The naive Bayes used in classification is superior to distributed variable bayes due to advantages of centralized processing in binary classification problems. However, through comprehensive factors and more iterative calculation, we notice that the results of distributed variational bayes processing are comparable to naive bayes. The variational bayes algorithm without distributed interaction among nodes is not as good as the distributed variational bayes algorithm in terms of accuracy, performance and misjudgment, except for the accelerated processing time.

3.4.4 Simulation on Mininet

In this section, we use the Mininet to emulate a virtual network environment to evaluate our architecture and algorithm. Mininet is a powerful network simulation platform, where the operator could easily simulate the network topology and monitor flow information in real time. We adopt the Floodlight as the controller of OVS switches. To monitor network traffic information, we use sFlow monitoring software, which consists of sFlow Agent and sFlow Collector. As a client, the sFlow Agent is embedded in the OpenVSwitch (OVS). By obtaining the interface statistics and data information on the device, agents package the information into sFlow messages and send them to the designated Collector. As a remote server, Collector is responsible for analyzing, summarizing, and generating traffic reports for sFlow messages.

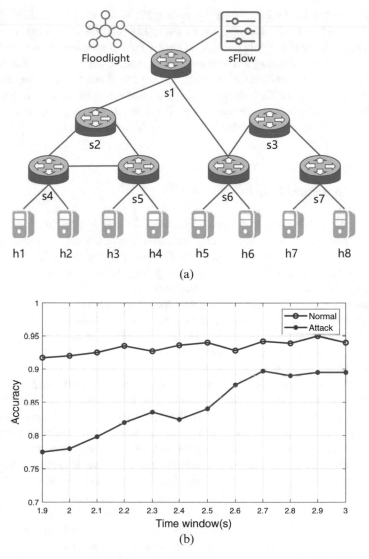

Fig. 3.34 Simulation analysis. (a) Simulation topology. (b) The relationship between classification accuracy and time window settings

3.4.4.1 Network Topology

In our simulation, we generalize the DDoS flood with random IP and a *timeout* command. We set the network topology as shown in Fig. 3.34a. The default port of IP is set as 6343 and sampling rate as every 10 packets. We set the IP of 8 hosts as 10.0.0.1–10.0.0.8, and each host can be a normal user also an attacker, randomly sending requests to switches during processing time. We set the process to run for

100 episodes and the length of each episode is 10. When we set the initial spoofing IP for a DDoS flood to be 10.1.1.1, which changes with different episodes to prevent it from being judged as abnormal traffic easily. After one episode, we use *timout* command to abort the *hping3* command, which matches the length of time window in the actual network architecture. During the process of simulations, we can use the *top* command to see the CPU usage, which is also a feature of DDoS attacks. Taking this vision one step further, we can modify the flow table to drop the specified flow, such as OpenFlowSwitch only drops the flow of ICMP without affecting the normal HTTP service.

3.4.4.2   Experiment Results

In this section, we evaluated the performance of our hybrid machine learning algorithm and OpenFlow based centralized naive Bayes detection algorithm in this topology. Additionally, we explored the performance fluctuation with the change of the time window ΔT, which is a set for aggregating statistics over multiple packets. For our hybrid machine learning algorithm, considering the time-consuming caused by iteration, we select the iteration times as 2000. Due to the cost of time-consuming deserves concern, we care about the time window in the simulation. Comparing the mislabeling and dropping of normal traffic, mislabeling attack traffic costs more. Therefore, we evaluated the relationship between classification accuracy and time window settings as shown in Fig. 3.34b.

We calculate the precision of classification both normal traffic and attack traffic using three methods and list the relatively best time window in Table 3.16. By comparing the results, we notice that the accuracy and efficiency of abnormal flow detection are significantly improved in the subsequent detection process for the labeled attack flows. Although centralized DDoS deployment using the naive Bayes classifier in the controller has shorter time windows, it increases the risk of mislabeling the traffic in edge switches. Traditional DDoS defense on SDN calls Floodlight's static flow entry pusher to discard DDoS attack packets for defense in the experiment. There is little difference in the precision of the three methods in labeling normal traffic, while naive Bayes classifier has an advantage in labeling attack traffic with a relatively short time window. Furthermore, the simulation experiment on Mininet can be transplanted to the real environment, which provides a guarantee for the realization and portability of the algorithm.

Table 3.16 IoT DDoS traffic classification

Methods	Precision (normal)	Precision (attack)	Time window
DVB	0.942	0.897	2.7 s
Naive Bayes	**0.953**	**0.917**	**1.2 s**
Flow-table	0.927	0.868	–

We highlight the highest accuracy and lowest latency of experimental results, which correspond to the method with better performance.

3.5 Summary

In this chapter, we discuss the main challenges of intelligence in the Internet of Things and introduce several network awareness algorithms based on machine learning. Network traffic information includes service-level information (such as QoS/QoE) and abnormal traffic detection information. To solve these problems, we first proposed an end-to-end IoT traffic classification method based on deep learning capsule network, aiming to form an efficient classification mechanism integrating feature extraction, feature selection, and classification model. Moreover, we propose a hybrid detection architecture to improve the security of the network and to reduce the bandwidth consumption in the meantime. In addition, we propose that the traffic flow is considered as time series and stored in the form of a matrix, which facilitates the use of deep learning models for classification. Moreover, attention based LSTM and HAN neural network architectures are constructed for traffic classification. Finally, we propose an in-network distributed machine learning-based DDoS detection architecture. The switches could collaboratively learn the DDoS detection policy through a centralized parameter synchronous platform.

References

1. H. Yao, P. Gao, J. Wang, P. Zhang, C. Jiang, Z. Han, Capsule network assisted IoT traffic classification mechanism for smart cities. IEEE Int. Things J. **6**(5), 7515–7525 (2019)
2. H. Yao, P. Gao, P. Zhang, J. Wang, C. Jiang, L. Lu, Hybrid intrusion detection system for edge-based iiot relying on machine-learning-aided detection. IEEE Netw. **33**(5), 75–81 (2019)
3. H. Yao, C. Liu, P. Zhang, S. Wu, C. Jiang, S. Yu, Identification of encrypted traffic through attention mechanism based long short term memory. IEEE Trans. Big Data **8**, 241–252 (2019)
4. W. He, Y. Liu, H. Yao, T. Mai, N. Zhang, F.R. Yu, Distributed variational bayes-based in-network security for the Internet of Things. IEEE Int. Things J. **8**(8), 6293–6304 (2020)
5. Y. Kawamoto, N. Yamada, H. Nishiyama, N. Kato, Y. Shimizu, Y. Zheng, A feedback control based crowd dynamics management in IoT system. IEEE Int. Things J. **4**, 1466–1476 (2017)
6. Y. Kawamoto, H. Nishiyama, N. Kato, Y. Shimizu, A. Takahara, T. Jiang, Effectively collecting data for the location-based authentication in internet of things. IEEE Syst. J. **11**, 1403–1411 (2017)
7. S. Verma, Y. Kawamoto, Z.M. Fadlullah, H. Nishiyama, N. Kato, A survey on network methodologies for real-time analytics of massive IoT data and open research issues. IEEE Commun. Surv. Tutor. **19**, 1457–1477 (2017)
8. T. Wang, J. Tan, W. Ding, Y. Zhang, F. Yang, J. Song, Z. Han, Inter-community detection scheme for social Internet of Things: a compressive sensing over graphs approach. IEEE Int. Things J. **5**, 1–1 (2018)
9. J. Ni, K. Zhang, X. Lin, X.S. Shen, Securing fog computing for internet of things applications: challenges and solutions. IEEE Commun. Surv. Tutor. **20**, 601–628 (2018)
10. C.L. Hsu, C.C. Lin, An empirical examination of consumer adoption of internet of things services: network externalities and concern for information privacy perspectives. Comput. Human Behav. **62**, 516–527 (2016)
11. D. Ventura, D. Casado-Mansilla, J. López-de Armentia, P. Garaizar, D. López-de Ipina, V. Catania, ARIIMA: a real IoT implementation of a machine-learning architecture for reducing energy consumption, in *International Conference on Ubiquitous Computing and Ambient Intelligence*, (Belfast, UK) (2014), pp. 444–451

12. T. Yonezawa, L. Gurgen, D. Pavia, M. Grella, H. Maeomichi, ClouT: leveraging cloud computing techniques for improving management of massive IoT data, in *IEEE International Conference on Service-Oriented Computing and Applications*, (Washington, DC) (2014), pp. 324–327
13. Y. Ma, J. Rao, W. Hu, X. Meng, X. Han, Y. Zhang, Y. Chai, C. Liu, An efficient index for massive IoT data in cloud environment, in *Proceedings of the 21st ACM International Conference on Information and Knowledge Management* (2012) pp. 2129–2133
14. Z. Ding, J. Xu, Q. Yang, SeaCloudDM: a database cluster framework for managing and querying massive heterogeneous sensor sampling data. J. Supercomput. **66**, 1260–1284 (2013)
15. J. Camhi, Former Cisco CEO John Chambers predicts 500 billion connected devices by 2025. Business Insider (2015)
16. J. Gubbi, R. Buyya, S. Marusic, M. Palaniswami, Internet of Things (IoT): a vision, architectural elements, and future directions. Future Gener. Comput. Syst. **29**, 1645–1660 (2013)
17. S. Chen, H. Xu, D. Liu, B. Hu, A vision of IoT: applications, challenges, and opportunities with China perspective. IEEE Int. Things J. **1**, 349–359 (2014)
18. D. Singh, G. Tripathi, A.J. Jara, A survey of internet-of-things: Future vision, architecture, challenges and services," in *IEEE World Forum on Internet of Things (WF-IoT)*, (Seoul, South Korea) (2014), pp. 287–292
19. J. Zheng, D. Simplot-Ryl, C. Bisdikian, H.T. Mouftah, The internet of things [guest editorial]. IEEE Commun. Mag. **49**, 30–31 (2011)
20. Y. Li, Q. Zhang, R. Gao, X. Xin, H. Yao, F. Tian, M. Guizani, An elastic resource allocation algorithm based on dispersion degree for hybrid requests in satellite optical networks. IEEE Int. Things J. **9**(9), 6536–6549 (2021)
21. W. Wang, M. Zhu, X. Zeng, X. Ye, Y. Sheng, Malware traffic classification using convolutional neural network for representation learning, in *IEEE International Conference on Information Networking*, (Da Nang, Vietnam) (2017), pp. 712–717
22. S. Sabour, N. Frosst, G.E. Hinton, Dynamic routing between capsules, in *Advances in Neural Information Processing Systems* (2017), pp. 3856–3866
23. J. Gong, X. Qiu, S. Wang, X. Huang, Information aggregation via dynamic routing for sequence encoding (2018). arXiv:1806.01501
24. Q. Liang, X. Wang, X. Tian, F. Wu, Q. Zhang, Two-dimensional route switching in cognitive radio networks: a game-theoretical framework. IEEE/ACM Trans. Netw. **23**, 1053–1066 (2015)
25. S. Hochreiter, J. Schmidhuber, Long short-term memory. Neural Comput. **9**, 1735–1780 (1997)
26. X. Tian, Y. Cheng, B. Liu, Design of a scalable multicast scheme with an application-network cross-layer approach. IEEE Trans. Multimedia **11**, 1160–1169 (2009)
27. J. Gu, Z. Wang, J. Kuen, L. Ma, A. Shahroudy, B. Shuai, T. Liu, X. Wang, G. Wang, J. Cai, et al., Recent advances in convolutional neural networks. Pattern Recogn. **77**, 354–377 (2018)
28. Y. Lecun, Y. Bengio, G. Hinton, Deep learning. Nature **521**, 436 (2015)
29. T. Mai, H. Yao, J. Xu, N. Zhang, Q. Liu, S. Guo, Automatic double-auction mechanism for federated learning service market in Internet of Things. IEEE Trans. Netw. Sci. Eng. **9**, 3123–3135 (2022)
30. F. Wang, H. Yao, Q. Zhang, J. Wang, R. Gao, D. Guo, M. Guizani, Dynamic distributed multi-path aided load balancing for optical data center networks. IEEE Trans. Netw. Ser. Manag. **19**, 991–1005 (2021)
31. S. Mukherjee, N. Sharma, Intrusion detection using naive Bayes classifier with feature reduction. Procedia Technol. **4**, 119–128 (2012)
32. B. Subba, S. Biswas, S. Karmakar, Intrusion detection systems using linear discriminant analysis and logistic regression, in *Annual IEEE India Conference (INDICON)*, (New Delhi, India) (2015), pp. 1–6
33. J. Zhang, M. Zulkernine, A hybrid network intrusion detection technique using random forests, in *The First International Conference on Availability, Reliability and Security*, vol. 2006, (Vienna, Austria) (2006), pp. 262–269

34. N. McLaughlin, J. Martinez del Rincon, B. Kang, S. Yerima, P. Miller, S. Sezer, Y. Safaei, E. Trickel, Z. Zhao, A. Doupe, et al., Deep android malware detection, in *The Seventh ACM on Conference on Data and Application Security and Privacy*, (Tempe, AZ) (2017), pp. 301–308
35. W. Wang, Y. Sheng, J. Wang, X. Zeng, X. Ye, Y. Huang, M. Zhu, HAST-IDS: learning hierarchical spatial-temporal features using deep neural networks to improve intrusion detection. IEEE Access **6**, 1792–1806 (2018)
36. A. Chawla, B. Lee, S. Fallon, P. Jacob, Host based intrusion detection system with combined CNN/RNN model, in *Proceedings of Second International Workshop on AI in Security* (2019), pp. 149–158
37. L. Xiao, X. Wan, C. Dai, X. Du, X. Chen, M. Guizani, Security in mobile edge caching with reinforcement learning. IEEE Wirel. Commun. **25**, 116–122 (2018)
38. M. Min, D. Xu, L. Xiao, Y. Tang, D. Wu, Learning-based computation offloading for IoT devices with energy harvesting. IEEE Trans. Vehic. Technol. **68**, 1930–1941 (2019)
39. D. Oh, D. Kim, W.W. Ro, A malicious pattern detection engine for embedded security systems in the internet of things. Sensors **14**, 24188–24211 (2014)
40. L. Wallgren, S. Raza, T. Voigt, Routing attacks and countermeasures in the RPL-based internet of things. Int. J. Distrib. Sensor Netw. **9**, 794326 (2013)
41. A. Le, J. Loo, K.K. Chai, M. Aiash, A specification-based IDS for detecting attacks on RPL-based network topology. Information **7**, 25 (2016)
42. K. Wang, M. Du, D. Yang, C. Zhu, J. Shen, Y. Zhang, Game-theory-based active defense for intrusion detection in cyber-physical embedded systems. ACM Trans. Embed. Comput. Syst. **16**, 18:1–18:21 (2016)
43. G. Giambene, S. Kota, P. Pillai, Satellite-5g integration: a network perspective. IEEE Netw. **32**, 25–31 (2018)
44. K. Wang, M. Du, Y. Sun, A. Vinel, Y. Zhang, Attack detection and distributed forensics in machine-to-machine networks. IEEE Netw. **30**, 49–55 (2016)
45. K. Wang, Y. Wang, Y. Sun, S. Guo, J. Wu, Green industrial Internet of Things architecture: an energy-efficient perspective. IEEE Commun. Mag. **54**, 48–54 (2016)
46. Z. Qin, H. Yao, T. Mai, D. Wu, N. Zhang, S. Guo, Multi-agent reinforcement learning aided computation offloading in aerial computing for the internet-of-things. IEEE Trans. Serv. Comput. (01), 1–12 (2022)
47. K. Gai, M. Qiu, H. Zhao, Privacy-preserving data encryption strategy for big data in mobile cloud computing. IEEE Transactions on Big Data **7**(4), 678–688 (2017)
48. Z. Cao, G. Xiong, Y. Zhao, Z. Li, L. Guo, A survey on encrypted traffic classification, in *International Conference on Applications and Techniques in Information Security*, (Berlin, Heidelberg) (2014), pp. 73–81
49. R. Alshammari, A.N. Zincir-Heywood, Investigating two different approaches for encrypted traffic classification, in *2008 Sixth Annual Conference on Privacy, Security and Trust* (2008), pp. 156–166
50. R. Alshammari, A.N. Zincir-Heywood, Machine learning based encrypted traffic classification: Identifying SSH and Skype, in *2009 IEEE Symposium on Computational Intelligence for Security and Defense Applications* (2009), pp. 1–8
51. M. Dusi, A. Este, F. Gringoli, L. Salgarelli, Using GMM and SVM-based techniques for the classification of SSH-encrypted traffic, in *2009 IEEE International Conference on Communications* (2009), pp. 1–6
52. Z. Wang, The applications of deep learning on traffic identification. BlackHat, USA (2015)
53. M. Lotfollahi, R.S.H. Zade, M.J. Siavoshani, M. Saberian, Deep packet: A novel approach for encrypted traffic classification using deep learning. Soft. Comput. **24**(3), 1999–2012 (2020)
54. W. Wang, M. Zhu, J. Wang, X. Zeng, Z. Yang, End-to-end encrypted traffic classification with one-dimensional convolution neural networks, in *2017 IEEE International Conference on Intelligence and Security Informatics (ISI)* (2017), pp. 43–48
55. M. Lopez-Martin, B. Carro, A. Sanchez-Esguevillas, J. Lloret, Network traffic classifier with convolutional and recurrent neural networks for Internet of Things. IEEE Access **5**, 18042–18050 (2017)

56. W. Wang, Y. Sheng, J. Wang, X. Zeng, X. Ye, Y. Huang, M. Zhu, Hast-ids: Learning hierarchical spatial-temporal features using deep neural networks to improve intrusion detection. IEEE Access **6**, 1792–1806 (2018)
57. A.H. Lashkari, G. Draper-Gil, M.S.I. Mamun, A.A. Ghorbani, Characterization of encrypted and VPN traffic using time-related features, in *Proceedings of the 2nd International Conference on Information Systems Security and Privacy (ICISSP)*, (Rome, Italy) (2016), pp. 407–414
58. G. Aceto, D. Ciuonzo, A. Montieri, A. Pescapè, Mobile encrypted traffic classification using deep learning, in *2018 Network Traffic Measurement and Analysis Conference (TMA)* (2018), pp. 1–8
59. R. Li, X. Xiao, S. Ni, H. Zheng, S. Xia, Byte segment neural network for network traffic classification, in *2018 IEEE/ACM 26th International Symposium on Quality of Service (IWQoS)* (2018), pp. 1–10
60. A. Graves, *Long Short-Term Memory* (Springer, Berlin, 2012), pp. 37–45
61. D. Bahdanau, K. Cho, Y. Bengio, Neural machine translation by jointly learning to align and translate (2014). arXiv preprint arXiv:1409.0473
62. Z. Yang, D. Yang, C. Dyer, X. He, A. Smola, E. Hovy, Hierarchical attention networks for document classification, in *Conference of the North American Chapter of the Association for Computational Linguistics: Human Language Technologies*, (San Diego, CA) (2016), pp. 1480–1489
63. Y.-A. Chung, H.-T. Lin, and S.-W. Yang, Cost-aware pretraining for multiclass cost-sensitive deep learning (2015). arXiv preprint arXiv:1511.09337
64. H. Yao, T. Mai, J. Wang, Z. Ji, C. Jiang, Y. Qian, Resource trading in blockchain-based industrial Internet of Things. IEEE Trans. Ind. Inf. **15**(6), 3602–3609 (2019)
65. C. Qiu, H. Yao, C. Jiang, S. Guo, F. Xu, Cloud computing assisted blockchain-enabled Internet of Things. IEEE Trans. Cloud Comput. **10**, 247–257 (2019)
66. R. Mahmoud, T. Yousuf, F. Aloul, and I. Zualkernan, Internet of things (IoT) security: Current status, challenges and prospective measures, in *2015 10th International Conference for Internet Technology and Secured Transactions (ICITST)* (IEEE, Piscataway, 2015), pp. 336–341
67. H. Yao, H. Liu, P. Zhang, S. Wu, C. Jiang, S. Guo, A learning-based approach to intra-domain qos routing. IEEE Trans. Veh. Technol. **69**(6), 6718–6730 (2020)
68. C. Qiu, H. Yao, F.R. Yu, F. Xu, C. Zhao, Deep q-learning aided networking, caching, and computing resources allocation in software-defined satellite-terrestrial networks. IEEE Trans. Vehic. Technol. **68**(6), 5871–5883 (2019)
69. Q. Yan, F.R. Yu, Q. Gong, J. Li, Software-defined networking (SDN) and distributed denial of service (DDoS) attacks in cloud computing environments: a survey, some research issues, and challenges. IEEE Commun. Surv. Tutor. **18**(1), 602–622 (2016)
70. F. Li, X. Xu, H. Yao, J. Wang, C. Jiang, S. Guo, Multi-controller resource management for software-defined wireless networks. IEEE Commun. Lett. **23**(3), 506–509 (2019)
71. F. Li, H. Yao, J. Du, C. Jiang, Y. Qian, Stackelberg game-based computation offloading in social and cognitive industrial Internet of Things. IEEE Trans. Ind. Inform. **16**(8), 5444–5455 (2019)
72. C. Qiu, F.R. Yu, H. Yao, C. Jiang, F. Xu, C. Zhao, Blockchain-based software-defined industrial Internet of Things: a dueling deep *Q*-learning approach. IEEE Int. Things J. **6**(3), 4627–4639 (2018)
73. H. Yao, T. Mai, X. Xu, P. Zhang, M. Li, Y. Liu, NetworkAI: An intelligent network architecture for self-learning control strategies in software defined networks. IEEE Int. Things J. **5**(6), 4319–4327 (2018)
74. H. Yao, S. Ma, J. Wang, P. Zhang, C. Jiang, S. Guo, A continuous-decision virtual network embedding scheme relying on reinforcement learning. IEEE Trans. Netw. Service Manag. **17**, 864–875 (2020)
75. R. Bifulco, G. Rétvári, A survey on the programmable data plane: abstractions, architectures, and open problems, in *2018 IEEE 19th International Conference on High Performance Switching and Routing (HPSR)* (IEEE, Piscataway, 2018), pp. 1–7

76. R. Harrison, Q. Cai, A. Gupta, J. Rexford, Network-wide heavy hitter detection with commodity switches, in *Proceedings of the Symposium on SDN Research* (2018), pp. 1–7
77. J. Liang, R. Liu, Stacked denoising autoencoder and dropout together to prevent overfitting in deep neural network, in *2015 8th International Congress on Image and Signal Processing (CISP)* (IEEE, Piscataway, 2015), pp. 697–701
78. G. Karatas, O. Demir, O.K. Sahingoz, Deep learning in intrusion detection systems, in *2018 International Congress on Big Data, Deep Learning and Fighting Cyber Terrorism (IBIGDELFT)* (IEEE, Piscataway, 2018), pp. 113–116
79. Y. Mirsky, T. Doitshman, Y. Elovici, A. Shabtai, Kitsune: an ensemble of autoencoders for online network intrusion detection (2018). Preprint arXiv:1802.09089
80. F.A. Khan, A. Gumaei, A. Derhab, A. Hussain, A novel two-stage deep learning model for efficient network intrusion detection. IEEE Access **7**, 30373–30385 (2019)
81. N. Shone, T.N. Ngoc, V.D. Phai, Q. Shi, A deep learning approach to network intrusion detection. IEEE Trans. Emerg. Topics Comput. Intell. **2**(1), 41–50 (2018)
82. D. Li, L. Deng, M. Lee, H. Wang, IoT data feature extraction and intrusion detection system for smart cities based on deep migration learning. Int. J. Inf. Manag. **49**, 533–545 (2019)

Chapter 4
Intelligent Traffic Control

Abstract Whether in a wired network or a wireless network, how to carry out effective and reliable traffic control is always discussed. However, most of the solutions to this problem rely heavily on manual processes. In order to solve this problem, in this chapter, we apply several artificial intelligence approaches to network traffic control. First, we introduce a social-based mechanism in the routing design of delay-tolerant network (DTN) and propose a cooperative multi-agent reinforcement learning (termed as QMIX) aided routing algorithm adopting centralized training and distributed execution learning paradigm. Then, in traditional network, we propose a new identity for networking routers—vectors, and a new routing principle based on these vectors and neural network is designed accordingly. In addition, we construct a jitter graph-based network model as well as a Poisson process-based traffic model in the context of 5G mobile networks and design a QoS-oriented adaptive routing scheme based on DRL. Finally, based on the SDN architecture, we propose a pair of machine learning aided load balance routing schemes considering the queue utilization (QU), which divide the routing process into three steps, namely the dimension reduction, and the QU prediction as well as the load balance routing. Extensive simulation results show that these traffic control methods have significant performance advantages.

Keywords Traffic control · Delay-tolerant network · Multi-agent reinforcement learning · QoS routing · Queue utilization

Whether in a wired network or a wireless network, how to carry out effective and reliable traffic control is always discussed. However, most of the solutions to this problem rely heavily on manual processes. In order to solve this problem, in this chapter, we apply several artificial intelligence approaches to network traffic control. First, we introduce a social-based mechanism in the routing design of delay-tolerant network (DTN) and propose a cooperative multi-agent reinforcement learning (termed as QMIX) aided routing algorithm adopting centralized training and distributed execution learning paradigm [1]. Then, in traditional network, we propose a new identity for networking routers—vectors, and a new routing principle based on these vectors and neural network is designed accordingly [2]. In addition, we construct a jitter graph-based network model as well as a Poisson process-based traffic model in the context of 5G mobile networks and design a QoS-oriented adaptive routing scheme based on DRL [3]. Finally, based on the SDN architecture, we propose a pair of machine learning aided load balance routing schemes considering the queue utilization (QU), which divide the routing process into three steps, namely the dimension reduction, and the QU prediction as well as the load balance routing [4]. Extensive simulation results show that these traffic control methods have significant performance advantages.

4.1 QMIX Aided Routing in Social-Based Delay-Tolerant Networks

Delay-tolerant network (DTN) is one of the emerging communication schemes that aim to operate effectively in extreme conditions (i.e., lack continuous network connectivity) and over very large distances [5]. In DTN, communication is challenged by sporadic and intermittent connections, and therefore traditional ad hoc routing protocols (e.g., ad hoc on-demand distance vector (AODV) and dynamic source routing (DSR)) are difficult to establish end-to-end routing path [6]. To overcome these challenges, DTN routing protocols adopt the "store and forward" mechanism. Packets are moved and stored incrementally in the whole network, hoping to reach the destination. DTN is originally developed for the initial interplanetary Internet (IPN) architecture to deal with message corruption and inevitable delays in deep-space communications. Recently, with the advance of 5G applications (e.g., Internet of Vehicles, intelligent communicable devices), DTN has also been introduced to these highly dynamic network scenarios [7]. As shown in Fig. 4.1, vehicles are connected through the roadside units (RSUs) to form the Internet of Vehicles, and mobile smart devices can also be interconnected. However, the highly dynamic topology and the limited node capacity in such network scenarios present great challenges to DTN routing [8].

Traditionally, DTN algorithms mostly rely on opportunity-based models, which are greedy schemes [9]. Most of them only rely on the comparisons between per-node metrics when encountering. For instance, in [10], Zhao et al. proposed a history-based forwarding message protocol that greedily transmits packets to all nodes whose delivery probability is greater than the current node. In [11], Dubois-Ferrière et al. proposed a FRESH scheme that forwards the message to the relay as long as it has encountered the destination more recently than the current node does. Although the greedy strategy is simple, however, this will not guarantee that the packet will eventually reach the destination. In addition, multiple copies of the same message may flood the network, which will eventually lead to routing failure and overload [12]. To overcome these insufficiencies, social-based methods are introduced into the routing design in DTN applications. Social-based DTN uses

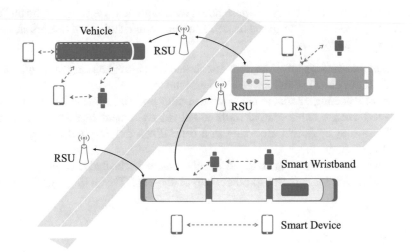

Fig. 4.1 The DTN scenario in 5G

the community and the centrality to increase the delivery rate of the network [13]. These attributes are usually long-term characteristics and less volatile than node mobility [14]. The community in DTN reflects the social relationship among wireless devices. A member in a community is usually more willing to communicate with members in the same community than with other members [15]. Therefore, through community detection, DTN nodes can more easily find the optimal next node. Centrality represents the "popularity" of the node, which means that the node is more willing to transmit data [12]. In this chapter, we introduce the social-based mechanism to our routing algorithm. We divide DTN nodes into countable communities. In addition, to speed up the convergence of the community detection algorithm, we set an iteration threshold. With that in mind, we propose a centralized and distributed hybrid architecture in the social-based DTN.

In a distributed system, it is worth asking how could the nodes cooperatively learn the routing policy while executing in a distributed fashion among multiple communities? Inspired by the recent success of the distributed reinforcement learning in multi-agent system control and the positive social attributes of DTN nodes [16, 17], we introduce a cooperative multi-agent reinforcement learning algorithm QMIX to our system, which helps nodes not only consider their own performance but also consider the undeniable connections of other nodes [18].

In general, for the positive social characteristics of DTN nodes, we describe the problem of selecting a reliable next-hop neighbor node as a cooperative Markov game, which can be seen as a Dec-POMDP model. Besides, we introduce the QMIX algorithm in a centralized training and decentralized execution way. Because the number of communities is far less than the number of nodes, the complexity of the algorithm is greatly reduced. What is more, to prevent the routing algorithm from transmitting a large amount of information to nodes with high social indicators without restrictions, which causes these nodes overburdened, we consider the buffer occupancy of nodes in a fine-grained way. The main contributions of this article can be summarized as follows:

- We adopt a centralized training and distributed execution learning paradigm and design a hierarchical social-based DTN architecture.
- We model selecting a reliable next-hop neighbor node as a Dec-POMDP model. Combined with a collaborative multi-agent reinforcement learning algorithm, QMIX, routing decisions are made in a centralized training and decentralized execution manner.
- Through a large number of in-depth simulations, it is verified that our proposed routing algorithm is superior to other state-of-the-art DTN routing algorithms.

4.1.1   System Model

In this section, we present a hierarchical social-based DTN architecture firstly. Then, we discuss the trade-off problem of social attributes and buffer congestion. Finally, we design two important evaluation metrics to evaluate routing protocols' performance.

4.1.1.1   Network Model

Consider a special ad hoc network with a weighted direct graph $G = \{V, \mathcal{E}\}$, consisting of a group of DTN nodes $V = \{1, ..., n\}$ and a set of directed links $\mathcal{E}$ in this model. We assume that this model lacks end-to-end connectivity. The weighted direct path is decided by the node pair's history connection records, which include the total connection times $M_{n_s}^{n_d}$ and two corresponding time stamps, T_{n_s,n_d}^{end} and T_{n_s,n_d}^{start}. Through the above parameters, we divide dozens of nodes into K communities $C = \{C_1, ..., C_k\}$ with the community detection algorithm in Sect. 4.4.

As shown in Fig. 4.2, we present a hierarchical social-based DTN architecture. In our architecture, it includes the community-aware clusters, the service units, and the computing center. Community-aware clusters are obtained through the clustering of our proposed community detection algorithm. They represent groups that are more willing to communicate with each other internally. Service units collect and analyze the social information of their respective communities and upload these attributes to the computing center. The computing center collects social attributes and issues routing strategies wisely in turn. Based on this architecture, we adopt the centralized training and distributed execution paradigm. The centralized training refers to the use of a joint function in the computing center to train the agents, while distributed execution means that each agent will only act according to the local observation. This paradigm cannot only avoid the large amount of computation cost of SDN centralized control but also tackle the problem that distributed single-

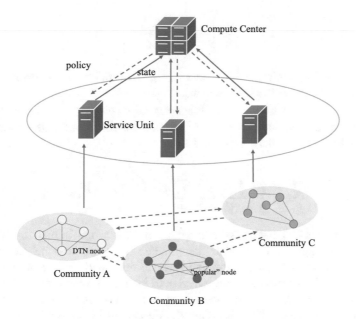

Fig. 4.2 The architecture of the social-based DTN

agent algorithm is likely to fall into a locally optimal solution. Note that we do not identify what the specific service unit is. At the mobile social DTN dominated by smart devices, the service unit could be the access point (ΛP) on the user side or just be a home-aware community model [13]. When it comes to the situation of vehicle delay-tolerant networks, the service unit may be the roadside unit (RSU) deployed along the road section [19].

4.1.1.2 Social Attribute Definitions and Problem Formulation

Next, we design two different centrality indexes at the community level, named the local centrality and the global centrality. The local centrality index $w_{C_i}^{local}$ of community C_i limits the source nodes and the destination nodes in C_i and sums up all the connection time considering the age constant. It can be expressed as

$$w_{C_i}^{local} = \sum_{n_s \in C_i} \sum_{n_d \in C_i} \sum_{m=1}^{M} \varphi^t \cdot \left(T_{n_s,n_d}^{end} - T_{n_s,n_d}^{start} \right), \tag{4.1}$$

where M indicates that the total connection times of n_s and n_d, and φ ($0 < \varphi < 1$) indicates the age constant calculated by time slices. Then, the time index t can be expressed as

$$t = \left\lfloor \frac{T_{now} - T_{n_s,n_d}^{end}}{T_{interval}} \right\rfloor, \tag{4.2}$$

where T_{now} is the simulation time and $T_{interval}$ is a variable time interval parameter that is selected for specific DTN scenarios.

Under the same restrictions, the global centrality index $w_{C_i}^{global}$ of community C_i also limits the source nodes in C_i, but it extends the scope of the destination nodes to the whole DTN. It can be expressed as

$$w_{C_i}^{global} = \sum_{n_s \in C_i} \sum_{n_d \in N} \sum_{m=1}^{M} \varphi^t \cdot \left(T_{n_s,n_d}^{end} - T_{n_s,n_d}^{start}\right). \tag{4.3}$$

Inspired by the fact that messages are more likely to be transmitted to destinations by passing through more "popular" communities [14], we take the centrality into consideration, and however, unlimited transmissions will lead to an overload of nodes with limited capacity. Thus, we should take the problem of buffer capacity limitation into account while transmitting messages cooperatively. Considering the bottleneck effect, $L_{C_i}^{buf}$ represents the maximum value buffer occupation in community C_i and can be calculated by

$$L_{C_i}^{buf} = \max\left(l_1^{buf}, ..., l_n^{buf}\right), n \in C_i, \tag{4.4}$$

where l_n^{buf} is the immediate buffer occupation in node n that belongs to C_i. It can be expressed as

$$l_n^{buf} = \begin{cases} \frac{\sum_{i=0}^{m} H_N^i}{n_{size}} & \text{if } l_n^{buf} \leqslant 1 \\ 1 & \text{if } l_n^{buf} > 1, \end{cases} \tag{4.5}$$

where n_{size} is the total buffer size of node n and is the number of bits of integer type, $\sum_{i=0}^{m} H_N^i$ represents the buffer occupation size in node n, m represents the number of messages, and H_N^i is the number of bits per message.

As a result, the optimization goal of each community can be formulated as

$$\max \alpha \cdot \Delta w_C - \beta \cdot L_C^{buf}, \tag{4.6}$$

where α and β ($0 < \alpha, \beta < 1$) are weight parameters. The optimization goal represents the improvement of the forwarding strategy [10], in which one node would consider any node as a relay as long as the latter node is more willing to forward messages to the destination node. We will specifically present the optimization objective of the design of rewards in Sect. 4.1.3. Here we list some important notations of this article in Table 4.1 for better comprehension.

Table 4.1 List of main notations in MARL routing protocol

Parameter	Definition
$\mathcal{G}$	Network model as a weighted direct graph
n_s, n_d	Source DTN node and destination DTN node in one message
C_i	Community of one DTN node
K	Number of communities (agents)
φ	Age constant
$w_{C_i}^{global}$	Global centrality index of community C_i
$w_{C_i}^{local}$	Local centrality index of community C_i
$L_{C_i}^{buf}$	Maximum value of the buffer occupation in community C_i
$P_{delivery}$	Delivery ratio of the social-based DTN routing protocol
$T_{latency}$	Mean latency of the delivered messages
$Q_{tot}(\boldsymbol{\tau}, \boldsymbol{a})$	Joint action-value function
s	State space, $s \in S$
a	Action space, $a \in A$
o	Observation for an agent, $o \in \Omega$
γ	Discount factor considering reducing the impact of earlier actions
τ_i	Action-observation history for each agent

4.1.1.3 Evaluation Metrics

We design three metrics to evaluate routing protocols' performance, namely the delivery ratio, the mean latency, and the overhead ratio. The delivery ratio is calculated by dividing the sum of data packets which is successfully transmitted to destinations by the total generated messages and can be described as

$$P_{delivery} = \frac{N_{delivered}}{N_{created}}. \tag{4.7}$$

It most intuitively reflects the delivery capacity of DTN routing strategies, while the mean latency can be calculated as

$$T_{latency} = \frac{\sum_{i=1}^{i=tot} T_{latency}^i}{N_{delivered}}. \tag{4.8}$$

Note that the main reason for the latency is the queueing delay of the node buffer. Whether it is not making good use of social attributes or unrestricted transmission of messages results in buffer overload, it will lead to poor performance in the mean delay. The overhead ratio of delivery result can be calculated as

$$P_{overhead} = \frac{N_{relayed} - N_{delivered}}{N_{delivered}}. \tag{4.9}$$

It represents the extra consumption rate in data delivery and reflects the routing transmission efficiency.

4.1.2 Community Detection

In this section, we present an improved heuristic successive iteration algorithm for convergence of modularity [20]. In our algorithm, the modularity function is introduced to measure the quality of community division. It can describe the cohesion within the community and the difference between the communities, and it can be expressed as

$$
Q = \frac{1}{2n} \sum_{i,j} \left[A_{ij} - \frac{s_i s_j}{2n} \right] \cdot \delta(C_i, C_j)
$$

$$
= \frac{1}{2n} \sum_{C} \left[\sum in - \frac{\left(\sum tot \right)^2}{2n} \right],
$$

(4.10)

A is the adjacency matrix. We regard the DTN topology as a weight graph, while A_{ij} represents the weight of the edge between node i and node j, specifically referring to the connection duration between the two. The sum of the weights of all edges s_i connected to node i can be expressed as $s_i = \sum_j A_{ij}$. The sum of weights of all edges n can be reformulated as $n = \frac{1}{2} \sum_{ij} A_{ij}$, which is the normalization function. $\delta(C_i, C_j)$ is 1 if node i and node j are in the same community. $\sum in$ represents the sum of the weights of the edges within the community and $\sum tot$ represents the sum of weights of edges connected to nodes in the community.

To maximize the modularity, we introduce the fast unfolding algorithm [21]. Rather than unrestricted multiple iterations, we set a threshold in the update of modularity value to speed up the convergence of the heuristic algorithm and avoid putting outlier nodes into communities. The modularity gain is a numerical index to evaluate the effect of an iteration, which is a heuristic optimization process. It can be calculated as follows:

$$
\Delta Q = \left[\frac{\sum_{in} + s_{i,in}}{2n} - \left(\frac{\sum_{tot} + s_i}{2n} \right)^2 \right] - \left[\frac{\sum_{in}}{2n} - \left(\frac{\sum_{tot}}{2n} \right)^2 - \left(\frac{s_i}{2n} \right)^2 \right].
$$

(4.11)

The first part means the modularity of a community after adding node n_i, while the second part means the modularity summed by node n_i independently and the independent community. Our improved algorithm iterates until ΔQ cannot reach a preset threshold. We can avoid outliers clustering into communities blindly and get faster convergence to accelerate our total routing algorithm in this way. The community detection algorithm is shown in Algorithm 4.1.

Algorithm 4.1 Community detection algorithm

1: Initialize:each node in the graph is regarded as an independent community
2: **while** The structure of the community stables **do**
3:    **for** node $i = 1, N$ **do**
4:        Get ΔQ_i by trying to assign node i to the community where each of its neighbors is located
5:        **if** $\Delta Q_i > Q_{thread}$ **then**
6:            assign node i to the neighbor's community
7:        **else**
8:            break
9:        **end if**
10:   **end for**
11:   Corresponding to each new node by each new community
12:   Update the weight of the ring of the new node
13:   Update the edge weight
14: **end while**

4.1.3 Dec-POMDP Model and Cooperative MARL Protocol

In this section, we introduce the cooperative Markov game and model the game as a decentralized partially observable Markov decision progress (Dec-POMDP) model [22]. Then we present a cooperative multi-agent reinforcement learning (MARL) algorithm, QMIX [18], to solve the social-based DTN routing problem.

4.1.3.1 Cooperative Markov Game

When multiple agents apply reinforcement learning in a shared environment at the same time, it may be beyond the scope of the Markov decision process (MDP) model because the optimal strategy of a single learner depends not only on the environment but also on the strategies adopted by other learners. MDP can be extended to the case of multi-agents by Markov games. Specifically speaking, as shown in Fig. 4.2, to obtain good routing performance, a community must send messages out on time when others have forwarded their messages to it. Otherwise, serious buffer blockage will occur. On the contrary, its enthusiasm as a relay is meaningless when little messages are delivered to it. This shows that routing decisions between communities must be at the same pace. Thus we describe the problem of choosing the next relay in social-based DTN as a collaborative Markov game.

The Markov game is a tuple $(n, S, A_{1,...,n}, R_{1,...,n}, T)$, where $T : S \times A_1 \times \cdots \times A_n \times \to [0, 1]$ is the transformation function. The Nash equilibrium strategy of the Markov game can be written as $(\pi_1^*, ..., \pi_n^*)$, and then for $\forall s \in S, i = 1, \cdots, n$ and for $\forall \pi_i, i = 1, \cdots, n$,

$$V_i \left(s, \pi_1^*, \cdots, \pi_i^*, \cdots, \pi_n^* \right) \geqslant V_i \left(s, \pi_1^*, \cdots, \pi_i, \cdots, \pi_n^* \right), \tag{4.12}$$

where $V_i(s, \pi_1^*, \cdots, \pi_i^*, \cdots, \pi_n^*)$ is the cumulative state value. According to the Bellman equation, the Nash equilibrium can be described by cumulative state–action value as

$$
\begin{aligned}
\sum_{a \in A_1 \times \cdots \times A_n} Q(s, a) \pi_1^*(s, a_1) \cdots \pi_1^*(s, a_i) \cdots \pi_1^*(s, a_n) \geqslant \\
\sum_{a \in A_1 \times \cdots \times A_n} Q(s, a) \pi_1^*(s, a_1) \cdots \pi_1(s, a_i) \cdots \pi_1^*(s, a_n).
\end{aligned}
\tag{4.13}
$$

Next, we will describe the cooperative Markov game as a Dec-POMDP model and then introduce the QMIX algorithm, to solve the social-based DTN routing problem.

4.1.3.2   Dec-POMDP Model

As mentioned above, we consider the social-based DTN routing strategy as a fully cooperative multi-agent task scenario, which can be described as a Dec-POMDP model. The Dec-POMDP model is the solution framework of the POMDP model under distributed conditions, and its solution process is more complex than the POMDP model. The fundamental reason is that the decision-making of agents under distributed conditions should not only consider the influence of the environment but also consider the strategies of other agents. The Dec-POMDP model is built in a discrete-time dynamic system, and there is usually a time upper limit, which is called steps or stages. In each step, the agent adopts an action, transfers to the next state according to the state transition function, and obtains the observation information in this state.

The Dec-POMDP model can be formalized as a tuple $G = \langle N, S, A, O, R, \Omega, \gamma \rangle$, where N represents the number of agents, $s_i \in S$ describes the true state of the environment, A is the action space, and Ω is the observation space. $O : A \times S \to \Omega$ is the observation function. Each agent has an action-observation history $\tau_i = (a_{i,0}, o_{i,1}, ..., a_{i,t-1}, o_{i,t})$. For each step, each agent conditions a stochastic policy $\pi(a|\tau_i)$, getting the cumulative reward $R_i = \sum_{t=0}^{T} \gamma^t r_i^t$, where γ is a discount factor. On the whole, the joint action-observation history is $\boldsymbol{\tau} = (\tau_1, ..., \tau_n)$ and $\boldsymbol{a} = (a_1, ..., a_n)$ means a joint action. The joint action-value function is $Q_{tot}(\boldsymbol{\tau}, \mathbf{a})$.

Specifically, in our scenario, the observation space of each agent refers to which communities forward messages to it within a time slice and its own bottleneck buffer occupancy. It can be expressed as

$$o_i = \left[U(C_{j_1}), ..., U(C_{j_m}), l_1^{buf}, ..., l_n^{buf} \right], \tag{4.14}$$

where the corresponding U function is 1 if C_j chooses forwarding messages to agent C_i.

The action space refers to whether C_i transferring messages to C_j before the next action coming. Notice that the packet can reach the destination node with less overhead and fewer hops once the packet's source node and destination node are in the same community [14]. So we mainly consider the action as the inter-community transmission. The joint observation and joint action of the multi-agent environment are all discrete $K - level$ tuples, where K is the total number of communities.

The reward of agent C_i can be described as

$$r_i = \begin{cases} \alpha_1 \cdot \Delta w - \beta_1 \cdot L_{C_i}^{buf} & \text{if } message.des \notin C_i \\ -\alpha_2 \cdot w_{C_i}^{local} - \beta_2 \cdot L_{C_i}^{buf} & \text{if } message.des \in C_i, \end{cases} \tag{4.15}$$

where Δw is

$$\Delta w = w_{C_j}^{global} - w_{C_i}^{global}. \tag{4.16}$$

If the message should be spread within this community but is forwarded to other communities or the message is forwarded to a community with a lower global centrality than the original community, the first item of the reward is negative. On the contrary, this action is beneficial to packet transmission. As shown in the second term of the reward function, we also regard the buffer occupancy as an important factor affecting message transmission.

A key stumbling block in a multi-agent environment is how to learn the joint action-value function. On the one hand, learning a central action-value function has a positive impact on the actions of agents. On the other hand, the action-value function is not easy to learn when there are many agents. Even if the central function can be learned, there is no obvious method to extract decentralized policies that allow each agent to select only an individual action based on the individual observation.

In the next section, we introduce QMIX aided routing algorithm to solve the above problem in a centralized training and distributed execution (CTDE) way.

Fig. 4.3 The collaborate
MARL structure

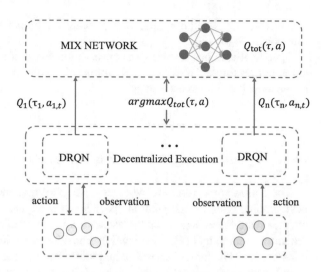

4.1.3.3   Cooperative Multi-agent Reinforcement Learning

The community detection could make messages forward between communities in a cooperative way. The centrality of a community means the willingness to forward packets. Emphasizing these positive cooperative social characteristics in social-based DTN, we introduce the QMIX algorithm to solve the selecting problem of the next relay in a centralized training and distributed execution way.

Figure 4.3 indicates the network structure of this cooperative multi-agent reinforcement learning algorithm. The lower layer shows that each agent uses a Deep Recurrent Q-Learning (DRQN) [23] to fit whose own Q value $Q_i(\tau_i, a_i; \theta_i)$. DRQN introduces RNN to deal with the partially observable problem, which is $Q(o, a|\theta) \neq Q(s, a|\theta)$ in DQN. DQN circularly inputs the current observation $o_{i,t}$ and the action $a_{i,t-1}$ of the previous time to obtain the Q value.

Nodes with positive social characteristics can extract decentralized strategies which are consistent with the centralization strategy. We only need to ensure that a global $argmax$ performed on Q_{tot} yields the same result as a set of individual $argmax$ operations which are performed on each Q_a. This means that the local optimal action chosen by each agent is just a part of the global optimal action and it can be expressed as

$$
\arg \max_{\boldsymbol{a}} Q_{tot}(\boldsymbol{\tau}, \boldsymbol{a}) = \begin{pmatrix} \arg \max_{a_1} Q_1(\tau_1, a_1) \\ \cdot \\ \cdot \\ \cdot \\ \arg \max_{a_n} Q_n(\tau_n, a_n) \end{pmatrix}, \tag{4.17}
$$

where

$$\frac{\partial Q_{tot}}{\partial Q_i} \geqslant 0, \forall i \in \{1, ..., n\}. \tag{4.18}$$

Equations (4.17) and (4.18) show that the monotonicity can be implemented by a constraint on the relationship between Q_{tot} and each Q_a. For each Q_a, agent a can execute distributed and greedy actions. So it is very easy to calculate $argmax_a Q_{tot}$. Conversely, the strategy of each agent can be explicitly extracted from Q_{tot}.

To achieve the above constraints, the mixing network takes the outputs of the agents' neural networks as inputs and mixes them monotonically, which can produce the values of Q_{tot}. To enforce the monotonicity constraint of Eq. (4.18), the weights (but not the biases) of the mixing network, which should be generated by a separate hyperparametric network, are restricted to be non-negative. Hence it can represent any joint action value function that could factor into a nonlinear monotonic combination of the agent's individual value functions.

In addition, there should be an appropriate order of the values of actions for the agent's individual value functions. It can be expressed as

$$Q_i(s_t, a_i) > Q_i(s_t, a_i') \rightleftharpoons$$
$$Q_{tot}(s_t, (\boldsymbol{a}_{-a_i}, a_i)) > Q_{tot}(s_t, (\boldsymbol{a}_{-a_i}, a_i')). \tag{4.19}$$

Equation (4.19) represents the ordering of the agent's actions in the joint action value function, while in Dec-POMDP, the observation of each agent could not observe the full state, which causes the disability of distinguishing the true state. If the ordering of the agent's value function is wrong depicted in Eq. (4.20), the mixing network would be unable to correctly represent Q_{tot} considering the monotonicity constraints.

$$Q_i(\tau_i, a_i) > Q_i(\tau_i, a_i') \text{ when}$$
$$Q_{tot}(s_t, (\boldsymbol{a}_{-a_i}, a_i)) < Q_{tot}(s_t, (\boldsymbol{a}_{-a_i}, a_i')). \tag{4.20}$$

The final cost function of QMIX can be expressed as

$$L(\theta) = \sum_{i=1}^{b} \left[\left(y_i^{tot} - Q_{tot}(\boldsymbol{\tau}, \boldsymbol{a}, s; \theta) \right)^2 \right], \tag{4.21}$$

where y^{tot} is the time difference target:

$$y^{tot} = r + \gamma \max_{a'} \overline{Q}_{tot}(\boldsymbol{\tau}', \boldsymbol{a}', s'; \overline{\theta}). \tag{4.22}$$

The update uses the way of gradient descent by the backpropagation. b represents the number of samples that are taken from empirical memory. $\overline{Q}(\boldsymbol{\tau}', \boldsymbol{a}', s'; \overline{\theta})$

Algorithm 4.2 Cooperative MARL algorithm for social-based DTN

1: Initialize replay buffer
2: **for** $episode = 1, M$ do **do**
3:    Initialize network environment
4:    **for** $step = 1, T$ do **do**
5:       Each agent obtains its observation o
6:       Execute the action to get each new observation o' and reward r of each
          agent
7:       Store (o, a, r, o') to replay buffer
8:    **end for**
9:    **for** agent $t = l, N$ do **do**
10:      Randomly extract a batch from replay buffer
11:      Calculate $Q_i(\tau_i, a_i; \theta_i)$ and $\max_{a_{i'}} \bar{Q}_i(\tau_i', a_i'; \theta_i')$ by DRQN
12:   **end for**
13:   Input all $Q_i(\tau, a, s; \theta)$ into the mixing network
14:   Set: $y^{tot} = r + \gamma \max_{a'} \bar{Q}(\tau', a', s'; \theta')$ by (1)
15:   Perform a gradient descend step on

$$L(\theta) = \sum_{i=1}^{b} [(y_i^{tot} - Q_{tot}(\tau, a, s; \theta))^2]$$

16: **end for**

---

indicates the target network. Under the action of Eq. (4.17), the computation of the maximum value of Q_{tot} can be solved linearly with an increase in the number of agents. The specific pseudo-code of the QMIX algorithm is shown in Algorithm 4.2.

4.1.3.4   Complexity Analysis

As for the time complexity of our proposed multi-agent routing algorithm, the time complexity of neural network training is $O(E * D/B * T)$, where E represents the epoch size, D represents the size of dataset, B denotes the batch size, and $O(T)$ denotes the time complexity of a single iteration. Under the restriction of Eq. (4.17), we consider $O(T) = O(n * T1)$, where $O(T1)$ is the time complexity of a single agent. Furthermore, the number of layers and the number of neurons in the agent are listed in Table 4.3.

4.1.4 Experiments and Simulation Results

In this section, we apply QMIX to real-life wireless datasets. We simulated in the Opportunistic Network Environment (ONE) simulator [24] based on JDK 1.8 and Pytorch 1.7.0 based on Python 3.6 in Windows 10. The hardware environment is Inter (R) Core (TM) i5-8500 CPU @ 3.00 GHz (6 CPUs), 3.0 GHz.

4.1.4.1 Experiment Configuration

We choose INFOCOM05 and INFOCOM06 dataset [25] for analysis. The information and some specific important configurations of using the datasets are listed in Table 4.2. First, we exploit the improved community detection algorithm with a threshold to aggregate a large number of nodes into countable communities. Figure 4.4 shows an iterative result graph in the process of maximizing the modularity in Gephi based on JDK1.8. The lines with color mean the connection of node pairs. The nodes of different colors belong to different communities which are formed after iteratively optimizing the modularity value.

Important hyperparameters of the multi-agent reinforcement learning are listed in Table 4.3. We choose Deep Q-learning (DQL) [26] published in Nature by Deep-mind as the baseline to prove the advantage of our proposed multi-agent reinforcement learning routing algorithm. DQL is a value-based and off-policy single-agent reinforcement learning algorithm, which updates by the time difference (TD) with ϵ-greedy choosing actions. It can be depicted as

$$a_i = \arg\max_a Q_i^{\mu}(o_i, a). \tag{4.23}$$

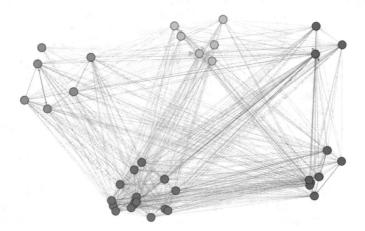

Fig. 4.4 An iterative graph of the community detection algorithm

Table 4.2 DataSet configuration

Parameter	Infocom5	Infocom6
Number of DTN nodes	41	98
Number of community	6	5
Simulation time	275,000 s	34,3000 s
Interface transmit speed	250 Kbps	250 Kbps
Node buffer occupation	5 M	5 M
Message interval	(300,400)	(300,400)
Single message size	(0.8 k,1 k)	(0.8 k,1 k)

Table 4.3 MARL and DQL configuration

Parameter	Value
Discount factor (γ)	0.95
Default learning rate (α)	0.005
Default batch size	60 episodes
Number of neuron cell	64
Number of network layers	3
Maximum episode duration	20 s
Maximum episodes	12,000
Use dueling	True

DQL also adopts the experience replay strategy to train the reinforcement learning process and sets the target network to deal with the TD bias in the time difference algorithm.

4.1.4.2   Training Performance

We evaluate the delivery ratio and mean latency with the lifetime of the packet as the independent variable. The convergence graph of reward is shown in Fig. 4.5a, b. From Fig. 4.5a, we test the effect of different learning rates on the training results. We can see that convergence can be achieved at about 3200 episodes when $learning Rate = 0.0005$ and $learning Rate = 0.005$. However, when the learning rate increases, convergence cannot occur. This can be explained that the gradient on Eq. (4.21) may oscillate back, forth near the minimum, and may not even converge. In the trade-off between accuracy and time cost, we use $learning Rate = 0.005$ for our following experiment.

We then compare the training effect between QMIX and the single-agent reinforcement learning, DQN. As depicted in Fig. 4.5c, our cooperative MARL routing algorithm gets a higher reward and can be more stable than the DQN trainer. Under the same training parameters, QMIX can reach a relatively convergent state under 3000 iterations, while DQN needs to spend more training time so that the reward can stabilize after 9000 iterations. The distributed training and distributed execution of DQN lack cooperation between agents, which results in a high reward for one single agent, but low rewards for other agents. In the specific routing

Fig. 4.5 Compare the convergence graph of reward. (**a**) Mean reward in different learning rate. (**b**) Mean reward in different batch size. (**c**) Mean reward in MARL and DQN

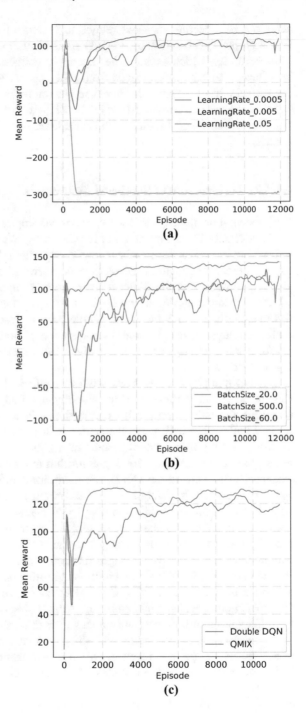

scenarios of selecting the next relays, to increase the value of the first term in Eq. (4.6), the agent will unrestrictedly transmit its messages to other agents with high centrality, which causes other agents to congest. In addition, the agent will be reluctant to receive messages from other agents in order to selfishly increase the second part of the reward. Therefore, the training curve cannot be stable and the training process performs poorly in DQN, while with the QMIX algorithm, excellent performance could get because of the consideration of the cooperation among multi-agents.

4.1.4.3   Performance of the Real-Life Datasets

With the final neural network model, nodes can implement their own policies separately after 12000 episodes of centralized training. We choose DQN as the baseline to outstand the performance of our routing algorithm. We also use BubbleRap [14] as the comparative method, which adopts greedy ideas to forward messages. We also experiment with the direct delivery routing algorithm as the benchmark, in which each node carries its message and moves continuously until it meets the destination node, which means the whole communication process never uses other nodes as relays. At last, in order to make the experiment more comprehensive, we list the result of the Epidemic routing algorithm with and without considering buffer capacity limitation.

As shown in Fig. 4.6a, the highest delivery ratio is the Epidemic routing protocol. Regardless of the buffer limit, the flooding method makes every node transmit messages through current valid connections. In actual situations, however, the limited buffer size will cause the buffer to fill up, thereby blocking the routing of data, resulting in poor performance of Epidemic when TTL increases. DQN is not as good as QMIX aided routing algorithm in terms of routing delivery rate. By extracted from Q_{tot} in Eq. (4.17), the centralized training in QMIX allows each agent to consider the overall reward according to the strategies of other agents when choosing actions. While adopting DQN, agents become selfish and competing with each other due to the lack of global constraints. Selfish communities will greedily transfer their own messages to the regions with higher centrality index and are reluctant to receive messages from other communities to increase their own reward in Eq. (4.15). QMIX aided routing algorithm outperforms the state-of-the-art algorithm in social-based DTN. BubbleRap chooses the relay node by the rank of "popular" nodes, which could cause the "first in first out" (FIFO) drop mechanism in buffer spaces of high centrality nodes over time. With the packet's time-to-live (TTL) increasing, it can reduce the situation that packets delete in their source node's buffer before they forward, which means there will be more opportunities to transfer messages between nodes. However, it can also cause the node buffer overflow so that the increased delivery ratio is not obvious. In our proposed routing algorithm, the design of the agent reward in Eq. (4.6) considers not only the positive social attributes but also the buffer occupation. With appropriate weight parameters

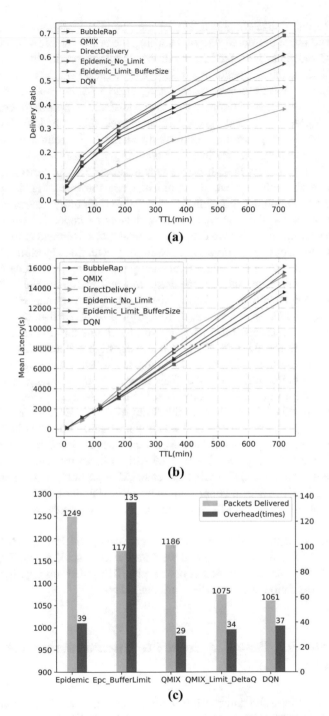

Fig. 4.6 Compare the convergence graph of reward. (**a**) Delivery Ratio in INFOCOM05. (**b**) Mean latency in INFOCOM05. (**c**) Packet delivered and overhead in INFOCOM05

α and β, we can utilize the advantage of the "popular" community's forwarding capability and avoid unlimited relaying causing the buffer overflow problems.

Figure 4.6b depicts the mean latency of the above five routing algorithms, taking messages' TTL as the independent variable. The highest average delay is the direct delivery routing algorithm because of being short of relay nodes. So messages can only passively wait in place for their destination nodes to communicate with them. Epidemic routing protocol also causes relatively high latency. This is because the flooding strategy requires messages to be routed to many relays before finally reaching the destination node. DQN is also inferior to QMIX in terms of delivery latency. Selfish nodes unrestrictedly transmit messages to other agents with high centrality, which would increase the queueing delay. Whether it is the intra-community transmission or inter-community transmission using BubbleRap, selecting the best relay among dozens of nodes is time-consuming. Instead, the routing algorithm we propose is calculated and spread among countable communities by adopting the improved community detection algorithm. In addition, strict control of the buffer size rather than greedy forwarding can reduce the message queueing delay. Therefore, the average delay was reduced compared to BubbleRap.

For better understanding, we list the histogram of delivery capacity in Fig. 4.6c when messages' TTL is 360 min and the buffer size of each node is 5 MB. We also compare the overhead ratio in different algorithms and different conditions. Compared with Epidemic, QMIX has a relatively high packet delivery rate and a much low overhead ratio. The second column result reflects the bad properties of Epidemic when buffer size is limited. This is because the flooding method causes the buffer to fill up quickly and blocks the transmission of packets. The fourth column is the result of decreasing the threshold of ΔQ in Eq. (4.11). It causes that the process of community clustering does not converge in time and clusters outliers into communities blindly. Then these outlier nodes cannot learn the routing strategy correctly in the training phase. Lastly, the fifth column reflects the single-agent reinforcement learning DQN routing scheme. After training convergence, agents trained under QMIX perform better than those trained under DQN. Compared with DQN, the data delivery rate is 12% higher and the overhead ratio is 21.5% lower in QMIX.

Similarly, we conduct the same experiment using the INFOCOM06 dataset. There are 98 DTN nodes recording and the duration time is 343,000 s. Figure 4.7a–c also indicates the effectiveness of our proposed QMIX aided routing algorithm with the distribute execution and centralized training way.

4.2   A Learning-Based Approach to Intra-domain QoS Routing

Nowadays, most intra-domain routing methods are based on distance vector algorithm or link state algorithm, such as RIPng, OSPFv3, EIGRP, and so on. These

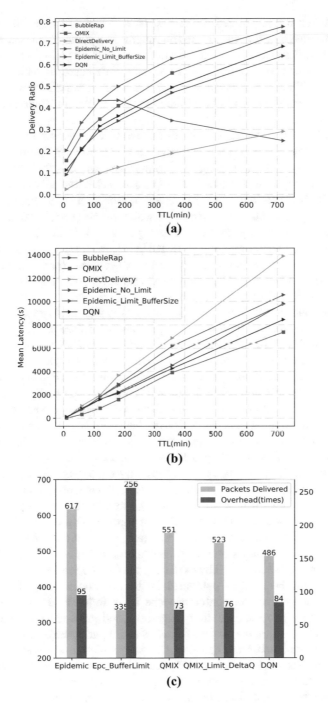

Fig. 4.7 Compare the convergence graph of reward. (**a**) Delivery Ratio in INFOCOM06. (**b**) Mean latency in INFOCOM06. (**c**) Packet delivered and overhead in INFOCOM06

routing methods are categorized as rule-based methods. As designed for specific network environments or for specific metric routing (e.g., delay), rule-based methods lack flexibility. For example, when extended to other metric (e.g., bandwidth), rule-based methods need to modify rules. In addition, rule-based methods cannot learn potential valuable information from massive network flows and network status, so their performance is limited because they cannot use such information to improve routing performance.

As a matter of fact, learning methods that feed on the variable networking data could be applied to advance the performance of network. Specifically, Fadlullah et al. [27] surveyed deep learning application in network traffic control systems. Meanwhile, deep learning has also been applied to packet transmission [28, 29], load balance [30], load prediction [31, 32], channel allocation [33], channel estimation [34], etc. J. Wang et al. [35] surveyed machine learning application in wireless network. Machine learning and related emerging methods are applied in wireless network [36–38], edge computing [39], ICN [40], IoT [41, 42], and routing [43]. Besides, some works focus on the application of reinforcement learning, such as virtual network embedding [44], resource allocation [45, 46], edge computing [47], data analytic [48], IoT [49], 5G [50], etc.

Recently, some works (Ref. [51–57]) focus on the graphical structure and encode vertices into the vectors with extracting features. Inspired by these works, we propose a new device identity, i.e., vectors, which can encode the network distance information and serve as universal patterns of network devices, such that it can adapt to various routing scenarios. On the one hand, unlike the IP address of devices, vectors can be directly applied to the routing computation. On the other hand, just like encoding words into vectors in natural language processing, vectors can provide a universal representation of network devices that could be put into neural networks. To obtain the optimal vectors, we apply the gradient descent method to minimize the proposed objective function of vectors. Based on these vectors, we design a basic routing algorithm that routers can independently choose the next hop by simply calculating the vectors between it and the relative nodes with a quite small amount of computation loads [58]. More importantly, the introduction of node vectors is not only for routing calculations but also for transforming networking units into appropriate input of neural networks. Therefore, our method is not limited to the basic routing, i.e., by using a neural network to process these vectors, different QoS demand routings can be devised by extracting the desired features.

The work has been early published in the conference [59]. The difference between this work and the conference one is as follows: (1) we modified the description of the proposed auxiliary algorithm and provided a theoretical proof for a special case, where the proposed vectors could guarantee delivery without the aid of the proposed auxiliary algorithm, (2) we provided time complexity analyses for our algorithms, such as "complexity analysis on the optimization of vectors", etc., (3) considering that multicast are indispensable functions of routing algorithm, we also provided the extension of the proposed method to the multicast scenario,

(4) we add a comparison with DBA (Mao et al. [28]) and the OSPF, and (5) we provided a demo of the special case for the vector-based routing.

4.2.1 The Basic Routing Based on Node Vectors

4.2.1.1 The Principle of Algorithm

In the shortest path routing, the next router (not the destination), where the packets are directly transported to, is the node which is closest to the destination node in the neighbors of the current router. If we model the network as $G = (V, E)$ (where V denotes the set of nodes and E denotes the set of links) and define the shortest distance from the node i to the other node j as D_{ij} ($i, j \in V$), this strategy could be formulated as

$$nextHop = \arg \min_i D_{id}$$
$$i \in \{x | D_{xd} < D_{cd}, e_{xc} \in E, \ e_{cd} \notin E\}$$

(4.24)

where c denotes the current node, and d denotes the destination.

Especially, if the distance D_{ij} is replaced by a distance function $f(v_i, v_j)$, our method for the shortest path routing (SP-RBNV) is proposed. The estimated value $\widetilde{D}_{ij}$ of the distance D_{ij} is obtained by calculating the proposed function of vectors $f(v_i, v_j)$: $\widetilde{D}_{ij} = f(v_i, v_j)$, where v_i denotes the represented vector of node i. For basic routing, delivering packets to the neighbor, which is the closest node to the destination, is not necessary. Any neighbor x that satisfies $D_{xd} < D_{cd}$ is an available candidate. Since the candidate is always closer to the destination than the current node, packets will be delivered to the target in a loop-free route.

The principle of our method could be summarized as follows: (1) the next hop must be closer to the target node than the current node, unless the target node is directly linked with the current one, and (2) the metric of distance is the proposed function $f(v_i, v_j)$. Specifically, to deliver packets, such values $f(v_{a_1}, v_d), ..., f(v_{a_k}, v_d)$ and $f(v_c, v_d)$ could be calculated in the current node, where $a_1, ..., a_k$ denotes these neighbors of the current one. Then, these nodes that satisfied the distance condition are the eligible candidates.

In this method, the routing is determined by the comparison of corresponding "distance function" between nodes and their neighbors. The comparison indicates the routing direction. Generally, there could be more than one neighbor that satisfies the condition. In such a case, multiple reliable paths could be selected for the routing. As the extension of Eq. (4.24), our strategy is formulated as follows:

$$nextHop = i$$
$$i \in \{x | f(v_x, v_d) < f(v_c, v_d), e_{xc} \in E, \ e_{cd} \notin E\}$$

(4.25)

4.2.1.2   Training of the Vectors

Assume there are n number of nodes in the network, each node of it is assigned an m-dimensional row vector. We could stack these vectors up as a matrix: $\boldsymbol{M} = (\boldsymbol{v}_1, \boldsymbol{v}_2, ..., \boldsymbol{v}_n)^T$. This matrix $\boldsymbol{M}$ is the target parameter that our learning method is going to optimize. A single vector $\boldsymbol{v}_i$ could be described by the matrix $\boldsymbol{M}$. For this purpose, we introduce an n-dimensional row $\boldsymbol{u}_i$, namely, the index vector. Any dimensional value of it is zero except the index of i, which is one. Then, the node vector $\boldsymbol{v}_i$ could be rewritten as a multiplication: $\boldsymbol{v}_i = \boldsymbol{u}_i \boldsymbol{M}$. In our method, the distance D_{ij} is replaced by a predicted value, $f(\boldsymbol{v}_i, \boldsymbol{v}_d)$. Thus, the objective of training these vectors is to minimize the difference between the distance and the function, which is formulated as follows:

$$M = \arg \min_{M} \sum_{i,j,D_{ij} \in S} \mathsf{loss}(|D_{ij} - f(\boldsymbol{u}_i \boldsymbol{M}, \boldsymbol{u}_j \boldsymbol{M})|), \qquad (4.26)$$

where S denotes training set that collects nodes and the distance between them, and $\mathsf{loss}(\cdot)$ denotes the object function.

Specially, if $S = \{i, j, D_{ij} | i, j \in V\}$, we stack all D_{ij} and corresponding $f(\boldsymbol{u}_i \boldsymbol{M}, \boldsymbol{u}_j \boldsymbol{M})$ as matrices, which we denote as $\boldsymbol{D}$ and $\boldsymbol{F}(\boldsymbol{M})$. Additionally, we assume $\mathsf{loss}(\cdot)$ is L1-norm, and then Eq. (4.26) could be rewritten as follows:

$$M = \arg \min_{M} ||\boldsymbol{D} - \boldsymbol{F}(\boldsymbol{M})||_1. \qquad (4.27)$$

Equation (4.27) shows the relation between the optimization and matrix manipulation. Considering the computational cost of the optimization, in general, simple function is the optimal selection for the object function $f(\cdot)$, e.g., the mean square function. While the optimization could be allocated to a certain node (e.g., the controller in an SDN network). Then, these vectors are distributed by this node to other net nodes. Considering the time-varying property of the net, the optimization should be processed periodically.

We apply the gradient descent to optimize Eq. (4.26), as shown in Algorithm 4.3. Meanwhile, other optimization algorithms are also able to solve Eq. (4.26) such as Adagrad (Adaptive Gradient [60]), RMSprop (Root Mean Square Prop [61]), Adam (Adaptive Moment Estimation [62]), etc. Nevertheless, the collection of training dataset is a tough problem, and thus two methods to collect training data are proposed in this work. One is to collect all pairs of nodes in the network and the distance between them into the training set, which requires the computing node to obtain the global topology. In such a case, the disadvantage of it is that additional algorithms are required to acquire the global topology information. On the other hand, the advantages are also obvious that the similarity between the predicted value $\widetilde{\boldsymbol{D}}$ of function $f(\cdot)$ and the true value $\boldsymbol{D}$ can be directly tested on the training set. The other is sampling based method. In [57], the random walks sampling method was proposed to construct a "text-like" dataset. In this work, a

---

Algorithm 4.3 The optimization on the vectors

Input: $\{(u_{s_1}, u_{d_1}, D_{s_1 d_1}), ..., (u_{s_k}, u_{d_k}, D_{s_k d_k})\}$
    notes as the training set S.
Output: The trained vectors: M
 1: Epoches: e, Batch size: bs, Learning rate: lr
 2: **Procedure** *gradient descent*
 3: Random initializes M
 4: **while** $turns < e$ **do**
 5:     Random shuffle S
 6:    **while** $i < k/bs$ **do**
 7:       $start = i * bs, end = (i + 1) * bs - 1$
 8:       $\Delta_j = D_{s_j, d_j} - f(u_{s_j} M, u_{d_j} M)$
 9. $l = \sum_{j=start}^{end} \text{loss}(|\Delta_j|)$
10:       $M = M - lr * \frac{\partial l}{\partial M}$
11:       $++i$
12:    **end while**
13:    $++turns$
14: **end while**

---

sample-based dataset is built based on the random walks approach as shown in Algorithm 4.4. Let $s_1, s_2, ..., s_L$ denote random walks sampled sequence of length L and $\{(s_1, s_2, D_{s_1 s_2}), ..., (s_{L-1}, s_L, D_{s_{L-1} s_L})\}$ denote the training set obtained by the sampling method. Compared with the method that collects all the node pairs into the training set, the sampling method based on random walks only collects neighboring nodes and the distances between them. The advantage of the sampling method is that it does not require the topology information of the entire domain, while the disadvantage is that the sampling sequences and distance information should be delivered to the computing node.

The value of D_{ij} may vary in a wide range, which could cause trouble in the optimization process. Since only the relative distance of $\widetilde{D}_{ij} = f(v_i, v_j)$ is concerned, rather than the absolute value of the distance, we utilize a regularization approach to D_{ij}:

$$D_{new} = w \cdot D_{old} + b, \tag{4.28}$$

where D_{old} denotes the absolute value of the distance (e.g., D_{ij}), w denotes the weight, and b denotes the bias. The purpose of the transformation is to limit the new distance D_{new} into a smaller range (e.g., $D_{new} \in (0, 1)$). If we scale D into $(0,1)$, the simplest one is that $w = 1/(D_{max} - D_{min})$ and $b = -D_{min}/(D_{max} - D_{min})$.

Algorithm 4.4 The random walks sampling

Input: The network $G = (V, E)$,
Input: Repeated times R_t,
Input: The length of the sampling sequence L.
Output: The sampled training set S
1: **for** node o in V **do**
2:     **while** $t < R_t$ **do**
3:         $cur = o$
4:         **while** $i < L$ **do**
5:             $Z = \sum_{(cur,x) \in E} Weight(cur, x)$
6:             $P(c_i = y | c_{i-1} = cur) = Weight(cur, y)/Z$
7:             Chooses y with $P(c_i = y | c_{i-1} = cur)$ from $cur.neighbors$
8:             Adds $(cur, y, Weight(cur, y))$ into S
9:             $cur = y$
10:            $++i$
11:        **end while**
12:        $++t$
13:    **end while**
14: **end for**
15: Random shuffle S

4.2.1.3 The Auxiliary Algorithm

The performance of the shortest path routing (SPR) depends on the training quality of M which cannot always guarantee $f(u_i M, u_j M) = D_{ij}$. Reliability of the routing relies on unbiased estimation of the shortest path distance D_{ij}. Thus, in practice, the proposed routing cannot provide true routing decision for all source node and destination node pairs. To ensure the reliability of the entire routing algorithm, we designed the auxiliary algorithm based on depth-first traversal algorithm. When SP-RBNV fails, the auxiliary routing algorithm (AuR) works to ensure the routing.

The auxiliary algorithm is based on the depth-first search (DFS) that roots on the source node. The principle of the algorithm is as follows: when the source node does not know how to deliver the packets to the destination node, it sends packets to inquiry all its neighboring nodes. Then, the neighboring node confirms whether the destination node is its neighbor. If not, the neighboring node sends packets to inquiry its neighboring nodes again. In the inquired packet, a list is maintained where visited nodes are recorded. The neighboring node that records in the list will not be inquired again in order to avoid the loop in the path. Finally, when the destination node is found by the inquiry mechanism, the inquired packet is sent back to the source node following the visited node list. Every node in the list will write the destination and the next hop into its table. As long as there are paths from the source node to the destination node, the route will eventually be found.

To reduce the cost of network communication, the auxiliary algorithm only looks for an available route and does not guarantee the shortest route. The complete pseudo-code of the auxiliary algorithm is shown in Algorithm 4.6.

In some special cases, the auxiliary algorithm is not an essential part of our method. We could theoretical prove that if the dimension of the vector d_v is equal to the size of network n and dot product is used as the distance function $(f(\boldsymbol{v}_i, \boldsymbol{u}_j) = \boldsymbol{v}_i \cdot \boldsymbol{u}_j^T)$, there is $\boldsymbol{M}$ that can guarantee $f(\boldsymbol{u}_i \boldsymbol{M}, \boldsymbol{u}_j \boldsymbol{M}) = D_{ij}$.

Assume the matrix A denotes the distance matrix of the network graph G. As condition $i = j$ is not used in the routing, the shortest distance between nodes i and j is as follows:

$$D_{ij} = \boldsymbol{u}_i(A + \alpha \boldsymbol{I})\boldsymbol{u}_j^T, (i \neq j), \tag{4.29}$$

where $\boldsymbol{I}$ denotes a unit diagonal matrix $(\boldsymbol{u}_i \boldsymbol{I} \boldsymbol{u}_j^T = 0, \boldsymbol{u}_i \boldsymbol{I} \boldsymbol{u}_i^T = 1)$, and $\alpha(\alpha > 0)$ denotes a coefficient. Obviously, the matrix $A + \alpha \boldsymbol{I}$ is a real symmetric matrix. When the coefficient α is large enough, the matrix $A + \alpha \boldsymbol{I}$ is a positive definite matrix. And its eigenvalue decomposition is performed as follows:

$$A + \alpha \boldsymbol{I} = \boldsymbol{P} \boldsymbol{\Lambda} \boldsymbol{P}^T. \tag{4.30}$$

The eigenvalues of $A + \alpha \boldsymbol{I}$ are all greater than zero. If the matrix $\sqrt{\boldsymbol{\Lambda}}$ satisfies $\boldsymbol{\Lambda} = \sqrt{\boldsymbol{\Lambda}}\sqrt{\boldsymbol{\Lambda}}^T$, $A + \alpha \boldsymbol{I}$ can be decomposed as follows:

$$A + \alpha \boldsymbol{I} = \boldsymbol{P}\sqrt{\boldsymbol{\Lambda}}(\boldsymbol{P}\sqrt{\boldsymbol{\Lambda}})^T. \tag{4.31}$$

If $\boldsymbol{M} = \boldsymbol{P}\sqrt{\boldsymbol{\Lambda}}$, we rewrite the distance D_{ij} as follows:

$$D_{ij} = \boldsymbol{u}_i \boldsymbol{M}(\boldsymbol{u}_j \boldsymbol{M})^T, (i \neq j). \tag{4.32}$$

In this case, the $\boldsymbol{M} = (\boldsymbol{v}_1, \boldsymbol{v}_2, ..., \boldsymbol{v}_n)^T$ optimized by our algorithm will converge to the $\boldsymbol{P}\sqrt{\boldsymbol{\Lambda}}$. The related demo is shown in Sect. 4.2.4.2.

However, the dimension of vectors is increasing with the expansion of the network scale. This scheme could be applied to small-scale networks, but not to the large-scale networks. There should be a trade-off between the application of the auxiliary algorithm and the expansion of dimension.

4.2.1.4 Complexity Analysis

Complexity Analysis on the Optimization of Vectors If we assume that the number of training data is $|S|$, the optimization algorithm iterates r times and the dimension of vectors is d_v. If the vectors are calculated following Eq. (4.27), where the objective function contains a vector multiplication, then the total cost is $O(r * d_v * |S|)$. The number of training data $|S|$ is decided by the collection method

---

Algorithm 4.5 The auxiliary algorithm based on depth-first traversal

---

Input: The current node c,
Input: The destination node d,
Input: The set of visited nodes $list$
Output: The next hop h, or null
1: *Procedure DFS(c, d, list)*
2: **if** d in $c.neighbor$ **then**
3:     **return** d
4: **end if**
5: **for** $node$ **in** $c.neighbor$ **if** $node$ **not in** $list$ **do**
6:     $list.add(node)$
7:     $next = DFS(node, d, list)$
8:     **if** $next! = null$ **then**
9:         $c.men[d] = node$
10:         **return** $node$
11:     **end if**
12:     $list.remove(node)$
13: **end for**
14: **return** null

---

of the training set: if the training set collects all pairs of nodes and the distance between them, the number of training data is $|S| = |V| * (|V| - 1)/2$, and if the training set is collected by sampling based method, $|S| = |V| * (L - 1) * R_t$.

Complexity Analysis on Random Walks Sampling As shown in Algorithm 4.4, random walks sampling contains three loops. For each node cur in V, repeated calculation for variable Z is not essential. It can only be calculated when node cur first occurred. Then, all operations of the innermost loop could be executed in $O(1)$. In conclusion, the total cost is $O(L * R_t * |V|)$.

Complexity Analysis on Choosing the Next Hop We assume that a node has at most De neighbor nodes and apply the greedy strategy. Therefore the times of vector multiplication and finding the next hop is $De + 1$. So, the worst cost is $O(d_v * De)$.

Complexity Analysis on the Auxiliary Algorithm The complexity analysis of auxiliary algorithm is kind of difficult, as the times of query are intractable. Nevertheless, the worst situation can be easily analyzed. If the number of nodes in the network is $|V|$, the query will be executed $|V| - 1$ times, and thus the worst cost is $O(|V|)$. Note that only the first lost package will activate the auxiliary algorithm, and then the path will be saved in the tables.

Algorithm 4.6 The extended shortest path routing (SP-RBNV + AuR)

Input: The source node s, the destination node d
Output: Route $list$, or null
1: *Procedure* $routing(s, d)$
2: $next = s$
3: $list.add(s)$
4: **while** $next! = d \&\& next! = null$ **do**
5:     $next = nextHop(next, d)$
6:     $list.add(next)$
7: **end while**
8: **if** $next! = null$ **then**
9:     **return** $list$
10: **else**
11:     **return** null
12: **end if**
13:
14: *Procedure:nextHop(c, d)*
15: **if** $c.mem.contain(d)$ **then**
16:     **return** $c.mem[d]$
17: **end if**
18: **if** $c == d$ **then**
19:     **return** c
20: **end if**
21: **if** d in $c.neighbor$ **then**
22:     **return** d
23: **end if**
24: $Benchmark = f(\boldsymbol{v}_c, \boldsymbol{v}_d)$
25: $list = []$
26: **for** a in $c.neighbor$ if $f(\boldsymbol{v}_a, \boldsymbol{v}_d) < f(\boldsymbol{v}_c, \boldsymbol{v}_d)$ **do**
27:     $list.add(a)$
28: **end for**
29: **if** $list$ is None **then**
30:     **return** $DFS(c, d, list)$
31: **else**
32:     Choose one from $list$ and return
33: **end if**

4.2.2   The Constrained Routing Based on Node Vectors

Except ensuring the proposed algorithm could guarantee the basic shortest path routing, we also explore the possibility of our algorithm for handling the constrained routing problem, which can be mathematically formulated as

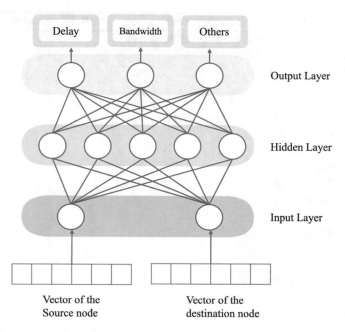

Fig. 4.8 The structure of the shallow neural network

$$z_{opt} = \min \sum_{x \in P_{sd}} cost(x)$$
$$s.t.\ g_1(x) \leq q_1,$$
$$...$$
$$g_m(x) \leq q_m,$$

(4.33)

where P_{sd}, $P_{sd} \subseteq \{0, 1\}^{|E|}$, denotes the set of paths from the source node to the destination node.

In this work, we propose a shallow neural network to learn the bandwidth and delay of links of the network as shown in Fig. 4.8. In this section, we reuse the node vectors that we train for the SP-RBNV. The input of the neural network is vectors of the source node s and destination node d, (v_s, v_d) which represent the virtual link (or physical link) between s and d. The output of it is the attributes of the link such as bandwidth, delay, packet loss rate, etc. The shallow neural network could be mathematically formulated as

$$output = \sigma \left(W_o \sigma \left(W_{hs} v_s^T + W_{hd} v_d^T + b_h \right)^T + b_o \right),$$

(4.34)

where $\sigma(x)$ is the sigmoid function ($\sigma(x) = \frac{1}{1+e^{-x}}$), which is a common activation function in ML tasks. $W_{hs}, W_{hd} \in R^{h \times m}$ denote weights of the hidden layer, $W_o \in R^{\kappa \times h}$ denotes weight of the output layer, h denotes the number of neuron in the hidden layer, and $b_h \in R^{h \times 1}$ and $b_o \in R^{\kappa \times 1}$ denote the bias. $output \in R^{\kappa \times 1}$ denotes κ predicted attributes of the input link. Since $\sigma(x) \in [0, 1]$, the values of the attributes are normalized into $[0, 1]$.

The proposed neural network is applied to extract fine-grained information from these vectors and networks, which maps from the input space (links) into the output space (attributes). To train it, we use cross-entropy as the objective function. The cross-entropy is a common one in such machine learning tasks, which is as follows:

$$loss_{nn} = \sum_{(v_s, v_d, q) \in S} \sum^{i} -q_i \log(output_i), \qquad (4.35)$$

where q_i denotes the i-th normalized attribute of the link between s and d, and the $output_i$ is the predicted counterpart of the q_i.

The neural network is optimized by back propagation (BP) [63, 64]. The BP algorithm applies the chain rule to calculate the gradients of the parameters as shown in Eq. (4.36). We consider two kinds of data feeding of the neural network: the one is that only the physical links are put into the network, and the other is that all pairs of node (virtual link and physical link) are put into the network. The attributes of virtual link between arbitrary two nodes could be taken as the attributes of the shortest path between them or the attribute expectations of all possible path. To make it simple, we adopt the attributes of the shortest path as the attributes of virtual link in the experiments.

$$
\begin{cases}
\frac{\partial loss}{\partial W_o} = \frac{\partial loss}{\partial ouput} \cdot \frac{\partial output}{\partial W_o}, \\
\frac{\partial loss}{\partial b_o} = \frac{\partial loss}{\partial ouput} \cdot \frac{\partial output}{\partial b_o}, \\
\frac{\partial loss}{\partial W_{hs}} = \frac{\partial loss}{\partial ouput} \cdot \frac{\partial output}{\partial h_1} \cdot \frac{\partial h_1}{\partial W_{hs}}, \\
\frac{\partial loss}{\partial W_{hd}} = \frac{\partial loss}{\partial ouput} \cdot \frac{\partial output}{\partial h_1} \cdot \frac{\partial h_1}{\partial W_{hd}}, \\
\frac{\partial loss}{\partial b_h} = \frac{\partial loss}{\partial ouput} \cdot \frac{\partial output}{\partial h_1} \cdot \frac{\partial h_1}{\partial b_h}.
\end{cases}
\qquad (4.36)
$$

The constrained routing decision is roughly the same as the basic routing. The difference of MC-RBNV is that only the neighbor node that the predicted attributes of the virtual link between it and the destination node are satisfied the constrained attributes are consideration in the decision of the next hop.

4.2.3   Extension to the Multicast Scenario

In the unicast routing, the calculation of the estimated distance $f(v_s, v_d)$ is needed to choose the next hop. As the shortest path distance is always a non-negative value, we propose a negative function $f(v_s, v_d) < 0$ for multicast routing. We only consider the situation that a single source node delivers packets to multiple destination (one to many), as many-to-many multicast can be achieved by one-to-many multicast. Suppose that node A is the source node and nodes C, F, G, H, and J are the destination nodes as shown in Fig. 4.9.

Let $f(\cdot)$ satisfy $f(k\boldsymbol{v}_s, \boldsymbol{v}_d) = kf(\boldsymbol{v}_s, \boldsymbol{v}_d)$. The procedure of building a multicast group (or new members join in) is as follows:

Step 1 The multicast destination nodes send CONNECTED packets to the source node.

Step 2 After receiving the CONNECTED packets, node A generates a random number $\gamma_A, (\gamma_A < 0)$ and calculates available routing paths to these destination nodes by the RBNV algorithm. Then, node A delivers the ESTABLISHED packets whose destination vector is $\gamma_A * \boldsymbol{v}_A$ to the next hops B and C.

Step 3 When the ESTABLISHED packet arrives at node B, node B is aware that it is a multicast packet, because $f(\gamma_A * \boldsymbol{v}_A, \boldsymbol{v}_B) < 0 \ (f(\boldsymbol{v}_A, \boldsymbol{v}_B) > 0)$. Then, node B searches its multicast table with the index $R_A = f(\gamma_A * \boldsymbol{v}_A, \boldsymbol{v}_B)$. If R_A does not exist in the table, node B creates a new rule in the table; otherwise, node B updates the existed rule and delivers the packet following the rule in the table.

Step 4 After a new rule is created, node B obtains the destination nodes F, G, H, and J from the packet and calculates the next hop for every target node. After that, node B generates a random number $\gamma_B, (\gamma_B < 0)$ and writes it into the new rule. The new rule contains the index R_a, the random number γ_B, and next hops.

Step 5 When node B delivers the packet to next hops D and E, the destination vector of the packet is replaced by $\gamma_B * \boldsymbol{v}_B$. The content of the packet is replaced by the destination nodes F and G (or H and J), which are reachable via hop D (or E).

Step 6 After the source node receives ACKNOWLEDGE from destination nodes, a multicast tree is built.

The intermediate nodes are periodically sending heartbeat packets to its adjacent nodes in the multicast tree to test connection status of the paths. The leaving procedure of members (or members miss connection) is as follows:

Leaving of Destination Node The destination node sends a DISCONNECTION packet to the source node. After receiving the packet, the source node sends an UPDATED packet back to update the subtree that the destination node locates on.

Destination Node or Intermediate Node Breaks Down The relative node sends a DISCONNECTION packet to the source node on behalf of the broken node, after it detects the breakdown. The next procedure is the same as the that of destination node leaving.

Source Node Breaks Down The intermediate nodes that directly connect with the source remove their relative rule and inform their multicast neighbors when they cannot connect with the source node. Recursively, the multicast tree will be removed from the network.

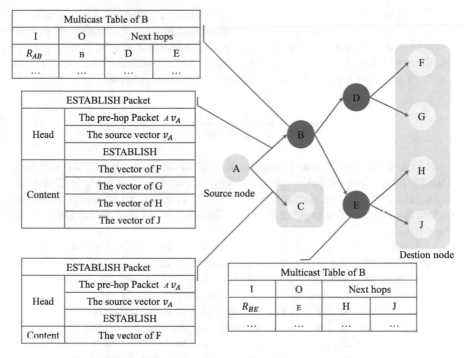

Multicast Table of B			
I	O	Next hops	
R_{AB}	B	D	E
...	...	...	...

ESTABLISH Packet	
Head	The pre-hop Packet $_A v_A$
	The source vector v_A
	ESTABLISH
Content	The vector of F
	The vector of G
	The vector of H
	The vector of J

ESTABLISH Packet	
Head	The pre-hop Packet $_A v_A$
	The source vector v_A
	ESTABLISH
Content	The vector of F

Multicast Table of B			
I	O	Next hops	
R_{BE}	E	H	J
...	...	...	...

Fig. 4.9 The muticast demo

Since every rule in multicast tables contains a unique value, i.e., γ, they could be retrieved in an effective manner by taking the γ as index. In addition, the QoS multicast routing can be provided by adding delay or other constraints into the routing paths, as the multicast routing is based on the proposed node vectors.

4.2.4   Experiments and Simulation Results

4.2.4.1   Simulations on the Basic Routing

There are four kinds of simulations in this part. The first one is testing the shortest path routing in a 100-node net without considering weights of link, in which we compared with different vectors (vectors from our method and vectors from node2vec [57]). Considering link costs of routing, the second one tests on net with weighted links. In the third one, the SP-RBNV was tested in a 1000-node network (the net without considering link weights) with the aim of verifying the availability of the routing in larger networks. In the fourth simulation, we randomly disconnected a link in a tiny net (a net with 34 nodes) to test the response of the RBNV to network transition.

The mean square error (MSE) is the objective function on the optimization process of vectors, and Eq. (4.37) is the estimated function. Two different methods are applied for the initialization of the matrix M: one is initializing M with random numbers, and the other is initialized by vectors from node2vec. The experimental topology was generated with graph generator algorithms [65, 66].

$$f(v_i, v_j) = k \left(1 - \frac{v_i \cdot v_j^T}{||v_i||_2 ||v_j||_2} \right). \tag{4.37}$$

Figure 4.10a shows the routing results of a randomly generated 100-node network graph. The methods that mentioned in Fig. 4.10a are listed in Table 4.4. In this table, M2-M5 are all proposed SP-RBNV methods. Additionally, the results on average training time cost are shown in Table 4.5 (trained on 4 core CPU virtual machines).

In these simulations, the paths generated by our method (SP-RBNV) are compared with the shortest paths. If the routes are the same as the shortest paths, we categorize it as the shortest route (SR), while if the routes are not the shortest one, we categorize it as non-shortest reachable route (NSR). In addition, if the route is not existed with the SP-RBNV, we categorize it as unreachable route (UR). In the routing model, all nodes in the network can generate and deliver packets [67]. All source–destination pairs in the net were test, except these whose source node is directly connected to its destination node (total 9408 pairs). Although the node2vec [57] is not designed for the routing tasks, we used the vectors learned by node2vec as the pre-trained vectors and learned our vectors by fine tuning these vectors (methods M3 and M5). To distinguish the node2vec method and our proposed method, the vectors of node2vec, as a kind of "routing" method, were testing on our simulation as well. Undoubtedly, our proposed method outperforms the node2vec method in the simulation. And the method that collects all node pairs terminally performs better than the sampling based one in Fig. 4.10a, which is reasonable.

Figure 4.10a shows that the initialized value of node vectors has influence on the performance of the SP-RBNV algorithm and specially has effort on the training time cost. Methods that initialized vectors by node2vec perform well. In these methods, the node2vec method serves as a pre-trained model, which is used for vector initialization. Then, our proposed method is applied to optimize these vectors, which is called fine-tuning.

In the follow-up experiments, we used the node2vec as a pre-trained model and fine tuned vectors by our proposed method, since pre-trained and fine-tuning mechanism has better performance.

L2 in Table 4.5 and Fig. 4.10b show the simulation results in a 100-node weighted net. The result shows that the auxiliary algorithm (AuR) guarantees the reliability of our method. The performance of these methods on the weighted net is inevitably worse than the un-weighted one. The cost ratio refers to the cost division between the NSR and the shortest path.

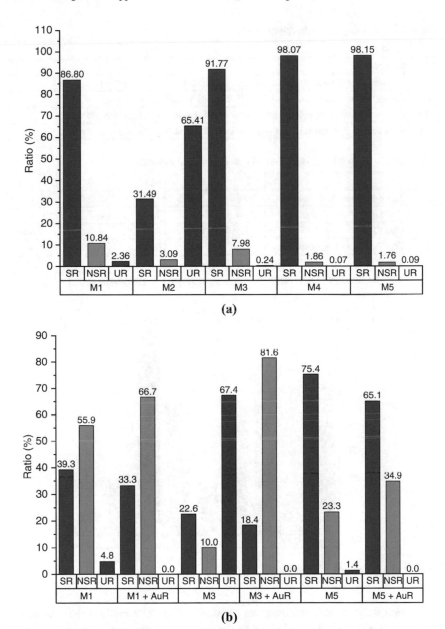

Fig. 4.10 Results analysis. (**a**) The results on 100-node unweighted net. (**b**) The results on 100-node weighted net

Similarly, L3 in Table 4.5 and Fig. 4.11 show the results in a 1000-node un-weighted net. Although the node2vec-trained vectors perform well in the 100-node unweighted net, experimental results in the weighted net and large networks have

Table 4.4 The methods on the experiments

	Method
M1	node2vec
M2	The sampling based method + random initialization of node vectors
M3	The sampling based method + initialized node vectors by node2vec
M4	The method that collects all node pairs + random initialization of node vectors
M5	The method that collects all node pairs + initialized node vectors by node2vec

Table 4.5 The average training time and cost ratio on the experiments

		M1	M2	M3	M4	M5
L1	Training time (unit:second)	29.8	588.2	57.9	762.0	66.4
L2	Training time (unit:second)	29.4	–	194.8	–	525.1
	Cost ratio	1.599	–	1.322	–	1.142
L3	Training time (unit:second)	661.3	–	823.3	–	4825.4
	Cost ratio	1.186	–	1.190	–	1.149

L1: 100-node unweighted net; L2: 100-node weighted net; L3: 1000-node net

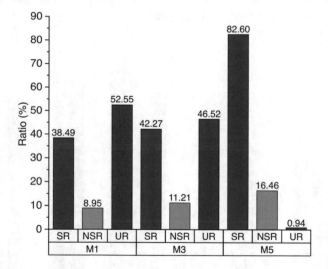

Fig. 4.11 The results on 1000-node net

proved that they are not ideal vectors for the routing. The method that trains node vectors on the sampling training set performs not that satisfied as well, despite the sampling method is more practical than the method of collecting the global topology information. Therefore, it is worthy of paying more attention on the improvement of the sampling method.

To test the response of SP-RBNV to the net transition, we randomly broke down a link in the experimental net and remained using the node vectors that trained in the original net to calculate the route in the broken net. We conducted ten sampling

Table 4.6 The results of coping to network mutation on 34-node net

	SR	NSR	UR	The number of test route
1	912	8	16	936
2	904	6	26	936
3	926	12	0	938
4	930	8	0	938
5	932	6	0	938
6	933	5	0	938
7	907	6	23	936
8	908	30	0	938
9	913	6	17	936
10	912	21	5	938
Baseline	934	2	0	936

Table 4.7 The node vectors of the demo network

	1-d	2-d	3-d	4-d	5-d	6-d
A	−1.542	−0.296	−0.168	1.117	−0.504	−0.064
B	−1.342	−0.202	0.443	−0.922	0.184	−1.038
C	−1.410	0.101	0.755	0.307	0.589	0.995
D	−1.426	0.297	0.019	−0.178	−1.342	−0.211
E	−1.356	0.214	−0.481	0.358	1.103	−0.735
F	−1.286	−0.100	−0.604	−0.894	0.092	1.0790

tests in a 34-node unweighted small-scale network, and the results are shown in
Table 4.6, in which the Baseline is the result of the original network (no broken
link). Table 4.6 shows that the SP-RBNV can still ensure reliable routes under most
of the circumstances. Specifically, in test cases 3, 4, 5, 6, and 8, the SP-RBNV
guarantees routes that are affected by the disconnected link rather than leading to
inaccessible results.

4.2.4.2 The Demo of Shortest Path Routing

Figure 4.12 shows the demo network, in which the black values represent the link
weights. In this demo, we choose dot product function as the distance estimated
function $f(\cdot)$:

$$f(v_i, v_j) = \begin{cases} v_i \cdot v_j^T & i \neq j \\ 0 & i = j. \end{cases} \tag{4.38}$$

Thus, we calculated vectors of all the nodes in this network, as shown in Table 4.7.
    We suppose that there are packets delivering from router B to the destination
router D. Blue numbers in Fig. 4.12a are the value of $f(v_i, v_D)(i = A, B, ..., F)$

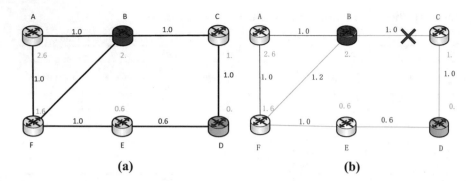

Fig. 4.12 The demo networks. (**a**) The original topology. (**b**) After the link between B and C break

we calculated. Obviously, among the neighboring nodes of B, nodes F and C satisfy the cosine function $f(v_B, v_D) < f(v_i, v_D)(i = F \ or \ C)$. Therefore, F and C are candidates that would transmit packets in the correct direction, and two paths can be chosen from B to D. Similarly, there are two paths available from A to D: $A - B \rightarrow D$ and $A - F \rightarrow D$.

Now we assume that the connection between router B and router C suddenly breaks, as shown in Fig. 4.12b. In other words, the original shortest path $B - C - D$ is disconnected. According to our proposed routing algorithm, an alternative solution can still be selected at router B, which is forwarded through router F. The time required for this decision is only the time cost for router B to detect the disconnection from router C. On the other hand, if the shortest path from router A to router D is broken at $E - D$ in $A - F - E - D$, router A can choose $A - B$ as an alternative route as long as the link-down message is feedback to router A via router F.

4.2.4.3 Simulations on the Constrained Routing

In this part, the MC-RBNV was tested on the DCLC problem. The resluts are shown in Fig. 4.13 with the EDSP [68], the H_MCOP [69], the Mixed_Metric [70], the DEB [71], and the Larac [72]. We assume that both costs and delays are independently selected from uniform distributions. The cost and delay of a link (u, v) are taken as $w_c(u, v) \sim uniform[1, 10]$, $w_d(u, v) \sim uniform[1, 10](ms)$. Since the source and destination of the requests may come from all possible node pairs in the net, the minimum hop-count between them is at least three. The EDSP [68] requires a predetermined parameter $x = coef * d_{s,t}$, where $coef$ is a given positive integer and $d_{s,t}$ is the distance from s to t. Following the suggestion of authors, we used $coef = 4$ and set $x = 4|V|$. Similarly, the Mixed_Metric [70] requires a constant ε. We set $\varepsilon = 5$, as the simulation in [70] shows that the Mixed_Metric performs best when $\epsilon = 5$. In these experiments, the cross-entropy

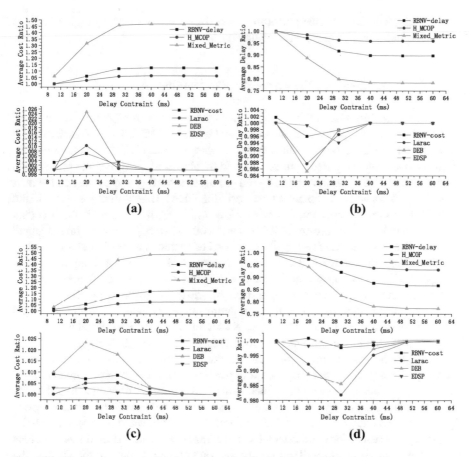

Fig. 4.13 The results of experiment on the constrained routing. (**a**) The average cost ratio in a 32-node network. (**b**) The average delay ratio in a 32-node network. (**c**) The average cost ratio in a 64-node network. (**d**) The average delay ratio in a 64-node network

is chosen as the objective function as mentioned in Eq. (4.35). Our method requires the shortest path from the source–destination pairs. Thus, two types of shortest paths were used in experiments (the cost shortest path, i.e., RBNV-cost, and the delay shortest path, i.e., RBNV-delay).

For better display, the results are drawn in two sub-layers. The cost ratio (delay ratio) refers to the cost (delay) division between the path provided by the model (i.e, RBNV-cost) and the actual optimal path. As shown in the figure, the metrics (the average cost ratio and the average delay ratio) of the proposed method is close to the best one (the EDSP and the Larac) with no distinct difference. The Mixed_Metric method seems to prefer to choose these low-delay paths. The performance of this method on average cost ratio is the worst. However, the Mixed_Metric is used to find a feasible path but not the least cost path and it performs the best on the average delay ratio. The RBNV-cost performs better than the RBNV-delay on the average cost ratio but worse on the average delay ratio.

4.2.4.4 Experiments on the Throughput of Routing

To test the performance of the proposed method on the throughput, latency, and signal overhead, we proceeded an experiment with the OMNET++. The simulation was conducted on the common network topology—the NTT backbone network. Our method was compared with the method of Mao et al. [28], which applies deep learning to optimize packet transmission of network routing and calculates the next hop for packets with a deep neural network (the DBA). In addition, we also add the OSPFv2 into the experiment, following Mao et al. [28]. In the experiment, the data rate and delay of links are set as 2 Gbps and 100 μs. Data generating rate changed from 1.74 to 3.48 Gbps. The signaling interval of methods is fixed at one second.

The detail of our method in the experiment is as follows: a computing node is specified for training the node vectors. All nodes check state of links between them and their neighbors and send state information to the computing node at every signaling interval. Then, the computing node cyclically distributes vectors to other nodes. The flood method (RBNV-flood) is applied to distribute vectors. Furthermore, to reduce the signaling overhead, we also apply the proposed multicast method (RBNV-multicast) to deliver vectors. As shown in Fig. 4.14a, compared with the flood method, the proposed multicast method reduces approximately 20% signaling overhead. The vectors trained in the next period are fine tuned based on the vectors of the previous period. In other words, vectors are initialized by the previous period one. As shown in Table 4.5, the training time of the vectors is greater than the signaling interval in the experiment. Even we train the vectors based on the previous period one, the time cost will not reduce into one second. However, the DBA in the work by Mao et al. requires less signaling interval. On the contrary, the proposed method requires greater signaling interval. To compare the signal overhead, in our method, the computing node distributes vectors to other nodes at every signaling interval. But the vectors are trained and updated every ten signaling intervals. For making full use of the advantages of the vector-based method, our method applied a simple load balancing policy to choose the next hop: all available candidates of the next hop that is chosen by the vector calculation have equal probability becoming the next hop. In other words, every time the routing decision that choosing the next hop is a sampling process from a discrete distribution. The results of the experiment are shown in Fig. 4.14. As shown in Fig. 4.14b, c, the proposed method with the simple load balancing policy outperforms other methods.

Fig. 4.14 The results of
experiment on the throughput
of routing. (**a**) The signaling
overhead. (**b**) The throughput.
(**c**) The average delay per hop

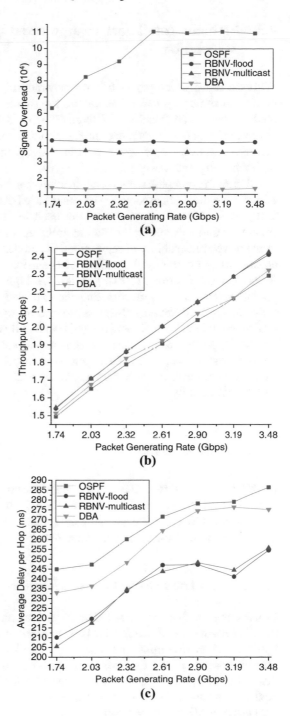

4.3 Artificial Intelligence Empowered QoS-oriented Network Association

With the rapid development of fifth generation (5G), the mobile networks will be able to share physical infrastructure to provide mobile applications, streaming media service, and Internet of Things [73–75]. Both the dimension and the category of information transmitted on the communication network have substantially increased. Hence, traffic services require high Quality of Service (QoS), such as high reliability and low queueing delay [76].

To meet the complex network application scenarios and diversified QoS requirements, many network association schemes have been proposed based on mathematical models [77]. However, it is still difficult to model the real network scenarios accurately and solve complex routing problems even with ideal assumptions. Due to the powerful capability of representation and decision-making, machine learning algorithms are currently the subject of extensive attention [78]. Especially, Deep Reinforcement Learning (DRL) [79, 80] can highlight how to take action relying on the environment by maximizing the expected reward function. Furthermore, Software-Defined Network (SDN) is proposed to simplify network management and enable innovation through network programmability so that a data-driven approach can be supported [81]. Hence, we aim for developing an efficient QoS-oriented network association, where DRL can learn to control a communication network from its experience, and an accurate mathematical mode can guarantee its reliability and interpretability.

4.3.1 System Model

In this section, we describe a jitter graph-based network model and a Poisson process-based traffic model. For a given traffic, a server and buffer space allocation problem constrained on the queueing delay and PLR is proposed to find the feasible path set and reduce the cost of allocated resources.

4.3.1.1 Data Transmission Framework in SDN

Considering the system model, a data transmission framework based on SDN in 5G mobile networks is shown in Fig. 4.15. In general, a data transmission framework based on SDN decouples the data and control planes. Specifically, the control plane is composed of a data collector to collect network information and a traffic dispatcher to manage the network. In the data plane, a set of connected network devices forward packets to the next hop according to the control logic. As for the data plane of 5G mobile networks, we consider a three-tier wireless heterogeneous network composed of Base Stations (BSs), relays, and users. In detail, a popular

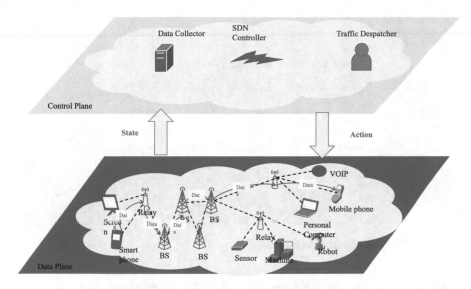

Fig. 4.15 Data transmission framework based on SDN in 5G mobile networks

approach for analyzing mobile networks is to use stochastic geometry and treat the location of BSs as points distributed according to a homogeneous Poisson Point Process (PPP) [82–84].

The data transmission based on SDN in 5G mobile networks can be summarized as the following steps. The BSs and relays in the data plane report the network states, such as the queue length and traffic distribution, to the data collector in the control plane. Then according to the overviews of the whole network and the states of each network device, the SDN controller generates a routing strategy. Finally, the specific action of routing strategy is deployed by the traffic dispatcher.

4.3.1.2 Network Model

The theoretical QoS values always consider the available resources, the traffic shape, and service disciplines. To determine the relationships between the QoS values and different resources inside a network, we propose a jitter graph-based network. Here, an undirected jitter graph [85] $G(V, E)$ is abstracted as a mobile network, where V denotes a set of jitter nodes representing routers and E denotes a set of jitter edges indicating communication links. Each edge $e(u, v)$ has a specific bandwidth BW_{uv}, which represents its available data transmission capability. Also, there are two fixed values of each node, which can be defined as C_u and B_u to represent the server space and buffer space of the router u, respectively. The relationship between our proposed jitter graph and the substrate network is shown in Fig. 4.16.

Specifically, the service discipline from router u to its neighbor v is composed of the jitter node u and the connection edge $e(u, v)$, which is characterized by (1)

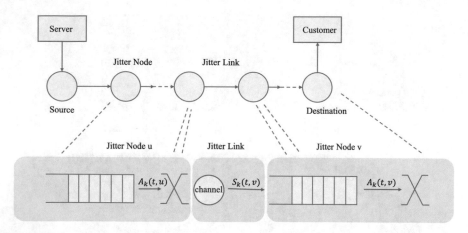

Fig. 4.16 Jitter graph-based network model

maximum delay D_{uv}, (2) minimum delay d_{uv}, (3) available bandwidth BW_{uv}, (4) packet loss rate $P_N(u)$, and (5) queueing delay $W_q(u)$. To elaborate a little further, the maximum delay D_{uv} models the end-to-end delay of the connection link $e(u, v)$ and the bounded delay of the node u. The minimum delay d_{uv} can be computed as the sum of the processing delay on the jitter node u and the propagation delay and transmission delay on the jitter edge $e(u, v)$. The available bandwidth BW_{uv} here defines the channel capacity in bits per second. Furthermore, PLR is defined as the failure of transmitted packets to reach their destination, which is mainly caused by congestion and mistakes in packet transmission. Assuming that $n_{tran}(u)$ is the packets transmitted from the router u and $n_{rec}(v)$ is the packets received by the router v, the PLR $P_N(u)$ on the edge $e(u, v)$ can be measured as $\frac{n_{tran}(u)}{n_{rec}(v)}$. The queueing delay $W_q(u)$ is defined as the time a packet waits in the router u's buffer until transmitted, which is a critical component of jitter caused by the difference in packets' delay.

4.3.1.3 Traffic Model

Each traffic f_k passes through the source s_k and the destination d_k in the communication networks, where the established path can be denoted by p_k. We use an $M/M/C/N$ queuing system [86] described below as a measurement tool for traffic policy. The arrival time of packets subjects to the Poisson distribution [87–89] and the service time of packets independently obeys the exponential distribution [90]. The considered queuing system is composed of C servers to handle incoming packets on the First-Come-First-Served (FCFS) discipline. The capacity of a queuing system is defined as $N(N \geq C)$, which contains both the packets in service and queue. If the number of packets in the system reaches N, the arrival packets will be discarded.

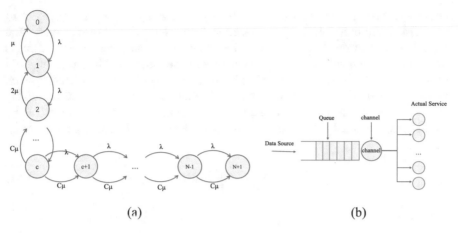

Fig. 4.17 Traffic and queueing models. (**a**) The state-transition-probability diagram, and (**b**) the traffic oriented queueing model

As the arrival time of packets subjects to the Poisson distribution, we can determine the following two characteristics for the packets arrival process. On the one hand, the number of arrival packets occurring in two non-overlapping intervals is a random variable that is independent of each other. On the other hand, for each time interval $(t, t + \tau]$, the number of arrival packets is a random variable that follows the Poisson distribution with associated parameter $\lambda \tau$ as

$$P[(N(t + \tau) - N(t)) = k] = \frac{e^{-\lambda \tau}(\lambda \tau)^k}{k!}, k = 0, 1, ... \tag{4.39}$$

Packet service time obeys an exponential distribution of parameter μ:

$$F(X; \mu) = \begin{cases} 1 - e^{-\mu x}, x \geqslant 0, \\ 0, x < 0, \end{cases} \tag{4.40}$$

and we write $X \sim \exp(\mu)$. If there are more than C packets, the packets will queue in the buffer. If the packets in the queuing system are less than C, some of the servers will be idle. As the corresponding state-transition-probability diagram shown in Fig. 4.17a, traffic intensity $\rho = \frac{\lambda}{C\mu}$ can be measured as the average occupancy of a server. If ρ is less than 1, there exists a stationary distribution of the system. Otherwise, the queue will be filled and the packets will be continuously discarded.

The routing strategy and resource allocation on the queueing model in Fig. 4.17b can affect the users' perceiving, so we aim to reduce the mismatch between the network available resources and the traffic distribution. In this work, a QoS routing strategy with resource allocation based on Poisson traffic is proposed to select a feasible path set and an adaptive routing algorithm is proposed to select an optimal path. Specifically, p_k is the available path from s_k to d_k composed of a sequence of nodes and links. Moreover, $V(p) = \{n | n \in p\}$ and $E(p) = \{e(u, v) | e(u, v) \in p\}$ are used to represent all the jitter nodes and jitter edges on p, respectively. The arrival rate of packets is determined by the traffic request, so its actual service rate is the minimum value of its arrival rate and routers' service rate. The path composed of multiple jitter nodes can be simplified as an $M/M/C/N$ queuing system, whose parameters are determined by the bottleneck router with the least available resources. Hence, the path p_k can be viewed as a queuing system characterized by the buffer space and the server space, which can be given by B_k and C_k, respectively, i.e.,

$$B_k = \min_{v \in p} B(v), \tag{4.41}$$

as well as

$$C_k = \min_{v \in p} C(v). \tag{4.42}$$

For an established path, the channel can be viewed as a simplified link characterized by the bottleneck bandwidth, which can be calculated as

$$R_k = \min_{e(u,v) \in p} BW_{uv}. \tag{4.43}$$

4.3.1.4 $M/M/C/N$ Queueing Model

We consider a problem involving the resource allocation of the bandwidth, the server, and buffer space, when establishing the available channel for the traffic f_k from s_k to d_k. As the mobile network always has security demands for low-jitter QoS and more bandwidth, we aim to maximize the available resources while satisfying the restrictions. The jitter node can be viewed as a truncated multi-channel queue with a general balk function. The steady-state distribution can be derived and the expected packet number in the queue can be obtained [86]. The corresponding mathematical derivation is given as follows.

The number of servers is C, the buffer space is B, and the system's capacity is $N = C + B$. Hence, the service rate μ_n on the state space $\{0, 1, 2, \ldots, C, \ldots, N\}$ can be obtained by

$$\mu_n = \begin{cases} n\mu & n = 1, 2, \cdots, C, \\ C\mu & C < n \leq N, \end{cases} \tag{4.44}$$

and the transition rate matrix Q can be calculated by

$$Q = \begin{bmatrix} -\lambda & \lambda & & & & \\ \mu & -(\mu+\lambda) & \lambda & & & \\ & & \ddots & & & \\ & & C\mu & -(C\mu+\lambda) & \lambda & \\ & & & & \ddots & \\ & & & & C\mu & -C\mu \end{bmatrix}. \tag{4.45}$$

The probabilities $P^t = \{P_0^t, P_1^t, \ldots, P_N^t\}$ of a state can be calculated by the following two equations:

$$P^{t+1} = Q P^t, \tag{4.46}$$

and

$$\sum_{n=1}^{N} P_n^t = 1, \tag{4.47}$$

where P_n^t is the probability that there are n packets at time t. After the long-term iteration, the system will be in a stable state satisfying $P^t = P^{t+1} = P = \{P_0, P_1, \ldots, P_N\}$.

For simplicity, $C_n (n = 1, 2, \cdots, N)$ are assigned to each queue state, which can be formulated by

$$C_n = \frac{\lambda_{n-1}\lambda_{n-2}\cdots\lambda_0}{\mu_{n-1}\mu_{n-2}\cdots\mu_0} = \begin{cases} \frac{1}{n!}(\frac{\lambda}{\mu})^n & n = 1, 2, \cdots, C, \\ \frac{1}{C!C^{n-C}}(\frac{\lambda}{\mu})^n & C < n \le N. \end{cases} \tag{4.48}$$

Therefore, the probability P_n can be obtained based on the traffic intensity $\rho = \frac{\lambda}{C\mu}$, which can be expressed as

$$P_n = C_n P_0 = \begin{cases} \frac{(C\rho)^n}{n!} P_0 & n = 1, 2, \cdots, C, \\ \frac{\rho^n C^C}{C!} P_0 & C < n \le N. \end{cases} \tag{4.49}$$

According to Eqs. (4.46–4.47), the probability that the queuing system is idle can be obtained by

$$P_0 = \frac{1}{\sum_{n=0}^{C-1} C_n + \sum_{n=C}^{N} C_n}$$

$$= \frac{1}{\sum_{n=0}^{C-1} \frac{1}{n!}\left(\frac{\lambda}{\mu}\right)^n + \frac{C^C}{C!}\frac{\rho(\rho^C - \rho^N)}{1-\rho}}. \tag{4.50}$$

Hence, the queueing length, which represents the number of packets waiting to be transmitted in the buffer, can be given by

$$
\begin{aligned}
L_q &= \sum_{n=C}^{N} (n - C) P_n, \\
&= \sum_{j=0}^{N-C} j P_{C+j}, \\
&= \frac{\rho (C\rho)^C P_0}{C!(1 - \rho)^2} \left[1 - \rho^{N-C} - (N - C)\rho^{N-C}(1 - \rho) \right].
\end{aligned}
\tag{4.51}
$$

Then, the average queueing delay proportional to L_q can be expressed as follows:

$$
W_q = \frac{L_q}{\lambda_{eff}} = \frac{L_q}{\lambda(1 - P_N)},
\tag{4.52}
$$

where the actual arrival rate of the packets λ_{eff} can be calculated as $\lambda_{eff} = \lambda(1 - P_N)$. Moreover, the packets will wait only when all the servers are busy and the buffer has free space. Thus, we can conclude that the queuing system is in the queueing state if the number of packets ranges from C to N.

4.3.2   QoS Routing with Resource Allocation

In this section, we define the problem of QoS routing with resource allocation relying on the jitter graph-based network model and the Poisson process-based traffic model, where the knowledge of the network elements implementation and the traffic shape inside the network can be fully incorporated. Also, a low computational complexity greedy algorithm is presented to solve the problem described above with the PLR and queueing delay satisfying the non-increasing conditions.

4.3.2.1   Problem Formulation

Consider a jitter graph-based network $G = (V, E)$ as well as the source and destination nodes $s_k, d_k \in V$ of the traffic f_k. We aim to find the feasible path set, which is composed of the jitter nodes connected in tandem. Then we can allocate the buffer space $b(v)$ satisfying $0 \le b(v) \le B(v)$ and the number of servers $c(v)$ satisfying $0 \le c(v) \le C(v)$ for the jitter nodes connected in tandem of p_k.

Definition 4.1 For a selected path p_k, the problem of QoS routing with resource allocation can be defined as

$$\max \quad Z(p_k) = aC(p_k) + bB(p_k)$$

$$\text{s.t.} \quad (15a) : P_N(p_k) \le PLR_k,$$

$$(15b) : W_q(p_k) \le delay_k, \qquad\qquad (4.53)$$

$$(15c) : C(p_k) = \min_{v \in p_k} c(v) \le \min_{v \in p_k} C(v),$$

$$(15d) : N(p_k) = b(v_{bn}) + c(v_{bn}).$$

Given traffic f_k, we aim for maximizing the available resources $Z(p_k)$ on the selected path so that the user can obtain better experience. Hence, we set the objective as a weighted sum of the server and buffer space, where a and b represent the corresponding coefficients. In Eqs. (4.53a) and (4.53b), the QoS metrics of p_k are composed of the PLR $P_N(p_k)$ as well as the queueing delay $W_q(p_k)$. Then, a feasible path set satisfying the QoS constraints $(PLR_k, delay_k)$ of the traffic f_k can be determined. Therefore, we can obtain the available server space of the queuing system $C(p_k)$ by Eq. (4.53c) and the maximum capacity $N(p_k)$ by Eq. (4.53d), which are constrained by the router v_{bn} with the minimum available server space $\min_{v \in p_k} c(v)$.

Unfortunately, the problem of Eq. (4.53) is non-convex and NP-hard, and thus it is hard to find the global optimal solution. To improve search efficiency, we simplify the problem based on two hypotheses, which can guide the search direction with some empirical suggestion proved in Sect. 4.3.4(B). We can obtain a solution toward the problem of QoS routing with resource allocation and assign numerical values to $b(v)$ and $c(v)$ for all jitter nodes $v \in p_k$, if the constraints can be satisfied simultaneously. Otherwise, increasing any other value $b(v)$ and $c(v)$ cannot affect the available resource $B(v_{bn})$ and $C(v_{bn})$ on the bottleneck router v_{bn}. In this case, we determine the available resources on p_k by $B(p_k) = B(v_{bn}), C(p_k) = C(v_{bn})$ and select the path by maximizing the objective to make $P_N(p_k)$ and $W_q(p_k)$ as low as possible.

Assumption The queuing system of p_k can be simplified with consideration of the bottleneck node v_{bn} whose allocated server space is minimum on p_k, because the number of servers plays a more important role in reducing the queueing delay and PLR compared with the buffer space [91].   □

Assumption As the key problem of queueing is always the available bottleneck resources [92], it is supposed that the number of servers of a path p_k can be viewed as a concave function $C(p_k) = \min_{v \in p_k} \{c(v)\}$ as depicted in Eq. (4.53c).   □

4.3.2.2 Optimality Conditions

In this subsection, the preconditions [93] are presented for the problem of QoS routing with resource allocation. We present the non-increasing QoS metrics of the queueing delay and PLR, which capture the routers' properties.

Definition 4.2 For the modeled jitter graph $G = (V, E)$, the operator $\oplus$ is used as a walk between two jitter nodes: $q = p_1 \oplus p_2$, where the last node in p_1 is the same as the first node in p_2. A sequence of jitter nodes q is constructed by concatenating p_2 to p_1. The function f is said to satisfy the non-increasing condition if

$$f(p_1(s, x)) \leq f(p_1(s, x) \oplus p_2(x, d)). \tag{4.54}$$

For a path $p_k(s, d) = < s, v_1, v_2, \ldots, v_n, d >$, the queueing delay and the PLR can be defined as $W_q(p_k)$ and $P_N(p_k)$, respectively, i.e.,

$$W_q(p_k) = W_q(s) + W_q(v_1) + \cdots + W_q(v_n) + W_q(d), \tag{4.55}$$

$$P_N(p_k) = 1 - [1 - P_N(s)] \cdot [1 - P_N(v_1)] \cdots \cdots [1 - P_N(d)]. \tag{4.56}$$

Because $W_q(v) \geq 0$ and $0 \leq P_N(v) \leq 1$, the queueing delay and PLR satisfy the non-increasing condition shown in Eq. (4.54), which can also be expressed as

$$W_q(p_k) \geq W_q(v_{bn}), \tag{4.57}$$

$$P_N(p_k) \geq P_N(v_{bn}). \tag{4.58}$$

According to Assumption 4.3.2.1, the resource on the routers is a concave function and we pay attention to the router with the least resource of p_k, namely the bottleneck router v_{bn}.

4.3.2.3 QoS Routing Strategy with Resource Allocation

We aim to select the feasible path set under constraints of the queueing delay and PLR while maximizing the available resources. The total latency can be obtained by

$$\begin{aligned} t_{tol} &= t_{pd} + t_{td} + t_{wat} + t_{ser} \\ &= \frac{dis}{c_l} + \frac{\lambda}{BW} + W_q(p_k) + \frac{1}{C \cdot \mu}, \end{aligned} \tag{4.59}$$

where the propagation delay t_{pd} is the proportion between the distance dis and the speed of light c_l, the transmission delay t_{td} is the proportion between the packet arrival rate λ and the bandwidth BW, t_{wat} is the queueing delay, and t_{ser} is the service delay.

For better user experience and network performance, our object is to maximize the available resources on p_k. According to Eqs. (4.53(a–b)) and (4.57–4.58), we have $P_N(v_{bn}) \leq P_N(p_k) \leq PLR_k$ and $W_q(v_{bn}) \leq W_q(p_k) \leq delay_k$. Thus, we determine the available server space $C(p_k)$ and maximum capacity $N(p_k)$ considering the bottleneck router v_{bn} selected by $\min_{v \in p_k} C(v)$. Then $C(p_k) = \min_{v \in p_k} C(v)$ and $N(p_k) = C(v_{bn}) + B(v_{bn})$ can be obtained from Eq. (4.53 (c–d)). Finally, the QoS routing strategy with resource allocation based on Eq. (4.53) can be redefined as

$$\max_{C,B} \ Z(p_k) = aC(p_k) + bB(p_k)$$

$$\text{s.t. } (22a) : P_N(v_{bn}) \leq P_N(p_k) \leq PLR_k,$$

$$(22b) : W_q(v_{bn}) \leq W_q(p_k) \leq delay_k, \qquad (4.60)$$

$$(22c) : C(p_k) = \min_{v \in p_k} C(v),$$

$$(22d) : N(p_k) = C(v_{bn}) + B(v_{bn}).$$

In the objective function, a and b are the corresponding coefficients for the server and buffer space. As described by Eqs. (4.60a) and (4.60b), a path p_k is feasible if the PLR $P_N(v_{bn}) = \frac{C^C \rho^N}{C!} \cdot P_0(v_{bn})$ and queueing delay $W_q(v_{bn}) = \frac{L_q}{\lambda_k[1-P_N(p_k)]}$ satisfy the corresponding constraints. Equation (4.60c) is used to select the bottleneck router v_{bn} according to the routers' available server space. Then, we can determine the capacity of p_k by Eq. (4.60d) based on v_{bn}'s available resource.

More specifically, the PLR of the established path p_k can be calculated as

$$P_N(p_k) \geq P_N(v)$$

$$= \frac{C^C \rho^N}{C!} \cdot \frac{1}{\sum_{n=0}^{C-1} \frac{1}{n!}(\frac{\lambda}{\mu})^n + \frac{C^C}{C!} \frac{\rho(\rho^C - \rho^N)}{1-\rho}}, \qquad (4.61)$$

and the queueing delay $W_q(p_k)$ can be obtained by

$$W_q(p_k) \geq W_q(v)$$

$$= \frac{1}{\lambda_k[1-P_N(p_k)]} \cdot \frac{\rho(C\rho)^C P_0}{C!(1-\rho)^2} \qquad (4.62)$$

$$\cdot \left[1 - \rho^{N-C} - (N-C)\rho^{N-C}(1-\rho)\right].$$

The number of servers $C(p_k)$, the buffer space $B(p_k)$, and the capability of queuing system $N(p_k) = C(p_k) + B(p_k)$ satisfy the following relationship:

$$C = C(p_k) = \min_{v \in p_k} c(v) = c(v_{bn}) \leq C(v_{bn}), \qquad (4.63)$$

as well as

$$N = N(p_k) = c(v_{bn}) + b(v_{bn}) \leq C(v_{bn}) + B(v_{bn}). \tag{4.64}$$

Hence, we can obtain an optimum solution by

$$
\begin{aligned}
(C^*, \ B^*) &= \arg\max_{C,B} Z(p_k) \\
&= \arg\max_{C,B} \{aC(p_k) + bB(p_k)\} \tag{4.65} \\
&= \left(c^*(v_{bn}), \ b^*(v_{bn}) \right).
\end{aligned}
$$

With the number of servers $C(p_k)$ and the buffer space $B(p_k)$ obtained, the available resources on p_k can be determined for all jitter nodes $v \in p_k$, including the number of servers and the buffer space, which can be defined as

$$c(v) = C(p_k) = c(v_{bn}), \tag{4.66}$$

and

$$b(v) = B(p_k) = b(v_{bn}). \tag{4.67}$$

The QoS routing with resource allocation problem in Eq. (4.60) can be divided into two cases:

Case (1) $P_N(v_{bn}) \leq PLR_k, W_q(v_{bn}) \leq delay_k.$
Case (2) $P_N(v_{bn}) > PLR_k \cup W_q(v_{bn}) > delay_k.$

Due to the non-increasing condition of latency and PLR, the path p_k can be determined as feasible only in the condition of case (1). Hence, QoS routing with resource allocation is proposed in algorithm 4.7 leveraging beam search [94]. We aim to maintain a feasible path set of K candidates at each step t:

$$\Psi_t = \left\{ \left(v_1^1, \ldots, v_t^1 \right), \ldots, \left(v_1^K, \ldots, v_t^K \right) \right\}. \tag{4.68}$$

Then the feasible path set at next time can be obtained by expanding Ψ_t and keeping the best K candidates by

$$\Psi_{t+1}^* = \left(\bigcup_{k=1}^{K} \Psi_{t+1}^k \right)^*, \tag{4.69}$$

where

$$\Psi_{t+1}^k = \left\{ \left(v_1^k, \ldots, v_t^k, v_1 \right), \ldots, \left(v_1^k, \ldots, v_t^k, v_{|V|} \right) \right\}. \tag{4.70}$$

Specifically, we have the valid outputs

$$\Psi := \{v_{src} \circ v \circ v_{dst} \,|\, v \in V\},\tag{4.71}$$

where $\circ$ is the string concatenation and V is the set of jitter nodes. Given the constraints $x : (PLR_k, delay_k)$, we define the probability distribution p_θ as the product of probability distributions

$$p_\theta(\Psi \,|\, x) = \prod_{t=1}^{|\Psi|} p_\theta(\Psi_t \,|\, x, \Psi_{<t}),\tag{4.72}$$

where each $p_\theta(\cdot \,|\, x, \Psi_{<t})$ is a distribution with support over $\Psi_{<1} = \Psi_0 := v_{src}$.
The objective for path generation aims to find the most probable hypothesis:

$$\Psi^* = \mathrm{argmax}\log p_\theta(\Psi \,|\, x).\tag{4.73}$$

For an elegant answer, we define the time-dependent surprising by log-likelihood equation to characterize the new information at time t:

$$u(\Psi_0) = 0,$$
$$u(\Psi) = -\sum_{t \geqslant 1} \log p_\theta(\Psi_t \,|\, x, \Psi_{<t}).\tag{4.74}$$

As minimally surprising means maximally probable, every local surprise u_t should be close to the minimally surprising choice.

Algorithm 4.7 QoS routing with resource allocation

Input: the number of path in the feasible path set K, the traffic $v_{src} \to v_{dst}$, its constraints $PLR_k, delay_k$ and its arrival rate λ_k
Output: the feasible path set
1: Initialize the current router $v_1^k = v_{src}$, the path set $\Psi = \{(v_1^1), \ldots, (v_1^K)\}$
2: **while** $v_t^k \neq v_{dst}$ **do**
3:    **for** any $i \in [1, K]$ **do**
4:       Select the available next hop v_i from v_t^k
5:       Compute the $P_N(v_i)$ and $W_q(v_i)$
6:       Expand Ψ_t as Eq. 4.70
7:       Keep the best K candidates as Eq. 4.69
8:    **end for**
9: **end while**
10: return $\Psi = \{(v_{src}^1, \ldots, v_{dst}^1), \ldots, (v_{src}^K, \ldots, v_{dst}^K)\}$

4.3.2.4   Computational Complexity Analysis

Proposition 4.1 *The time complexity of the QoS routing with resource allocation is $O((K^2(logN + logC))^{len})$, where K is the number of paths in the feasible path set, C is the number of available servers, $N = B + C$ is the capacity, and len is the average number of iterations required to search the feasible paths.*

Proof Given a traffic request, we need to iteratively calculate PLR and queueing delay to determine feasible paths that meet the constraints. We simplify the established path as an $M/M/C/N$ queuing system based on the bottleneck router, so each router has to satisfy given constraints. For the path set Ψ of the traffic request f_k, we need to calculate the QoS value of all possible next hops to determine the path feasibility with complexity of $O(K^2NC)$. To improve the efficiency of calculation, we turn an enumerated problem into a divide-and-conquer process considering the monotonicity. Thanks to the strategy of binary search, the time complexity for determining the feasible path set Ψ at each step is $O(K^2(logN + logC))$. The iteration stops when f_k reaches the destination, so the computational complexity is $O((K^2(logN + logC))^{len})$. The average number of iterations len is determined by the network size and the search size.   □

4.3.3   Deep Reinforcement Learning for QoS-oriented Routing

The problem formulated in Eq. (4.53) targets to only one path. However, there are always multiple traffics arriving at the same time and each pair's path decision impacts others. Even if the packet arrival at every source node follows a Poisson distribution, packet arrivals at intermediate nodes may not. In this section, we propose a QoS-oriented adaptive routing strategy based on deep reinforcement learning techniques.

4.3.3.1   Deep Reinforcement Learning Framework

Reinforcement learning is a kind of reward-guided algorithm, where the agent learns in a trial-and-error manner to maximize the reward through the interaction with the environment. Boyan et al. [95] propose the Q-routing algorithm for packet routing, where the Q-learning is first applied to the routing algorithm. v_{cur} estimates the delivering time $Q_{v_{cur}}(v_{dst}, v_{nxt})$ from v_{cur}'s neighbor v_{nxt} to v_{dst} and select the next hop with minimum delivering time by

$$a_{nxt} = \min_{v_{nxt} \in \text{neighbors of } v_{cur}} Q_{v_{cur}}(v_{dst}, v_{nxt}). \tag{4.75}$$

As for updating Q-table, v_{cur} immediately gets back v_{nxt}'s estimate for the time remaining t. Then v_{cur} can revise its estimate as follows:

$$\Delta Q_{v_{cur}}(v_{dst}, v_{nxt}) = \alpha(\overbrace{q + s + t}^{\text{new estimate}} - \overbrace{Q_{v_{cur}}(v_{dst}, v_{nxt})}^{\text{old estimate}}), \tag{4.76}$$

where α is a "learning rate" parameter, q is the queueing delay, and s is the transmission time. To search for an incentive solution, a trade-off by ϵ-greedy policy is considered to balance the exploration and exploitation, where the agent takes a random action with probability ϵ and the action of the highest Q value otherwise.

According to [96], deep reinforcement learning is a strategy, which approximates some component of reinforcement learning with deep neural networks. DDPG is a deep reinforcement learning strategy that can learn a good and specific QoS-oriented routing strategy with low-dimensional observations [97]. The agent selects the optimal action simply relying on the reward function. Specifically,

- State Space: The state is denoted by the traffic request, which is composed of the parameters of queuing systems, the position messages of packets, and the requirements of traffic. The parameters of queuing systems illustrate their character by the arrival rate and the service rate. The position messages demonstrate the source, the current position, and the destination in packet-level simulation. The requirement of traffic is used to make a routing strategy satisfied QoS constraints. The state vector can be formulated as $s = [\lambda, \mu, v_{src}, v_{cur}, v_{dst}, PLR_k, delay_k]$.
- Action Space: The action is used to select next hop from the located router of traffic and can be represented as the probability of each router to be the next hop. We formulate the action vector as $a = [a^1, a^2, \cdots, a^V]$, where a^i represents the probability of v_i selected as the next hop satisfying $\sum_{i=1}^{V} a^i = 1$.
- Reward Space: The reward is defined considering the distance difference and available resource.

As shown in Fig. 4.18a, the framework of DDPG is composed of the primary network and the target network, both of which are actor-critic-based (ac-based) framework. The actor neural network updated by policy gradient aims for specifying the optimal action founded on the current network status. Hence, its input is the real-time state s_t, while the output is the selected probability of each router $a = \pi(s; \theta^\mu)$ as shown in Fig. 4.18b. In the final output layer, we employ the softmax [98] as the activation function to ensure that the sum of output equals one. The critic neural network aims to predict the value function generated by the state and corresponding QoS routing action as Fig. 4.18c. We formulate the critic network as $Q(s, a) = Q(s, a; \theta^Q)$ and perform parameter update using deep Q-Network [99].

4.3.3.2   DDPG Model for QoS-Oriented Routing

The key parts of a DDPG model are the primary actor network and primary critic network. The so-called routing policy is actually a series of actions. In this subsection, we will elaborate on the update of the two networks as well as the definition of reward.

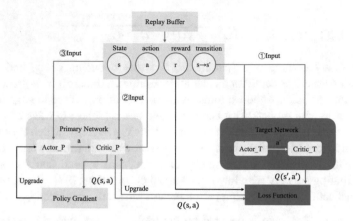

(a)Deep Deterministic Policy Gradient

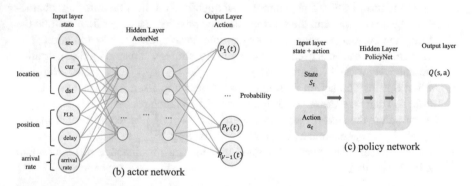

Fig. 4.18 DDPG-based QoS routing: (**a**) the framework of DDPG, (**b**) the architecture of actor network, and (**c**) the architecture of policy network

Primary Actor Network (Actor_P) is used to find a good policy, which is a mapping from a traffic state to a probability distribution, i.e., action. The action a for the current state s can be determined as

$$a = \pi(s; \theta^{\mu}), \tag{4.77}$$

where s is the normalized metrics referred in state space. a represents the probability of each router to be the next hop, which can be selected from the neighbor routers of packets' current position with the largest probability. The objection of *Actor_P* is to maximize the expectation of long-term cumulative reward with the given state and the objective function can be predicted through *primary critic network (Critic_P)*:

$$\max_{\theta^\mu} J(\theta^\mu) = \max E_{\theta^\mu}[r(1) + \gamma r(2) + \gamma^2 r(3) + \cdots]$$

$$= \max E_{\theta^\mu}[Q(s, a; \theta^Q)].$$

(4.78)

In Eq. (4.78), $r(t)$ represents the instant reward obtained by executing the policy $a = \pi(s; \theta^\mu)$ with the discount γ at time t. The prediction of reward is the output of $Critic\_P$ represented by $Q(s, a; \theta^Q)$. Also, the gradient of the objective function with respect to θ^μ is equivalent to the value function [97]. Therefore, parameters of $Actor\_P$ can be updated in the direction where the objective function increases based on its gradient

$$\frac{\partial J(\theta^\mu)}{\partial \theta^\mu} = E_s\left[\frac{\partial Q(s, a|\theta^\mu)}{\partial \theta^\mu}\right]$$

$$= E_s\left[\frac{\partial Q(s, a|\theta^Q)}{\partial a}\frac{\partial \pi(s|\theta^\mu)}{\partial \theta^\mu}\right].$$

(4.79)

Primary Critic Network ($Critic\_P$) aims to obtain the value function for updating $Actor\_P$. As for the update of $Critic\_P$, it is similar to DQN. The loss function of DQN is defined as $loss = (Q_{target} - Q(s, a))^2$, where $Q_{target} = r + \gamma max_{a'} Q(s', a'; \theta')$. Instead of traversing the action space to obtain $Q(s', a'; \theta')$, the DDPG algorithm only needs to use the *target critic network* ($Critic\_T$) to evaluate the prediction $\pi'(s'; \theta^\mu)$ of *target actor network* ($Actor\_T$). Hence, the $Critic\_P$ can be updated by minimizing the loss function

$$loss = \left[r + \gamma Q'(s', \pi'(s'; \theta^{\mu'}); \theta^{Q'}) - Q(s, a; \theta^Q)\right]^2.$$

(4.80)

Target Network is used to generate the target value according to the input transformed state to train $Critic\_P$. To solve the problem of unstable convergence of the ac-based framework, DDPG updates and modifies the parameters using "soft" target updates unlike DQN. Rather than a simple snapshot of the earlier primary network, the parameters of *Target Network* approach the primary network parameters with a small amount in each iteration by

$$\theta^{Q'} \leftarrow \tau\theta^Q + (1 - \tau)\theta^{Q'}$$

$$\theta^{\mu'} \leftarrow \tau\theta^\mu + (1 - \tau)\theta^{\mu'}.$$

(4.81)

The coefficient τ for updating *Target Network* is generally smaller due to the minor environmental changes in an ultra-low time.

Reward is the feedback from the environment based on the current state and action. From the 2G to 5G era, people's demand for network communication has become more diverse as in Table 4.8.

To meet the QoS requirements for different use cases, the reward function can be defined as

Table 4.8 QoS requirements with respect to traffic requests

Application	QoS requirements
Telnet connection	Delay, packet loss rate
Simple web page	Delay
Heavy web page	Throughput
STMP/POP3/IMAP	Packet loss rate
FTP data connection	Throughput
Data with Telnet	Packet loss rate
Real-time multimedia	Delay, throughput, jitter
Control message	Delay

$$R_t = R(i \rightarrow j|s_t, a_t)$$
$$= \omega_1 f(C, B) + \omega_2 f(W_q) + \omega_3 f(P_N) \qquad (4.82)$$
$$+ \omega_4 f(BW) + \omega_5 f(dis),$$

where the environment takes action a_t at state s_t. In Eq. (4.82), $\omega_1, \omega_2, \omega_3, \omega_4, \omega_5 \in [0, 1)$ represent the weights of corresponding QoS requirements. For the sake of explicitly evaluating our proposed routing method, we normalize abovementioned QoS-oriented benchmarks as follows:

• $f(C, B)$ is the normalized capacity of the queuing system on the router j:

$$f(C, B) = -\frac{4}{3\pi} arc\tan\left(\frac{1}{C(v_j)}\right) - \frac{2}{3\pi} arc\tan\left(\frac{1}{B(v_j)}\right). \qquad (4.83)$$

• $f(W_q)$ is the normalized function of the queueing delay $W_q(v_j)$ for the traffic arriving at the router j:

$$f(W_q) = -\frac{2}{\pi} arc\tan W_q(v_j). \qquad (4.84)$$

• $f(P_N)$ is the normalized function of PLR $P_N(v_j)$ for the traffic arriving at the router j:

$$f(P_N) = 1 - 2P_N(v_j). \qquad (4.85)$$

• $f(BW)$ is the normalized function of available bandwidth BW_{ij}^A on the edge $e(i, j)$ and we have

$$f(BW) = \frac{2}{\pi} arc\tan\left(0.01 BW_{ij}^A\right). \qquad (4.86)$$

• $f(dis)$ is the distance difference to destination:

$$f(dis) = dis_{id} - dis_{jd}, \qquad (4.87)$$

where dis_{id} and dis_{jd} represent the number of hops from the router i and j to the destination d, respectively.

According to Eqs. (4.83–4.87), all the above functions normalize the QoS metrics within $[-1, 1]$. The selection for next hop is preferred if the value of $f(\cdot)$ is close to 1, while the action is likely to be rejected with the value of $f(\cdot)$ close to -1.

4.3.3.3   DDPG Aided QoS-Oriented Routing

A randomized algorithm is proposed to trade-off exploration and exploitation. Specifically, we take action $a + \epsilon \cdot N$ with $1 - \epsilon$ probability where $a = \pi(s; \theta^\mu)$ is the output of $Actor\_P$. To explore a hopefully optimal policy, we derive action $u_{base} + \epsilon \cdot N$, where a_{base} is a base routing strategy. We chose N to suit the environment and used an Ornstein–Uhlenbeck process proposed by Uhlenbeck and Ornstein as

$$N = \theta(\mu - a) + \sigma W,$$
$$N = \theta(\mu - a_{base}) + \sigma W, \tag{4.88}$$

where W is the velocity of a Brownian particle with friction. As for parameters in Eq. (4.88), $\mu = \frac{1}{V}$ is the mean, θ is the weight of noise, and σ is the rate of mean regression.

Motivated by the proposed DDPG model, a QoS-oriented routing strategy is described in Algorithm 4.8.

In the data plane, the source forwards its generated traffics regularly according to the QoS-oriented routing strategy, which is trained and updated by the control plane. For training of control plane, the agent initializes $Actor\_P : \pi(\cdot)$ and $Critic\_P : Q(\cdot)$ with parameters θ^μ and θ^Q generated randomly (line 1). $Target\ Network$ is initialized in the same way as the primary network whose parameters $\pi'(\cdot)$, $Q'(\cdot)$ are snapshot of $\pi(\cdot)$, $Q(\cdot)$ (line 2). As for its update, "soft" target updates are used with the control of hyperparameter τ (line 21).

We define the hyperparameter ϵ to balance the exploration of feasible path and the exploitation of recommended path (lines 7–13), which is updated according to ϵ_{decay}, ϵ_{min} (lines 22–24). We connect the selected next hop with the path and observe the environment for the reward and next state. Then we store the transition sample s, a, r, s' into the replay buffer R (line 15). We sample a random minibatch of M transitions to update the DDPG network (lines 16–22), so that the relation between transitions sampled sequentially can be destroyed and the agent can learn with fewer oscillations and less divergence (line 17). We determine the next hop for the transformed state s' from $Actor\_T : \pi'(\cdot)$ and Q_{target} for training $Critic\_P$ (lines 18–19); In addition, we update $Actor\_P$ with its parameters θ^μ (line20) by applying the chain rule to the expected return of $Critic\_P$.

Algorithm 4.8 QoS-oriented adaptive routing

Input: the initialized hyperparameter γ, τ, M, ϵ, ϵ_{decay}, ϵ_{min}, the traffic v_{src}, v_{dst} and constraints PLR_k, $delay_k$

Output: the established path
 1: Randomly initialize $Actor\_P$ and $Critic\_P$ with weights θ^μ and θ^Q respectively
 2: Initialize $Target\ Network$ with weights $\theta^{\mu'} = \theta^\mu$, $\theta^{Q'} = \theta^Q$
 3: Initialize replay buffer R, $count = 0$
 4: Receive the initial observed state $s = [\lambda, \mu, v_{src}, v_{cur}, v_{dst}, PLR_k, delay_k]$
 5: Initialize the path $p_k = [v_{src}]$
 6: **while** ($v_{cur}! = v_{dst}$) **do**
 7:     Find the feasible set A according to algorithm 4.7
 8:     $z \leftarrow$ uniform random number $[0, 1]$
 9:     **if** $z < \epsilon$ **then**
10:         select the action $v_{next} \in A$ according to a base solution and exploration noise $a = a_{base} + \epsilon \cdot N$
11:     **else**
12:         select v_{next} according to the current policy and exploration noise $a = \pi(s; \theta^\mu) + \epsilon \cdot N$
13:     **end if**
14:     Update $p_k = p_k + [v_{next}]$
15:     Take action by $v_{cur} = v_{next}$, obtain r and s'
16:     Store transition (s, a, r, s') in R and $count + = 1$
17:     **if** $count > M$ **then**
18:         Sample a random minibatch of M transitions (s_i, a_i, r_i, s_i') from R
19:         Set $Q_{target} = r_i + \gamma * Q(s_i', \pi'(s_i'; \theta^{\mu'}); \theta^{Q'})$
20:         Update $Critic\_P$ by minimizing the loss $L = \frac{1}{M}\sum_i \left(Q_{target} - Q(s_i, a_i; \theta^Q)\right)^2$
21:         Update $Actor\_P$ by sampled policy gradient $\nabla_{\theta^\mu} J \approx \frac{1}{M}\sum_i \nabla_{a_i} Q(s_i, a_i; \theta^Q)\nabla_{\theta^\mu}\pi(s_i; \theta^\mu)$
22:         Update $Target\ Network$ by Eq. (4.81)
23:         **if** $\epsilon > \epsilon_{min}$ **then**
24:             $\epsilon = \epsilon \cdot \epsilon_{decay}$
25:         **end if**
26:     **end if**
27: **end while**
28: Return the established path p_k for traffic f_k

There are some hyperparameters in the proposed DDPG aided QoS-oriented routing. A comprehensive empirical study is made to find the best structure for optimal performances. Regarding the neural network architecture of $Actor\_P$, it is composed of four fully connected layers as shown in Fig. 4.18b: three hidden layers with 32, 32, and 64 neurons, respectively, activated by the Rectifier Linear Unit

(ReLU), as well as output layer with V neurons using softmax as activation function. As for the $Critic\_P$, we use a 4-layer fully connected neural network as shown in Fig. 4.18c, which is composed of three hidden layers with 50, 50, and 30 neurons each, all using the ReLU activation, and the output layer with one neuron using the standard linear transfer function. Furthermore, dropout is added before the output layer to prevent $Actor\_P$, $Critic\_P$ from overfitting by letting the activation value of a certain neuron stop working with probability 0.5. Moreover, we find the best settings for the other hyperparameters: $\gamma = 0.95$, $\tau = 0.01$, $M = 128$, $\epsilon = 1.0$, $\epsilon_{decay} = 0.995$, and $\epsilon_{min} = 0.2$ according to the empirical study.

4.3.3.4 Computational Complexity Analysis

Proposition 4.2 *The time complexity of the "DDPG Aided QoS-Oriented Routing" on an established channel is $O((K^2(logN + logC))^{len} \cdot \epsilon_{min} + (l_{Input} \cdot l_1 + l_1 \cdot l_2 + l_2 \cdot l_3 + l_3 \cdot l_{Output})^{len} \cdot (1 - \epsilon_{min}))$, and the time complexity for updating and sampling is $O(M \cdot (l_{Input} \cdot l_1 + l_1 \cdot l_2 + l_2 \cdot l_3 + l_3 \cdot l_{Output}))$.*

Proof In the process of Algorithm 4.7, the QoS routing with resource allocation on an established channel is $O((K^2(logN + logC))^{len})$, where K is the number of paths in the feasible path set, C is the number of available servers, and $N - B + C$ is the bottleneck router's capacity. For the neural network agent, the time complexity of the feedforward and backward propagation for one sample is $O(l_{Input} \cdot l_1 + l_1 \cdot l_2 + l_2 \cdot l_3 + l_3 \cdot l_{Output})$, where l_{Input} and l_{Output} represent the units' number of the input and output layers, while l_i ($i \in [1, n]$) represents the units' number of the ith hidden layer. It can be noticed that we leverage Algorithm 4.7 as the baseline during exploration with probability ϵ_{min}. Otherwise, we employ a new action $a = \pi(s; \theta^{\mu})$ with careful consideration for both actor and critic networks. As for the iteration time, len is determined by the convergence and performance of routing strategy. Moreover, the actor and critic networks are updated according to the real-time network status with a random minibatch of M transitions, so its time complexity is $O(M \cdot (l_{Input} \cdot l_1 + l_1 \cdot l_2 + l_2 \cdot l_3 + l_3 \cdot l_{Output}))$. $\qquad \square$

4.3.4 Experiments and Simulation Results

We utilize the random regular graph generator algorithm to generate a substrate network as a simulation environment [65]. Our proposed QoS-oriented routing strategy is tested on an ER random graph with 25 nodes. The performance on the PLR, queueing delay, and path length can be evaluated based on the formula in Sect. 4.3.1. We first depict the settings of the jitter graph-based network and traffic pattern in detail. Then we prove the correctness of the mathematical derivation in Sect. 4.3.1 and the availability of QoS routing strategy with resource allocation in Sect. 4.3.2. Furthermore, we analyze the performance of our proposed strategy, which combines the QoS routing strategy for a feasible path set and adaptive routing strategy for the optimal routing strategy. Finally, we make an expansion experiment on another ER random network with 50 nodes to interpret the accuracy of our proposed strategy more deeply.

4.3.4.1   Database

In this work, an ER random graph is generated to provide a jitter graph-based network model for routing and corresponding resource allocation of different traffic requests. There are 25 jitter nodes and 76 jitter links in our generated network, where each pair of nodes is connected with the probability of 0.25. The time for every packet service is a Poisson process obeying the independent and identical distribution with an average rate of $\mu = 3$, which represents the number of packets that can be processed per unit time. Each router can be viewed as a queuing system with $C = 20$ servers to provide the service for the packets in order of their arrivals (First-Come-First-Server). The available buffer space is $B = 10$ and the capacity of the queuing system is set as $N = B + C = 30$.

In practical mobile networks, the interference between routers cannot be accurately modeled as collisions [100]. A popular approach for analyzing mobile networks is to use stochastic geometry and treat the location of BS as PPP [82–84]. Furthermore, we concentrate on the traffic with fair contention access period [101]. Hence, we model the packet arrival at the source node of each traffic request as a Poisson process (note that the packet arrivals at intermediate nodes may not follow a PPP) in our simulator. To model the dynamic traffic requests, we set the traffic pattern uniformly distributed in a window with a size of 6 requests per unit time. Considering the available resources of our generated network, we set arrival rate $\lambda = 15$ for each traffic, which can be characterized by the source, destination, starting time, and departure time. The duration time of each traffic is a random number from 0 to 1.

We implement the SDN-based data transmission framework of 5G mobile networks and set up the environment for packet-level simulation using Python 3.7. Due to the light weight of our design, we could easily run and train the proposed framework on a regular desktop with an Intel Quad-Core 2.6Ghz CPU with 8GB memory.

4.3.4.2   QoS Routing with Resource Allocation

For simplification of the QoS routing with resource allocation, we analyze the queueing length, queueing delay, and PLR versus the different server and buffer space. We calculate the mathematical derivation of QoS metrics. As shown in Fig. 4.19, the buffer space has a subtle effect on the queueing delay only when the number of servers is limited and has almost no effect on the PLR. Assumption 1 can be proved that the number of servers is the key factor for reducing the queueing

Fig. 4.19 The mathematical derivation of QoS metrics versus the different capability of server and buffer. (**a**) Packet loss rate. (**b**) Queueing length. (**c**) Queueing delay

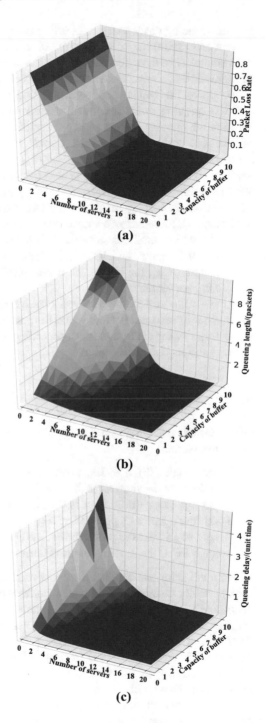

length, queueing delay, and PLR. Furthermore, we simulate an $M/M/C/N$ queuing system and record the simulation result in Fig. 4.20 by plotting scatter plots. The high similarity of the calculated and simulated QoS metrics witnesses the correction of mathematical derivation depicted in Sect. 4.3.1.4.

To prove the concave of concatenating path in Assumption 2, we simulate multiple queuing systems with different routers' resources. As for the parameter settings, the queuing system p_k is connected by five routers with a different number of available servers and the same capacity of buffer 10. Hence, the path can be represented by $p_k(C(v_1) - C(v_2) - C(v_3) - C(v_4) - C(v_5))$, where $C(v_i)$ is the available buffer space of router v_i and the simulation result is shown in Fig. 4.21. We can observe from the performance of p_2, p_3, and p_4 that the available resources of p_k are determined by the router with the least available resources on p_k, namely the bottleneck router. Moreover, we can also observe that the QoS metrics of p_1, p_2, and p_5 have high similarity because their bottleneck routers have the same available resources of 6 servers and 10 buffers. It can be concluded that compared with the other routers of p_k, the bottleneck router plays a more critical part in determining the routing performance.

Above all, we select the bottleneck router on the given channel only considering the number of servers. Assuming that the constraint of traffic f_k is $(PLR_k, delay_k)$, we aim for finding the feasible path according to the bottleneck router for the routing strategy and resource allocation.

4.3.4.3 Adaptive Routing Strategy

We compare our proposed QoS-oriented adaptive routing strategy with three experiments, i.e.,

- Shortest Path (SP): a widely used baseline solution where every traffic delivers all packets on the shortest path.
- Q-routing: every traffic load is distributed to multiple paths obtained from the Q-learning agent with probability $\epsilon = 0.1$ for exploration.
- Q-base: to address the problem that the agent does not know how to explore, we leverage SP as the baseline method during exploration.

Both the actor and critic networks constructed by the Tensorflow are initialized with normally distributed parameters. Some hyperparameter settings are determined according to the empirical study. To guarantee the speed and convergence of the training process, we use Adam to update neural networks with the learning rate of 0.0001 for $Actor\_P$ and 0.001 for $Critic\_P$. For Q value, we use a discount factor of $\gamma = 0.95$. We use $\tau = 0.01$ to control the soft target update rate. We train with minibatch sizes of $M = 64$ and a replay buffer size of 1000. In addition, we let the parameter ϵ decay with decision epoch at rate ϵ_{decay}. We need to adjust parameter ϵ appropriately according to ϵ_{decay} and ϵ_{min}. To find the best strategy after enough exploration, we use a decay rate of $\epsilon_{decay} = 0.995$. According to the environmental

Fig. 4.20 The simulation result of QoS metrics versus the different capability of server and buffer. (**a**) Packet loss rate. (**b**) Queueing length. (**c**) Queueing delay

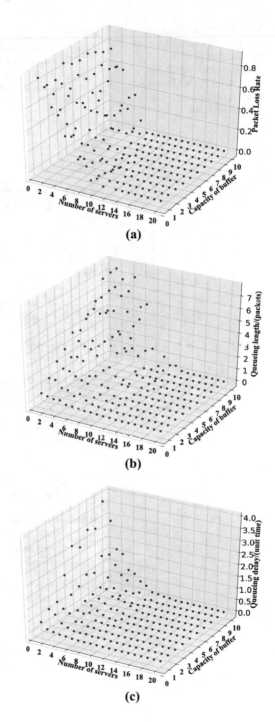

Fig. 4.21 The simulation result of QoS metrics versus the different path concatenated by multiple routers. (**a**) Packet loss rate. (**b**) Queueing length. (**c**) Queueing delay

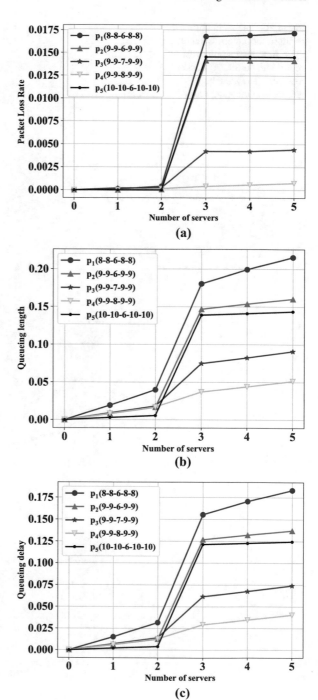

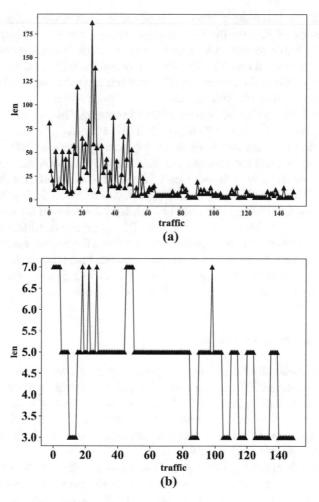

Fig. 4.22 The change of path length in first 150 traffics. (**a**) Shows the convergence with the random exploration, and (**b**) shows the convergence with the specific exploration

dynamics, we set ϵ value after fully explored as $\epsilon_{min} = 0.2$. For the exploration noise process, we use an Ornstein–Uhlenbeck process with $\theta = 0.15$ and $\sigma = 0.2$.

Compared with the supervised learning process, the reinforcement training agent must interact with the environment continuously to perceive the real-time state. To show the convergence, we set a_{base} as a random selection and record the change of path length in first 150 traffics as shown in Fig. 4.22a. The path length decreases when more traffics arrive, because the random selection of next hop allows the model to explore all possibilities of routing. To accelerate the convergence rate, the specific simple solution such as SP can help to reduce the path length intuitively, so our proposed QoS-oriented adaptive routing unarguably has superior performance as Fig. 4.22b. To show the performance of our proposed QoS-oriented adaptive routing strategy relying on the QoS routing strategy for a feasible path set, we generate 1000 traffic requests and record the average QoS metrics versus a different set of traffic patterns in Fig. 4.23. For the steady results, we record and average the results of 100-1000 traffics, when the RL agent has learned a lot about the network topology from the previous 100 traffics.

As shown in Fig. 4.23a, the path lengths for Q-routing, Q-base, and QoS-oriented adaptive routing have almost no change as traffic intensity grows. Hence, we can conclude that the RL agents have learnt a lot from the previous 100 traffics. With the guide from exploration process, the path lengths of Q-base and QoS-oriented adaptive routing are only 20% more than SP, which is acceptable for the guarantee on QoS including the PLR and queueing delay. As shown in Fig. 4.23b, c, the queueing delay and PLR grow as traffic intensity grows due to the fixed available resource. Compared with the SP strategy, the RL agents can distribute the traffic on multiple paths according to its experiences. Therefore, the two RL routing strategies accelerated by SP strategy have a better performance than the others. Furthermore, the QoS-oriented adaptive routing performs better than Q-base and our explanation is that the DDPG agent can learn better from the experience.

4.3.4.4 Expansion Experiment

To verify the generalization of our proposed algorithm, we make an expansion experiment on another ER random network topology with 50 nodes and 301 links. The router parameters are the same as the ER random network of 25 nodes including the service rate $\mu = 3$ and the server and buffer space $C = 20$ and $B = 10$. We generate traffics from 0 to 12 per unit time and record the QoS performance versus different traffic patterns in Fig. 4.24.

Since the objective of the RL agent is to minimize the path length, PLR, and queueing delay, we can see that our proposed QoS-oriented adaptive routing strategy performs as expected. On the one hand, the DDPG agent learns well about the network topology to find a relatively short path as shown in Fig. 4.24a. On the other hand, our proposed QoS-oriented adaptive routing strategy outperforms the others in terms of PLR and queueing delay as Fig. 4.24b, c. Our explanation is that the QoS-oriented adaptive routing strategy can learn better and distribute the traffic requests according to its reward function more efficiently. The above analysis proves the effectiveness and generalization of our proposed strategy, where the path length, PLR, and queueing delay are considered.

Fig. 4.23 Performance on routing of all the methods over the ER random topology with 25 nodes. (**a**) Path length. (**b**) Packet loss rate. (**c**) Queueing delay

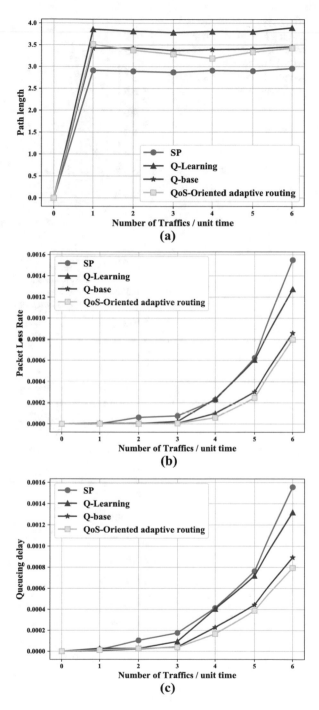

Fig. 4.24 Performance on routing of all the methods over the ER random topology with 50 nodes. (**a**) Path length. (**b**) Packet loss rate. (**c**) Queueing delay

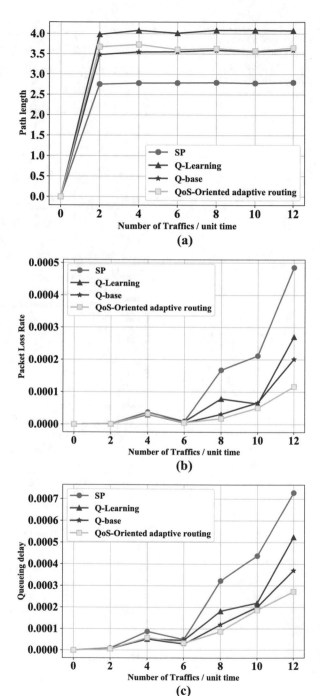

4.4   Machine Learning Aided Load Balance Routing Scheme

In the packet-switched networks, data traffic experiences a sharp growth because of the rapid development of digital video services such as Internet video, mobile streaming media, and IPTV [27]. Considering the bursty nature of the packet traffic, conventional routing algorithms cannot effectively avoid network congestion [102]. In order to achieve the load balance of the network's routers, it is necessary to invoke an intelligent routing scheme.

Traditional link state-based routing algorithms, such as the Bellman–Ford algorithm [103, 104], the link state algorithms, and the Dijkstra algorithm [105], just to name a few, require each router to know the entire topology information. By contrast, the distributed routing algorithms iteratively find the best path to the destination relying on all the neighbor nodes [106].

Multiple Constrained Path (MCP) selection is a popular member of the combinatorial optimization family, which is often used to find available paths that satisfy multiple constraints [107]. It is proven to be an NP-complete problem [108, 109] if these constraints are mutually independent [106]. A range of work has been conducted for addressing Quality of Service (QoS) routing to achieve the reliability and stability of networks [110, 111]. The majority of QoS routing algorithms take multiple constraints into account relying on intuitive or empirical construction. Hence, the feasible solution can be obtained within an acceptable computational complexity [112].

However, heuristic algorithms still have a slow convergence speed when dealing with large-scale problems. Also, their solutions based on a series of ideal assumptions may be out of physical reality. Additionally, owing to the diversity of QoS constraints, the feasible path for one constraint may not be available for the others [113]. Motivated by these issues, in this section we propose a machine learning aided routing scheme to address the QoS routing process. The performance comparison of several routing algorithms involved in this section is shown in Table 4.9.

4.4.1   System Model

In this section, we formulate the load balance routing problem and introduce the PCA algorithm to reduce the substrate network's dimension. The symbols in this section is shown in Table 4.10.

4.4.1.1   Packets Detection in the Dataplane

Recently, SDRs have appeared along with programming languages. SDRs offer the possibility to gather and export important packets' meta-data, while the packets are being processed [114]. As shown in Fig. 4.25, we propose a hierarchical architecture based on SDR to detect the packet-switched network and achieve the load balance.

Local routers monitor the real-time status, process the arrival packets, and select the next hop for them. Meanwhile, central routers detect the QU and the traffic pattern of all the substrate routers. Then central routers predict and distribute the predicted QU vector based on their detection and neural network framework.

4.4.1.2 Routing Scheme

Assuming that $G = (V; E)$ is the model of the network, we can know that V is the set of nodes and $n = |V|$ indicates the number of nodes; E is the set of links and $m = |E|$ indicates the number of edges. Each link $e \in E$ can be characterized by the value $h_i(j)$, which represents the number of routing hops between routers i and j. The BF does not consider the queue state and the topology, which have a great impact on the packets' transmission. In order to achieve load balance of the network, the MLQU and DLQU routing schemes consider both the current and predicted QU. The MLQU and DLQU routing schemes decide the next hop for the packets in the buffer when the current router is not the destination. The packet is mainly composed of destination IP address, source IP address, and load data [107]. When a router receives data packets, it selects the next hop intelligently according to multi-metrics, such as the QU and topology. In order to get the destination as soon as possible, BF may choose the shortest path for routing. However, the MLQU and DLQU routing schemes will select the next hop intelligently to avoid the loss of packets.

Table 4.9 Comparison among multiple routing algorithms

Algorithm name	Problem solve	Complexity	Delay	Topology	Traffic pattern
Dijkstra	Shortest path (SP)	$O(n^2)$	\	\	\
BF	Shortest path (SP)	$O(n \cdot e)$	\	\	\
CBF	Constrained shortest path (CSP)	$O((deg - 1)^{e-deg-3})$	✓	\	\
LARAC	Constrained shortest path (CSP)	$O(e \cdot n^2)$	✓	\	\
QUBF	Constrained shortest path (CSP)	$O(n \cdot d \cdot e)$	✓	\	\
MLQU, DLQU	Constrained shortest path (CSP)	$O(n \cdot d \cdot e)$	✓	✓	✓

Table 4.10 Symbols

Notations	Definitions
s	The source router
d	The destination router
$h_i(j)$	The number of routing hops between router i and j
$X_{N \times N}$	The adjacency matrix
x_{ij}	The connection of routers i and j
$p(i)$	The topology of the router i
$t(i)$	The traffic pattern of the router i
$q(i)$	The QU of the router i
$q^n(k)$	The QU of the router k's neighbors
$q_i(k)$	The QU of the router k during the ith time interval
$RANK(k)$	A single metric of router k
$RANK_i(k)$	A single metric of router k during i time interval
$N(k)$	The neighbor set of router k
$H(k)$	The candidate set of router k's next hop
$\widehat{h}$	The selected next hop with best resource

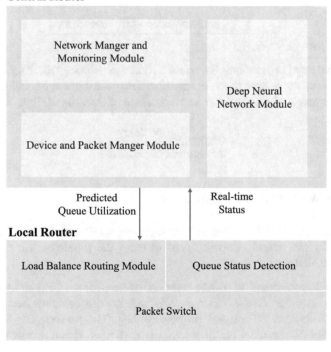

Fig. 4.25 Hierarchical design of the control plane

4.4.1.3   Metrics

The main goal of load balance routing is to process packets as many as possible while reducing and avoiding network congestion. Through traffic control, we can easily achieve load balance while providing a high-quality routing service and a great user experience. This work uses the three evaluation metrics, i.e., the packet loss rate, the worst throughput, and the average delay, to measure the QoS of routing requests.

Packet Loss Rate (PLR) [115] is a phenomenon leading to the loss of data transmitted in the network and queue loss is the main reason for the increase of PLR. In this work, we assume that all packages contain the same size of data, and then *packet loss rate* can be defined as follows:

$$\frac{\text{Number of lost packets in queue of router } k}{\text{Number of arrival packets in queue of router } k}. \tag{4.89}$$

Throughput [28] refers to the maximum data transmission and reception capability of network devices. The throughput is mainly determined by the hardware efficiency and program algorithms. In this work, we take the number of packets successfully transmitted by the router per unit time as *throughput*:

$$\lim_{T \to \infty} \frac{\sum_{t=0}^{T} \text{packets successfully transmitted}}{T}. \tag{4.90}$$

Average delay [116] reflects the transmission time, during which the packets are forwarded from the source to the destination. The achievement of load balance can lead to an increase of *throughput* and a decrease of *average delay*:

$$\frac{\sum_{packets} \text{the delay}}{\text{the number of packets}}. \tag{4.91}$$

4.4.1.4   PCA-Based Feature Extraction

In the MLQU and DLQU routing schemes, the connection of the substrate network is critical for the neural network agent to learn the topology. We consider the following attributes of routers:

Computing Resources (CPU)   The basic components of the CPU include operator, buffer, and controller. And its main processing functions contain receiving instructions, performing an action, and processing data. The routers' CPU determines its availability. When the CPU of a router is sufficient, it can accept more router requests.

The Length of Queue Buffer   The length of the router's buffer reflects its ability to store packets. As for routers, the longer the buffer, the lower the lost packets.

Degree The degree is the most direct measure of node centrality in the substrate network. The router with a higher degree is more important in the substrate network and can accept more routing requests.

Assuming that the router's CPU and buffer length are the fixed values, the topology of the router can be provided by PCA as depicted in Algorithm 4.9. The obtained topology represents the nodes' connectivity in the substrate packet-switched network better than the degree.

Algorithm 4.9 PCA aided feature extraction

Input: $X_{N \times N} = \begin{bmatrix} X_1, X_2, \dots, X_N \end{bmatrix}$, Low dimensional space dimension: 1

Output: $P_{N \times 1} = P_1 = l_1^T X = l_{11} X_1 + l_{12} X_2 + \cdots + l_{1N} X_N$

1: $\overline{x_j} = \frac{\sum_{i=1}^{N} x_{ij}}{N}$

2: $\sigma_j^2 = \frac{\sum_{i=1}^{N} (x_{ij} - \overline{x_j})^2}{N}$

3: **for** $j = 1 \rightarrow N$ **do**

4:      **for** $i = 1 \rightarrow N$ **do**

5:          $x_{ij}^* = \frac{x_{ij} - \overline{x_j}}{\sigma_j}$

6:      **end for**

7: **end for**

8: Calculate the covariance matrix $\Sigma_{N \times N} = \frac{1}{N-1} X^* (X^*)^T$

9: Calculate the eigenvalues and eigenvectors $\Sigma_{N \times N} l_i = \lambda_i l_i$

10: sort the eigenvalues $\lambda_1 \geq \lambda_2 \geq \cdots \geq \lambda_N$

11: sort the eigenvectors $l_1, l_2, \dots, l_N$ according to the eigenvalues

12: return $P_{N \times 1} = l_1^T X$

PCA is the most commonly used dimension reduction and feature extraction techniques in multivariate statistical analysis. It converts multiple related variables into a few unrelated feature variables, which contain most of the information provided by the original variables. Therefore, we can perform PCA on the adjacency matrix to obtain the information of the substrate network connection. In this way, we can obtain the topology of each router by reducing its dimension and extracting its main variation characteristics. PCA is a process of eliminating redundancy and overlap of related information and it can help the SDRs to find the next hop with a good connection.

In multivariate statistical analysis, the amount of information contained in a set of data can be characterized by its variance. The core idea of PCA is to convert the original related variables (M variables) into a set of new unrelated variables (still M), namely the principal components. PCA searches a set of orthogonal linear changes between the original variables. In the transformation process, the first principal component has the largest variance value in the linear combination, and the second principal component has the second largest variance value in the linear combination and is orthogonal to the first principal component. Thus, after transformation, the first few principal components (k principal components) retain the most variance and information of the original variable sets. So the first few principal components can represent the main features of the original variables and the latter principal components can be discarded.

As the original input, the adjacency matrix X (or $X_{N \times N}$) contains the connection information of routers:

$$X_{N \times N} = \begin{bmatrix} x_{11} & x_{12} & \cdots & x_{1N} \\ x_{21} & x_{22} & \cdots & x_{2N} \\ \vdots & \vdots & \vdots & \vdots \\ x_{N1} & x_{N2} & \cdots & x_{NN} \end{bmatrix} = \begin{bmatrix} X_1, & X_2, & \cdots, & X_N \end{bmatrix}^T. \tag{4.92}$$

When there is a link between routers i and j, $x_{ij} = 1$; otherwise, $x_{ij} = 0$. Moreover, X_i reflects the router i's connection to the others. Let $\Sigma_{N \times N}$ be the covariance matrix calculated by the vector $\begin{bmatrix} X_1, & X_2, & \cdots, & X_N \end{bmatrix}^T$:

$$\Sigma_{N \times N} = \begin{bmatrix} cov(X_1, X_1) & cov(X_1, X_2) & \cdots & cov(X_1, X_N) \\ cov(X_2, X_1) & cov(X_2, X_2) & \cdots & cov(X_2, X_N) \\ \vdots & \vdots & \vdots & \vdots \\ cov(X_N, X_1) & cov(X_N, X_2) & \cdots & cov(X_N, X_N) \end{bmatrix}, \tag{4.93}$$

where $cov(X_i, X_j), i, j = 1, 2, \ldots, N$, represents the covariance of the variables X_i and X_j. According to the definition of PCA, all the principal components are a linear combination of the input variables. And they can be represented by the following formulation:

$$\begin{cases} P_1 = l_1^T X = l_{11} X_1 + l_{12} X_2 + \cdots + l_{1M} X_M, \\ P_2 = l_2^T X = l_{21} X_1 + l_{22} X_2 + \cdots + l_{2M} X_M, \\ \vdots \\ P_M = l_M^T X = l_{M1} X_1 + l_{M2} X_2 + \cdots + l_{MM} X_M, \end{cases} \tag{4.94}$$

where $l_{ij}, (i, j = 1, 2, \ldots, M)$ is the transform coefficient.

According to the knowledge of multivariate statistical analysis, the variance of the principal component P_i is $Var(P_i) = Var(l_i^T X) = l_i^T \Sigma_{M \times M} l_i$, ($i = 1, 2, \ldots, M$). We assume that l_i is a feature vector of the covariance matrix $\Sigma_{M \times M}$. And according to the definition of the eigenvalue, we know that $\Sigma_{M \times M} l_i = \lambda_i l_i$, where λ_i is the eigenvalue corresponding to the feature vector l_i. Therefore, we can derive the variance of the principal components by $Var(P_i) = l_i^T \Sigma_{M \times M} l_i = l_i^T \lambda_i l_i = \lambda_i l_i^T l_i$. To ensure that the variance of each principal component is a finite value, we assume that the feature vector l_i has a unit length, i.e., $l_i^T l_i = 1$. Hence, there is $Var(P_i) = \lambda_i$. Obviously, the variance $Var(P_i)$ of the principal component corresponds to the eigenvalue λ_i of the covariance matrix $\Sigma_{M \times M}$ and the eigenvector l_i subordinate to the corresponding eigenvalue λ_i. Since the first principal component of all principal components has the largest variance, it corresponds to the largest eigenvalue of the covariance matrix $\Sigma_{M \times M}$. On the other hand, since the covariance matrix $\Sigma_{M \times M}$ is a real symmetric matrix, the eigenvectors corresponding to different eigenvalues are orthogonal to each other. In other words, $cov(P_i, P_j) = l_i^T \Sigma_{M \times M} l_j = \lambda_i l_i^T l_j = 0, (i \neq j)$, so P_i and P_j are irrelevant.

4.4.2 Network Modeling

In this section, we focus on the prediction of the routers' QU based on machine learning. And we can divide the prediction procedure into initializing, training, and running phases. An ideal output value close to the routers' QU at the next time slot can be obtained by an accurate prediction.

4.4.2.1 Input and Output Design

In order to simply describe our research problem, we consider a simple wireless network backbone composed of some routers. In the packet-switched network, we select the next hop for packets in the routers' buffer according to their current position and destination. This mechanism can be called the routing strategy.

As the routing strategy can be expressed as a classic combinatorial optimization problem, it can be defined as a shortest path problem with multi-constraints. However, traditional routing strategies are not intelligent and they cannot learn from the occurred invalid routing decisions. Therefore, they make the same routing decisions for similar congestion scenarios, such as the bursty traffic. For example, as shown in Fig. 4.26, source routers R0, R3, and R6 receive input packets and send them to destination router R5. The traditional routing method chooses R4 to forward the packets to R5. However, the increasing load leads to congestion at R4. Faced with this network congestion problem, packets can be forwarded through alternate paths (R1 or R7) to alleviate too much burden on R4. However, when this happens again, the traditional routing method is "unintelligent" and always makes the same decision without considering the substrate topology, the routers' resources, and the traffic pattern. By contrast, the machine learning system can collect ineffective routing decisions to predict and avoid possible congestion triggered by load imbalance.

As machine learning has been used to many complicated nonlinear problems to learn the features of input, we adopt a neural network for QU prediction.

Topology As proved in [115], the queue loss ratio is roughly proportional to the size of the subtree, which depicts the routers' connectivity. Thus, the topology $P_{1 \times N} = [p(1), p(2), \ldots p(N)]$ can indicate the routers' connectivity and determine the router's availability. p_i indicates the topology in the substrate network of router i. The higher a router's topology is, the more likely the router is selected as an intermediate router.

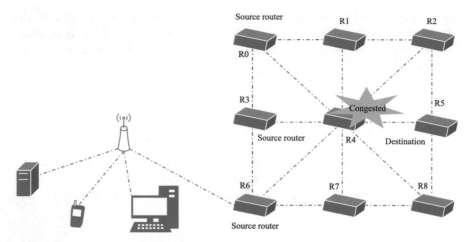

Fig. 4.26 The unintelligence of the traditional algorithm

Traffic Pattern Traffic pattern is the generation rate of data packets on each router. The traffic pattern can indicate its current situation, such as burst traffic and steady traffic. Hence, the traffic pattern can be adopted as the input of our neural network [28]. $T_{1 \times N} = [t(1), t(2), \ldots t(N)]$ indicates the normalized number of packets arriving at the substrate routers during the previous time interval. And we define traffic pattern $t(i)$ as the normalized number of packets in the buffer of router i.

Queue Utilization Queue loss is the main reason for packet loss in high traffic scenario. $Q_{1 \times N} = [q(1), q(2), \ldots q(N)]$ represents the current queue status of the substrate network and $q(i)$ is the normal number of packets in the queue of router i. The router with the fewer packets in its queue has more available storage resources, so it is often selected as the intermediate router.

4.4.2.2 Intialization Phase

In the initialization phase, the training data should be obtained to train the parameters of our proposed neural network. As demonstrated in Fig. 4.27a, the topology, the traffic pattern, and the queue state are served as the input and the neural network is supposed to process for the QU at the next time slot. Also, since the traffic pattern indicates the number of arrival packets in the routers' buffer, it can be added to the QU. However, if the summary of the traffic pattern and the QU is larger than 1, it means that the router is overflowing with packets and we record the summary as 1. Thus, we can use a 2N-dimensional vector x as the input and an N-dimensional vector y to represent the output of the neural network model. x and y are given as follows:

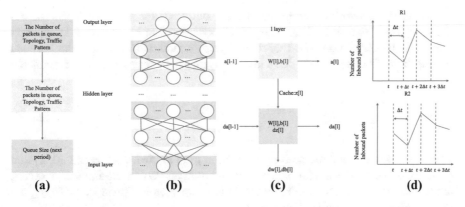

Fig. 4.27 The model of our proposed QU prediction system based on neural network. (**a**) Characterized input and output. (**b**) The structure of deep neural network. (**c**) The building block of deep neural network. (**d**) Considering traffic patterns at each routers as input

$$x_{1\times 2N} = [Q_{1\times N} + T_{1\times N}, P_{1\times N}],\qquad(4.95)$$

$$y_{1\times N} = [Q_{1\times N}].\qquad(4.96)$$

To obtain training data corresponding to the above formula, we can approach some dataset and extract information on topology, traffic pattern, and QU. In addition, we can also run the traditional routing algorithm according to the given traffic pattern and record the routers' QU of every time interval. We use the recorded QU of the BF algorithm to build a training set and then train the neural network by a machine learning system on the central router.

4.4.2.3   Training Phase

The training of the neural network could be processed on a certain computing router of the substrate network (e.g., the central router in the SDN). In the training phase, the obtained data by BF is used as features and labels to train the neural network. The training process can be divided into two steps: initializing the parameters $\theta(\boldsymbol{\omega}, \boldsymbol{b})$ randomly and tuning them with Adam, an algorithm based on adaptive estimates of lower order moments. The values of $\theta(\boldsymbol{\omega}, \boldsymbol{b})$ can be obtained by the training phase and the details are depicted as follows.

As depicted in Fig. 4.27b, the neural network model is composed of L layers, including the input $\boldsymbol{x}$, output $\boldsymbol{y}$, and $(L-2)$ hidden layers located in the middle. As Fig. 4.27a shows, the input layer contains three metric, where $Q_{1\times N} = [q(1), q(2), \ldots q(N)]$ represents the routers' QU, $T_{1\times N} = [t(1), t(2), \ldots t(N)]$ represents the routers' traffic pattern, and $P_{1\times N} = [p(1), p(2), \ldots p(N)]$ represents the routers' topology in the substrate network.

The neural network is composed of many building blocks of two layers' neural network. The detail of each building block is depicted in Fig. 4.27c. It is apparent that each block is composed of two layers' neural network and the forward propagation of the building block with two layers can be mathematically formulated as

$$
\begin{aligned}
A^{[l]} &= g^{[l]}(Z^{[l]}), \\
Z^{[l]} &= W^{[l]}X + b^{[l]},
\end{aligned}
\tag{4.97}
$$

where we have defined the activation function $g(x)$ as $ReLu(x) = \max(0, x)$ and $\sigma(x) = \frac{1}{1+e^{-x}}$ in the hidden layers and output layer. When the input is a vector, activation functions ($ReLu(x)$ and $\sigma(x)$) are also vectors. Since $\sigma_x \in [0, 1]$, the output values are normalized into [0,1].

The parameter dimensions of the neural network can be described as

$$
\begin{aligned}
&W^{[l]} \ (n^{[l]}, n^{[l-1]}), \\
&b^{[l]} \ (n^{[l]}, 1),
\end{aligned}
\tag{4.98}
$$

where $l = 1, \ldots L$. Moreover, $n^{[l]}$ represents the units' number for the l and $n^{[l-1]}$ for the $l - 1$ layer. $n^{[0]} = n_x$ indicates the number of input features.

In supervised ANNs, training samples include features and labels, which are served as the input and output of the network model. After the corresponding calculations with features, the network model will obtain the predicted values. As shown in Fig. 4.27c, the neural network is a mapping from the topology, the traffic pattern, and the current QU to the QU at the next time slot. We use the neural network to extract fine-grained information from the input and output. The parameters need to be tuned to mine the relation among the topology, traffic pattern, and QU. The mapping is determined by the parameters and the activation functions of the neural network.

The closer the predicted value and the true label are, the better the training of the neural network is. The distance between the true and predicted label is often used as the loss function. We choose the mean square error as the stochastic objective function of the neural network, which can be formulated as

$$
loss_{mn} = \sum_{Q,P,T \in S} \sum_i (y_i - output_i)^2,
\tag{4.99}
$$

where y_i represents the router i's QU at the next time slot and $output_i$ is the predicted value of y_i.

The neural network is optimized by Adam, a stochastic gradient-based optimization [117]. The gradient is calculated based on the principle as follows:

$$dZ^{[l]} = dA^{[l]} \cdot g^{[l]'}(Z^{[l]}),$$
$$dW^{[l]} = \frac{1}{m}dZ^{[l]} \cdot A^{[l-1]T},$$
$$db^{[l]} = \frac{1}{m}\sum_i dZ^{[l]}(i),$$
$$dA^{[l-1]} = W^{[l]^T} \cdot dZ^{[l]}.$$
(4.100)

Hence, Eq. (4.101) can be obtained to reflect the recursive relationship between $dZ^{[l+1]}$ and $dZ^{[l]}$, which is expressed as

$$dZ^{[l]} = W^{[l+1]^T} \cdot dZ^{[l+1]} \cdot g^{[l]'}(Z^{[l]}).$$
(4.101)

As shown in Fig. 4.27c, the Adam algorithm is depicted in Algorithm 4.10.

Algorithm 4.10 Adam, an algorithm for first-order gradient-based optimization of stochastic objective functions, based on adaptive estimates of lower order moments

---

Input: training set$(x, y) = \{(x^{(t)}, y^{(t)} | t = 1, \ldots, m\}$
Output: $\theta = (\omega, b)$
Require: $\beta_1, \beta_2 \in [0, 1)$: Exponential decay rates for the moment estimates
  1: epoches: e, batch size bs, learning rate η
  2: **procedure** *Adam*
  3: random initialize parameters $\theta = (\omega, b) \in (0, 1)$
  4: $m_0 \leftarrow 0$(Intialize 1st moment vector)
  5: $v_0 \leftarrow 0$(Intialize 2nd moment vector)
  6: $t \leftarrow 0$(Intialize timestep)
  7: **while** θ_t not converged **do**
  8:   $t \leftarrow t + 1$
  9:   $g_t \leftarrow \nabla_\theta f_t(\theta_{t-1})$
     //Get gradients w.r.t. stochastic objective Eq.(11) at timestep t by Eq.(12),(13)

 10:   $m_t = \beta_1 \cdot m_{t-1} + \beta_1 \cdot g_t$
 11:   $v_t = \beta_2 \cdot v_{t-1} + \beta_2 \cdot g_t^2$
     //Update biasd moment estimate
 12:   $\widehat{m}_t \leftarrow m_t/(1 - \beta_1^t)$
 13:   $\widehat{v}_t \leftarrow v_t/(1 - \beta_2^t)$
     //Compute bias-corrected moment estimate
 14:   $\theta_t \leftarrow \theta_{t-1} - \alpha \widehat{m}_t/(\sqrt{\widehat{v}_t} + \varepsilon)$
     //Update parameters
 15: **end while**
 16: return θ_t (Resulting parameters)

4.4.2.4 Running Phase

The time interval to record stacked data packets is Δt as in Fig. 4.27d and the traffic pattern is described as the recorded number of stacked packets in the last time interval $\beta \cdot \Delta t$ (β is any positive integer). At the discrete time $\beta \cdot \Delta t$, the SDRs make routing decisions for the packets in the buffer. Given an input x, the agent tries to predict the QU at the next time slot of each router based on the input. The running phase is a process of forwarding propagation as depicted in Eq. (4.97) to predict the QU at the next time slot. After the predicted QU being calculated, the central router distributes the QU vector to each SDR.

We consider a substrate network with N routers and M links. For each router, we can adopt the topology, the traffic pattern, and the current QU as input. After the running phase and the information distribution by central routers, each SDR can make the routing decision based on the output of the neural network to achieve the load balance routing scheme.

4.4.3 Routing Based on Queue Utilization

Actually, the SDRs mainly work on the load balance routing. The detailed procedures of the routing are described in this section. Here, we propose the MLQU and DLQU algorithms that consider not only the number of routing hops but also the current and predicted QU. Hence, MLQU and DLQU perform better in load balance, especially in the scenario depicted in Fig. 4.29.

4.4.3.1 The Representation and Update of Queue Utilization

It has been proved in [115] that queue loss is the main reason for the growth of PLR in high traffic scenario. Due to the limited storage capacity of the buffer, it is inevitable that the packets will be discarded if the buffer is filled. To avoid the condition that some routers' buffers are idle while some are busy, we also choose QU at each router k as an impact on next hop:

$$q(k) = \frac{\text{Number of packets in queue of node } k}{\text{Total queue size of node } k}. \tag{4.102}$$

Figure 4.29a shows that the MLQU and DLQU routing schemes will select R6 for routing to avoid the loss of packets on R2 whose QU is much larger. Based on the $q(k)$ calculated by the above formula, we obtain a $1 \times N$-dimensional matrix $Q_{1 \times N}$. Then, we update the $q(k)$ of each router according to the matrix $Q_{1 \times N}$ by

$$q(k) = \max \left\{ \overline{q_n(k)} - \lambda, q(k) \right\}, \tag{4.103}$$

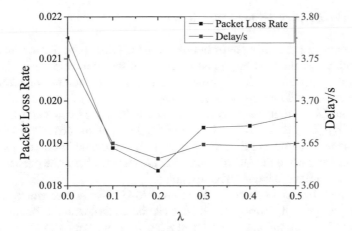

Fig. 4.28 The impact of λ on the PLR and the delay

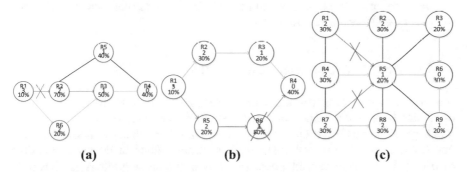

Fig. 4.29 The deep learning aided load balance routing scheme relying on queue utilization. (**a**) The routing scheme considering the QU (source: R1, destination: R4). (**b**) The routing scheme considering the neighbors' QU (source: R1, destination: R4). (**c**) The routing scheme considering the predicted QU (source: R1, R4, R7, destination: R6)

where $q(k)$ is the QU obtained from Eq. (4.102). $q_n(k)$ represents the QU of the router k's neighbors and $\overline{q_n(k)}$ means the average of all the $q_n(k)$. We have also analyzed the specific experiment for selecting the threshold that can trigger QU adjustment and Fig. 4.28 shows the relationship between the network performance and λ. When λ = 0.20, both the PLR and the delay achieve their optimal values.

The intuition behind this adjustment comes from a situation, as depicted in Fig. 4.29b, where $q(k)$ might be small even when the router k's neighbors are severely congested. In this case, although the buffer of R5 is more available, it is better not to be selected as the next hop. As R5 forwards all the packets received to its neighbors, the packets are more likely to be discarded at R5's neighbors. Thus, we perform the QU adjustment Eq. (4.102) to reduce the possibility of a router whose neighbors have a poor queue status to be selected as an intermediate router. Consequently, we set R2 as the next hop.

4.4.3.2   The Preprocessing of Metric Data

$h_i(j)$: the number of routing hops between routers i and j. To deliver data packets to the destination, the next hop must be closer to the destination than the located router, or the next hop is the destination. Therefore, the number of routing hops should be considered into the routing strategy.

$q(k)$: the QU of router k. According to Eq. (4.102), we know that $q(k)$ satisfies the equation $q(k) \in [0, 1]$. When the router's QU is 1, all the arrival packets will be dropped since its buffer is full. From Eq. (4.103) and the definition that $\lambda = 0.20$, we conclude that when the QU of router k is greater than 0.80, the packets arriving at the router k are more likely to be discarded.

$p(k)$: the topology of router k. In addition to QU, the degree which means the routers' topology in the substrate network can be a consideration for load balance network [115]. From the aforementioned Sect. 4.4.1, we get the routers' topology, $P_{1 \times N}$, by PCA dimension reduction method. Since the value processed by PCA is a normal distribution, the topology can be used directly as the neural network's input to predict the QU at the next time slot.

4.4.3.3   The Selection Mechanism of Next Hop

In our deployment scenario, there are three metrics. Among them, the number of routing hops is destination-oriented, the QU is the dynamic attribute, and the topology is the static attribute. Our proposed routing schemes choose the next hop based on a single metric aggregated as a combination of the above multiple metrics [118]. The proposed queue utilization routing algorithm is shown in Algorithm 4.11.

We propose a queue utilization routing algorithm, where the QU at the current and next time slots is used for the calculation of RANK and select the next hop based on a single metric (RANK). Moreover, we defined RANK of router k as

$$\text{RANK}(k) = \alpha \cdot q_i(k) + \beta \cdot q_{i+1}(k), \tag{4.104}$$

where α and β are the coefficients which control the weight given to the metrics, $q_i(k)$ and $q_{i+1}(k)$, respectively. α and β satisfy that $\alpha + \beta = 1$ and we can select the next hop considering both the current QU and the predicted. For further analysis, we set traffic pattern as 30pps (packets per second) and record the network performance versus different weighting coefficients. Figure 4.32 shows the impact of considering the QU at the next time slot which can reflect both the network topology and the traffic pattern. According to the experimental results, the network performance is improved with the increase of the QU at the next time slot until $cur/pred < 7/3$. The network performance begins to degrade when the QU at the next time slot is more than 0.3. Thus, we set the coefficients α to 0.7 and β to 0.3.

Each router recognizes its neighbor routers by metric messages. Router k determines the next hop candidate set $\boldsymbol{H}_k$ according to its neighbors $N(k)$ as

---

Algorithm 4.11 Queue utilization routing algorithm

Input: the pkt(src,dist), the network graph, the routers' current queue, the the
routers' predicted queue

Output: the routers' queue utilization at the next time slot route list, or null

1: **procedure** $:routing(pkt(s, d), graph, current\ queue, predicted\ queue)$
2: $next = s$
3: $s.buffer.push(pkt)$
4: $list.add(s)$
5: **while** $next! = d\&\&next! = null$ **do**
6:      $next = nextHop(next, d, current\ queue, predicted\ queue))$
7:      $list.add(next)$
8:      $s.buffer.pop(pkt)$
9:      $next.buffer.push(pkt)$
10: **end while**
11: **if** $next! = null$ **then**
12:     return $list, future\ queue$
13: **else**
14:     return null, $future\ queue$
15: **end if**
16: **procedure** $:nextHop(c, d, pkt, current\ queue, predicted\ queue)$
17: **if** $c == d$ **then**
18:     return c
19: **end if**
20: **if** d in $c.neighbor$ **then**
21:     return c, d
22: **end if**
23: list=$[c]$
24: Benchmark=$Hop(c, d)$
25: **for** a in $c.neighbor$ if $Hop(a, d) \leq Hop(c, d)$ **do**
26:     $list.add(a)$
27: **end for**
28: **if** list is None **then**
29:     return Null
30: **else**
31:     choose one from list as $nexthop$
32:     $RANK = current\ queue + predicted\ queue$
33:     $nexthop = argmin(list[].RANK)$
34:     $c.buffer.pop(pkt)$
35:     $nexthop.buffer.push(pkt)$
36:     return
37: **end if**

$$H_k = \{n \in N(k) | h_n(d) \le h_k(d), q_i(n), q_{i+1}(n+1) < 1\}, \qquad (4.105)$$

where $h_n(d) \le h_k(d)$ means that router n is as close as or closer to the destination than k, and $q_i(n), q_{i+1}(n+1) < 1$ indicates that there are resources available for the buffer of router n. Each router performs the next hop selection process at discrete time $\beta \cdot \Delta t$. The strategy to select the next hop is as follows:

$$\widehat{h} = \arg \min_{n \in N_k} \{\text{RANK}(n)\}. \qquad (4.106)$$

Thus, the MLQU and DLQU routing schemes allow each router to select the next hop which is as close as or closer to the destination with lower QU. Although the selected next hop may not satisfy the minimum number of routing hops, load balance can be achieved considering the QU. In this way, the substrate network resources are utilized more efficiently and the QoS routing is achieved.

As shown in Fig. 4.29, source routers R1, R4, and R7 receive the input packets and send them to destination R6. Since the MLQU and DLQU routing schemes consider not only the current but also the QU at the next time slot, R1 and R7 can avoid choosing R5 as the next hop and choose R2 and R8 instead. Therefore, we can achieve load balance of the network and guarantee the QoS of the routing.

4.4.3.4 Time Complexity of the Algorithm

Proposition 4.3 *The time complexity of the machine learning aided routing scheme is $O(n^3 + m(l_{Input} \cdot l_1 + l_1 \cdot l_2 + \cdots + l_n \cdot l_{Output}) + n^2 \cdot e)$, where n is the number of substrate routers, e is the number of substrate edges, d is the average degree of routers, and m is the number of training samples. l_{Input} and l_{Output} represent the units' number of the input and output layers, respectively, while l_i ($i \in [1, n]$) represents the units' number of the ith hidden layer.*

Proof *(Proof)* In the process of PCA, the time complexity to calculate the covariance matrix $\Sigma_{M \times M}$ of a square matrix $X_{N \times N}$ is $O(n^2)$, while the time complexity to calculate all the eigenvectors is $O(n^3)$. For the neural network agent, the time complexity of the feedforward and backward propagation for one sample is $O(l_{Input} \cdot l_1 + l_1 \cdot l_2 + \cdots + l_n \cdot l_{Output})$. For the BF algorithm, the time complexity is $O(n \cdot e)$. After every time interval, the QUBF algorithm is used to construct the routing table for each router and make the adjustments with the time complexity of $O(n^2 \cdot e)$." $\qquad \qquad \square$

4.4.4   Experiments and Simulation Results

4.4.4.1   Datasets

This work utilizes the random regular graph generator algorithms to generate a substrate network as a simulation environment for routing algorithms [65, 66]. And we suppose a simulation environment with 30 nodes and 45 links, where each node's degree is set as 3. Here, we consider all the substrate nodes as routers that can generate, forward, and process the arrival packets. The CPU of each router is 5 units (packets per seconds) and the bandwidth resources of each link are assumed to be large enough. We set the buffer length as 8, while the packets are randomly discarded when its buffer is filled.

In the conducted simulation, the generating rate of packets is set as 30pps. And the time interval of QU updating is set as 1 s during which packets are forwarded and processed. Considering the initial phase, all the routers choose the shortest path to send packets by BF. Then, the recorded traffic pattern and QU are used as the training set of our proposed neural network to predict the QU at the next time slot. Finally, the routing decisions are made based on the predicted QU. For comparison of our proposed routing scheme, BF is used as the benchmark method.

To guarantee the generalization of the QU prediction, we recorded the routers' QU from 100 to 500 s. The number of substrate physical nodes is 30 in our proposed simulation environment of the training set and the QU before and after each time interval are recorded as input and output. The test set contains the QU information of 500 s based on a substrate network with 30 nodes. First, we trained the network model and the network parameters can be improved by the training phase. Then we use the obtained network parameters to make QU prediction on the test set. Consequently, the generalization of the training model on the test set can be observed by our experimental results.

4.4.4.2   Experiments Settings

The hidden layers of the shallow and deep neural networks are set as one and two, respectively. The specific features of the two neural networks are shown in Table 4.11. We use the Python 3.7 programming environment to construct a neural network model through TensorFlow and initialize it with normal distribution parameters.

In Sect. 4.4.2, when optimizing neural networks with Adam, many hyperparameters have been fixed at certain values. As for selection of network parameters, the output is the routers' QU from 0 to 1 and the type of activation function is determined as $ReLu(x) = \max(0, x)$ for hidden layers and $\sigma(x) = \frac{1}{1+e^{-x}}$ for the output layer. Then we select the mean square error as the loss function to reduce the difference between the predicted values and true labels. According to the broad strategy, we construct a simple structure by determining the number of hidden layers

Table 4.11 Neural network parameters

Shallow neural network						
Input layer		Full connection layer			Output layer	
Width	60	Layer	1		Node	30
		Node	50		Connection	Full
Width	60	Active	Relu		Active	Sigmoid
		Initialize	Normal		Initialize	Normal
Deep neural network						
Input layer		Full connection layer			Output layer	
Width	60	Layer	1	2	Node	30
		Node	50	50	Connection	Full
Width	60	Active	Relu	Relu	Active	Sigmoid
		Initialize	Normal	Normal	Initialize	Normal

and their corresponding number of neurons. Given a random value of the remaining hyperparameters, the learning rate is adjusted to obtain a suitable value. Then the number of epochs can be determined by a whole observation based on the given hyperparameters. According to the training results as in Fig. 4.30, we set the learning rate as 0.025 and the epoch as 5000.

Four experiments were designed to test the performance of the proposed DLQU and MLQU routing schemes. The first two test the application of the MLQU routing schemes in a 30-node net. In the experiments, Fig. 4.30 shows the training performance of routing based on the QU prediction, where the predicted QUs are trained by the shallow neural network (MLQU) and the deep neural network (DLQU). The third one (QUBF) tests the application of BF considering the current QU to prove the critical importance of QU in load balance routing. In the fourth experiment, the BF algorithm was tested in a 30-node network, which only considers the number of routing hops as a benchmark. Both BF and QUBF algorithms aim to solve the single source shortest path problem. Since each router needs to find the optimal next hop for its packets, their computational complexity is $O(n^2 \cdot e)$ in each time interval. In addition, as described in Section V-D, the time complexity of our proposed routing schemes is $O(n^3 + m(l_{Input} \cdot l_1 + l_1 \cdot l_2 + \cdots + l_n \cdot l_{Output}) + n^2 \cdot e)$. Apparently, the computational complexity of MLQU and DLQU can be similar to BF and QUBF with the appropriate number of the hidden layer units.

4.4.4.3   Training Results

Mean square error (MSE) is used as a value to measure the performance of our proposed neural network model. It is important to select the appropriate learning rate and epoch because the convergence of the supervised learning process is decided by them.

Fig. 4.30 Performance on both training set and testing set. (**a**) Shows the trend of loss versus learning rate in 6000 epoch in the shallow neural network. (**b**) Shows the trend of loss versus epoch at the trained learning rate (0.025) in the shallow neural network. (**c**) Shows the trend of loss versus epoch of the shallow and deep neural network at the trained learning rate and epoch

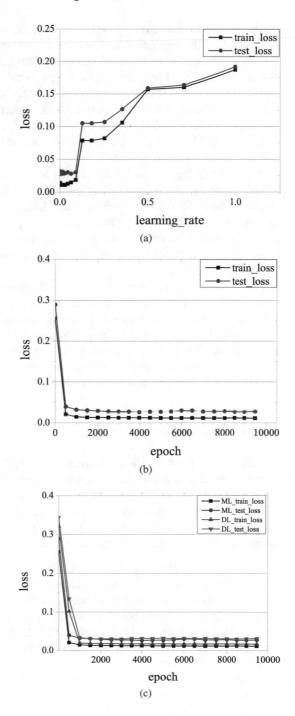

Figure 4.30a shows the trend of loss to learning rate in 6000 epochs. The gradient is updated by the learning rate α, which determines the convergence of objective function. When α is set too small, the convergence process will become very slow. When α is set too large, the gradient may oscillate around the minimum value and may not even converge. An appropriate α can enable the objective function to converge to a minimum value in a suitable time. According to the result, we choose the learning rate as 0.025, where the loss of both training and testing sets can be reduced well. Hence, the learning agent can adjust its embedding matrix according to the gradient and learning rate.

As shown in Fig. 4.30a, the loss (MSE) is constantly decreasing with the trained learning rate in the shallow neural network. As the number of epoch increases, the curve becomes over-fitting from under-fitting. The diversity of data affects the number of the suitable epoch and we set the epoch as 5000 based on Fig. 4.30a. If training loss is declining and the testing loss is declining, the network is still learning, and if training loss is decreasing and testing loss tends to be constant, the network is over-fitting.

As shown in Fig. 4.30c, the shallow neural network performs better than the deep neural network on reducing the loss between the output and label. Our interpretation is that the number of routers is small and our dataset is simple. The more the number of network layers is, the better the neural network learns. Deep neural networks are more suitable for the large amounts of data such as complex large-scale problems. Hence, there is no need to predict by the deep neural network, which is too complex to ignore the random noise and memory the general trend in the training process.

4.4.4.4   Performance Evaluation

The Routing Scheme with Consideration of QU We compare the performance of our proposal and the benchmarks versus different input traffic evenly distributed from 10 pps to 40 pps. It can be concluded that the current QU affects the load balance of the network because the packets can be avoided to forward by the busy router. The comparison result of three metrics among the two benchmarks and our proposed schemes is demonstrated in Fig. 4.31.

Figure 4.31a depicts the PLR (i.e., dropped packets divided by injected packets into the IP layer) with different traffic patterns. The core purpose of the packets-switched network is to transmit as many packets as possible. And we evaluate our proposed routing scheme based on the average PLR. It can be observed that our proposed routing schemes have a lower PLR. It is easy to explain that considering the router's queue states, the packets can be well avoided transmitting to the router with fewer buffer resources.

Figure 4.31b shows the throughput (i.e., the number of successfully transmitted packets per unit time) with varying traffic pattern. The mean throughput value means nothing to the evaluation of load balance, so we record the worst throughput of the recommended four algorithms. The higher the worst throughput is, the better the

Fig. 4.31 Performance on routing schemes. (**a**) Shows packet loss rate at routers' queue versus traffic pattern. (**b**) Shows throughput versus traffic pattern. (**c**) Shows average delay per packet versus traffic pattern

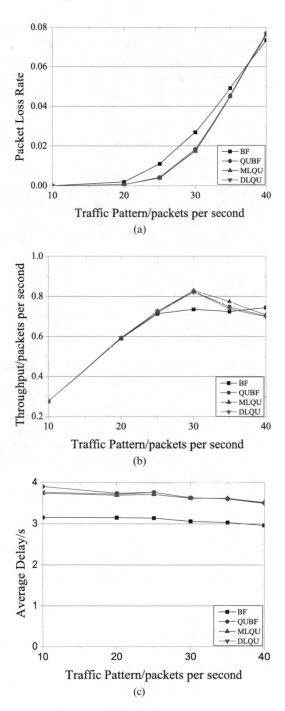

(a)

(b)

(c)

load balance network performs. As Fig. 4.31b shows, the MLQU and DLQU have the highest throughput. Since MLQU and DLQU can forward the packets with both the current QU and the predicted, the load balance routing scheme is achieved.

Figure 4.31c depicts the average delay with varying traffic pattern. Delay is one of the important metrics of packet-switched network performance, and the smaller the delay is, the better the user experience. Thus, we record mean values to evaluate the performance on the network load balance of our proposal. As Fig. 4.31c shows, the delay of MLQU and DLQU is 20% more than the shortest routing scheme (BF). As a trade-off for the good performance of the PLR and the throughput, the slightly lower delay is acceptable. On the one hand, BF chooses the shortest path and the packet's transmission time is relatively short. On the other hand, the MLQU and DLQU routing schemes consider the buffer length to reduce the queue delay. Compared with BF, our proposal can avoid selecting the router with fewer buffer resources as the next hop and the queuing delay can be lowered.

The Routing Scheme with Prediction of QU To illustrate the success of QU prediction by the neural network in the routing process, three kinds of experiments were designed to examine our proposed MLQU and DLQU routing schemes. Figure 4.32 depicts their performance versus the different proportion of the QU at the current and next time slot (cur/pred), where the QU at the next time slot is replaced by the real value, the predicted value by shallow and deep neural networks.

On the one hand, the packet loss rate is constantly decreasing and the throughput is constantly increasing at the beginning. It can be indicated that the QUBF always deals with bursty traffic unintelligently. While our proposed algorithm can achieve a better global convergence and more effective optimization of the network resources. The neural network has the following three advantages: strong learning ability, robustness to noisy data, and association with approximate nonlinear relationship. Therefore, our proposal based on the neural network can predict and avoid the same problem at a later instant.

On the other hand, performances of MLQU and DLQU routing schemes are better after $cur/pred < 6/4$, where the QU at the next time slot is predicted by the neural network. Our interpretation is that the neural network prediction is obtained by analyzing the current QU, so it includes more information about the current QU.

The above analysis proves the effectiveness of our proposed scheme, where the predicted QU of the next interval is considered.

4.5   Summary

In this chapter, we discuss the problem of traditional traffic control methods and introduce several intelligent traffic control methods. We first propose a collaborative multi-agent reinforcement learning aided routing algorithm. Then, we design a vector-based routing principle. What is more, we design a QoS-oriented adaptive routing scheme using deep reinforcement learning in 5G mobile network envi-

Fig. 4.32 Performance on
the prediction of routing
schemes. (**a**) Shows loss rate
at packet queue versus
different proportion of the
current QU and the predicted
(cur/pred). (**b**) Shows
throughput versus different
cur/pred. (**c**) Shows average
delay per packet versus
different cur/pred

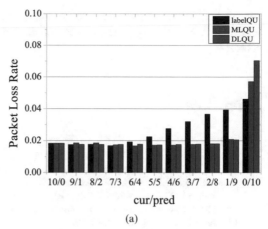

(a)

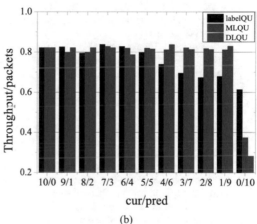

(b)

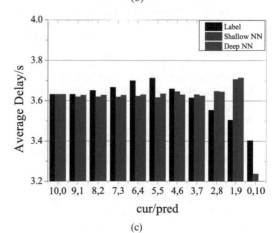

(c)

ronment. Finally, we propose a load balancing routing scheme aided by machine learning. Extensive experimental results show that these methods are feasible and effective.

References

1. C. Han, H. Yao, T. Mai, N. Zhang, M. Guizani, QMIX aided routing in social-based delay-tolerant networks. IEEE Trans. Veh. Technol. **71**(2), 1952–1963 (2021)
2. H. Yao, H. Liu, P. Zhang, S. Wu, C. Jiang, S. Guo, A learning-based approach to intra-domain QoS routing. IEEE Trans. Veh. Technol. **69**(6), 6718–6730 (2020)
3. X. Yuan, H. Yao, J. Wang, T. Mai, M. Guizani, Artificial intelligence empowered QoS-oriented network association for next-generation mobile networks. IEEE Trans. Cogn. Commun. Netw. **7**(3), 856–870 (2021)
4. H. Yao, X. Yuan, P. Zhang, J. Wang, C. Jiang, M. Guizani, Machine learning aided load balance routing scheme considering queue utilization. IEEE Trans. Veh. Technol. **68**(8), 7987–7999 (2019)
5. L. Rashidi, R. Entezari-Maleki, D. Chatzopoulos, P. Hui, K.S. Trivedi, A. Movaghar, Performance evaluation of epidemic content retrieval in DTNs with restricted mobility. IEEE Trans. Netw. Serv. Manag. **16**(2), 701–714 (2019)
6. C.E. Perkins, E.M. Royer, Ad-hoc on-demand distance vector routing, in *Proceedings WMCSA'99 of the Second IEEE Workshop on Mobile Computing Systems and Applications* (1999), pp. 90–100
7. S.M. Tornell, C.T. Calafate, J. Cano, P. Manzoni, DTN protocols for vehicular networks: an application oriented overview. IEEE Commun. Surv. Tutorials **17**(2), 868–887 (2015)
8. H. Yao, T. Mai, C. Jiang, L. Kuang, S. Guo, AI routers & network mind: a hybrid machine learning paradigm for packet routing. IEEE Comput. Intell. Mag. **14**(4), 21–30 (2019)
9. T. Spyropoulos, K. Psounis, C. Raghavendra, Spray and wait: an efficient routing scheme for intermittently connected mobile networks, in *Proceedings of ACM SIGCOMM Workshop on Delay-Tolerant Networking* (2005)
10. A. Lindgren, A. Doria, O. Schelen, Probabilistic routing in intermittently connected networks, in *1st International Workshop on Service Assurance with Partial Intermittent Resources, Fortaleza, BRAZIL, AUG 01-06, (2004)*, vol. 3126 (2004), pp. 239–254
11. H. Dubois-Ferrière, M. Grossglauser, M. Vetterli, Age matters: efficient route discovery in mobile Ad hoc networks using encounter ages, in *MobiHoc '03* (2003)
12. Y. Zhu, B. Xu, X. Shi, Y. Wang, A survey of social-based routing in delay tolerant networks: positive and negative social effects. IEEE Commun. Surv. Tutorials **15**(1), 387–401 (2013)
13. M. Xiao, J. Wu, L. Huang, Community-aware opportunistic routing in mobile social networks. IEEE Trans. Comput. **63**(7), 1682–1695 (2014)
14. P. Hui, J. Crowcroft, E. Yoneki, Bubble rap: social-based forwarding in delay-tolerant networks. IEEE Trans. Mob. Comput. **10**(11), 1576–1589 (2011)
15. J. Tao, H. Wu, S. Shi, J. Hu, Y. Gao, Contacts-aware opportunistic forwarding in mobile social networks: a community perspective, in *2018 IEEE Wireless Communications and Networking Conference (WCNC)* (2018), pp. 1–6
16. P. Sunehag, G. Lever, A. Gruslys, W.M. Czarnecki, V.F. Zambaldi, M. Jaderberg, M. Lanctot, N. Sonnerat, J.Z. Leibo, K. Tuyls et al., Value-decomposition networks for cooperative multi-agent learning based on team reward, in *AAMAS* (2018), pp. 2085–2087
17. J. Foerster, G. Farquhar, T. Afouras, N. Nardelli, S. Whiteson, Counterfactual multi-agent policy gradients, in *Proceedings of the AAAI Conference on Artificial Intelligence*, vol. 32 (2018)
18. T. Rashid et al., Monotonic value function factorisation for deep multi-agent reinforcement learning. The. J. Mach. Learn. Res. **21**(1), 7234–7284 (2020)

19. S. Correia, A. Boukerche, R.I. Meneguette, An architecture for hierarchical software-defined vehicular networks. IEEE Commun. Mag. **55**(7), 80–86 (2017)
20. M.E.J. Newman, Modularity and community structure in networks. Proc. Natl. Acad. Sci. **103**(23), 8577–8582 (2006)
21. V.D. Blondel, et al., Fast unfolding of communities in large networks. J. Stat. Mech. Theory Exp. **2008**(10), P10008 (2008)
22. Oliehoek, A. Frans, Amato, Christopher, *A Concise Introduction to Decentralized POMDPs* (Springer International Publishing, Berlin, 2016)
23. M.J. Hausknecht, P. Stone, Deep recurrent q-learning for partially observable MDPs, in *AAAI Fall Symposia* (2015)
24. A. Keränen, J. Ott, T. Kärkkäinen, The ONE simulator for DTN protocol evaluation, in *SIMUTools '09: Proceedings of the 2nd International Conference on Simulation Tools and Techniques*, New York (ICST, 2009)
25. J. Scott, R. Gass, J. Crowcroft, P. Hui, C. Diot, A, Chaintreau, CRAWDAD dataset cambridge/haggle (v. 2009). Downloaded from https://crawdad.org/cambridge/haggle/2009 (2009)
26. V. Mnih, et al., Human-level control through deep reinforcement learning. Nature **518**(7540), 529–533 (2015)
27. Z.M. Fadlullah, F. Tang, B. Mao, N. Kato, O. Akashi, T. Inoue, K. Mizutani, State-of-the-art deep learning: evolving machine intelligence toward tomorrows intelligent network traffic control systems. IEEE Commun. Surv. Tutorials **19**(4), 2432–2455 (2017)
28. B. Mao, Z.M. Fadlullah, F. Tang, N. Kato, O. Akashi, T. Inoue, K. Mizutani, Routing or computing? The paradigm shift towards intelligent computer network packet transmission based on deep learning. IEEE Trans. Comput. **66**(11), 1946–1960 (2017)
29. B. Mao, Z.M. Fadlullah, F. Tang, N. Kato, O. Akashi, T. Inoue, K. Mizutani, A tensor based deep learning technique for intelligent packet routing, in *GLOBECOM 2017 – 2017 IEEE Global Communications Conference* (2017), pp. 1–6
30. H. Yao, X. Yuan, P. Zhang, J. Wang, J. Chunxiao, M. Guizani, Machine learning aided load balance routing scheme considering queue utilization. IEEE Trans. Veh. Technol. 1–1 (2019)
31. F. Tang, et al., An intelligent traffic load prediction-based adaptive channel assignment algorithm in SDN-IoT: a deep learning approach. IEEE Internet Things J. **5**(6), 5141–5154 (2018)
32. F. Zhao, G. Zeng, K. Lu, EnLSTM-WPEO: short-term traffic flow prediction by ensemble LSTM, NNCT weight integration, and population extremal optimization. IEEE Trans. Veh. Technol. **69**(1), 101–113 (2020)
33. F. Tang, B. Mao, Z.M. Fadlullah, N. Kato, Deep spatiotemporal partially overlapping channel allocation: joint CNN and activity vector approach, in *2018 IEEE Global Communications Conference (GLOBECOM)* (2018), pp. 1–6
34. Z. Zhang, Y. Li, C. Huang, Q. Guo, C. Yuen, Y. L. Guan, DNN-aided block sparse bayesian learning for user activity detection and channel estimation in grant-free non-orthogonal random access. IEEE Trans. Veh. Technol. **68**(12), 12 000–12 012 (2019)
35. J. Wang, C. Jiang, H. Zhang, Y. Ren, K. Chen, L. Hanzo, Thirty years of machine learning: the road to pareto-optimal wireless networks. IEEE Commun. Surv. Tutorials 1–1 (2020)
36. Q. Yang, T. Jiang, N.C. Beaulieu, J. Wang, C. Jiang, S. Mumtaz, Z. Zhou, Heterogeneous semi-blind interference alignment in finite-SNR networks with fairness consideration. IEEE Trans. Wirel. Commun. 1–1 (2020)
37. C. Jiang, H. Zhang, Y. Ren, Z. Han, K. Chen, L. Hanzo, Machine learning paradigms for next-generation wireless networks. IEEE Wirel. Commun. **24**(2), 98–105 (2017)
38. K. Gai, M. Qiu, H. Zhao, Energy-aware task assignment for mobile cyber-enabled applications in heterogeneous cloud computing. J. Parallel Distrib. Comput. **111**, 126–135 (2018)
39. H. Liao, Z. Zhou, X. Zhao, L. Zhang, S. Mumtaz, A. Jolfaei, S.H. Ahmed, A.K. Bashir, Learning-based context-aware resource allocation for edge computing-empowered industrial IoT. IEEE Internet Things J. 1–1 (2019)

40. H. Yao, M. Li, J. Du, P. Zhang, C. Jiang, Z. Han, Artificial intelligence for information-centric networks. IEEE Commun. Mag. **57**(6), 47–53 (2019)
41. Z. Xiong, Y. Zhang, N.C. Luong, D. Niyato, P. Wang, N. Guizani, The best of both worlds: a general architecture for data management in blockchain-enabled internet-of-things. IEEE Netw. **34**(1), 166–173 (2020)
42. H. Yao, T. Mai, J. Wang, Z. Ji, C. Jiang, Y. Qian, Resource trading in blockchain-based industrial internet of things. IEEE Trans. Ind. Inform. **15**(6), 3602–3609 (2019)
43. H. Yao, T. Mai, C. Jiang, L. Kuang, S. Guo, AI routers network mind: a hybrid machine learning paradigm for packet routing. IEEE Comput. Intell. Mag. **14**(4), 21–30 (2019)
44. H. Yao, X. Chen, M. Li, P. Zhang, L. Wang, A novel reinforcement learning algorithm for virtual network embedding. Neurocomputing **284**, 1–9 (2018)
45. X. Liao et al., A model-driven deep reinforcement learning heuristic algorithm for resource allocation in ultra-dense cellular networks. IEEE Trans. Veh. Technol. **69**(1), 983–997 (2019)
46. K. Gai, M. Qiu, Reinforcement learning-based content-centric services in mobile sensing. IEEE Netw. **32**(4), 34–39 (2018)
47. K. Gai, K. Xu, Z. Lu, M. Qiu, L. Zhu, Fusion of cognitive wireless networks and edge computing. IEEE Wirel. Commun. **26**(3), 69–75 (2019)
48. C. Xu, K. Wang, P. Li, R. Xia, S. Guo, M. Guo, Renewable energy-aware big data analytics in geo-distributed data centers with reinforcement learning. IEEE Trans. Netw. Sci. Eng. 1–1 (2018)
49. J. Wang, C. Jiang, K. Zhang, X. Hou, Y. Ren, Y. Qian, Distributed q-learning aided heterogeneous network association for energy-efficient IIoT. IEEE Trans. Ind. Inform. **16**(4), 2756–2764 (2020)
50. Z. Xiong, Y. Zhang, D. Niyato, R. Deng, P. Wang, L. Wang, Deep reinforcement learning for mobile 5G and beyond: fundamentals, applications, and challenges. IEEE Veh. Technol. Mag. **14**(2), 44–52 (2019)
51. W.L. Hamilton, R. Ying, J. Leskovec, Representation learning on graphs: methods and applications. Preprint. arXiv:1709.05584 (2017)
52. S. Cao, W. Lu, Q. Xu, Deep neural networks for learning graph representations, in *Thirtieth AAAI Conference on Artificial Intelligence* (2016), pp. 1145–1152
53. W. Hamilton, Z. Ying, J. Leskovec, Inductive representation learning on large graphs, in *Advances in Neural Information Processing Systems* (2017), pp. 1025–1035
54. M. Niepert, M. Ahmed, K. Kutzkov, Learning convolutional neural networks for graphs, in *International Conference on Machine Learning* (2016), pp. 2014–2023
55. T.N. Kipf, M. Welling, Semi-supervised classification with graph convolutional networks. Preprint. arXiv:1609.02907 (2016)
56. B. Perozzi, R. Al-Rfou, S. Skiena, Deepwalk: online learning of social representations, in *Proceedings of the 20th ACM SIGKDD International Conference on Knowledge Discovery and Data Mining* (ACM, New York, 2014), pp. 701–710
57. A. Grover, J. Leskovec, node2vec: scalable feature learning for networks, in *ACM Sigkdd International Conference on Knowledge Discovery & Data Mining* (2016), pp. 855–864
58. H. Pan, H. Yao, T. Mai, N. Zhang, Y. Liu, Scalable traffic control using programmable data planes in a space information network. IEEE Netw. **35**(4), 35–41 (2021)
59. H. Yao, H. Liu, P. Zhang, S. Wu, C. Jiang, S. Guo, An intelligent approach to energy efficient transportation and QoS routing, in *ICC 2019 – 2019 IEEE International Conference on Communications (ICC)* (2019), pp. 1–6
60. J. Duchi, E. Hazan, Y. Singer, Adaptive subgradient methods for online learning and stochastic optimization. J. Mach. Learn. Res. **12**(Jul), 2121–2159 (2011)
61. T. Tieleman, G. Hinton, Lecture 6.5-rmsprop: divide the gradient by a running average of its recent magnitude. COURSERA Neural Netw. Mach. Learn. **4**(2), 26–31 (2012)
62. D.P. Kingma, J. Ba, Adam: a method for stochastic optimization. Preprint. arXiv:1412.6980 (2014)
63. D.E. Rumelhart, G.E. Hinton, R.J. Williams, Learning representations by back-propagating errors. Nature **323**(6088), 533 (1986)

64. S. Suresh, S.N. Omkar, V. Mani, Parallel implementation of back-propagation algorithm in networks of workstations. IEEE Trans. Parallel Distrib. Syst. **16**(1), 24–34 (2005)
65. A. Steger, N.C. Wormald, Generating random regular graphs quickly. Comb. Probab. Comput. **8**(4), 377–396 (1999)
66. J.H. Kim, V.H. Vu, Generating random regular graphs, in *ACM Symposium on Theory of Computing* (2003), pp. 213–222
67. C. Jiang, Y. Chen, Y. Ren, K.J.R. Liu, Maximizing network capacity with optimal source selection: a network science perspective. IEEE Signal Process Lett. **22**(7), 938–942 (2015)
68. S.S. Chen, K. Nahrstedt, On finding multi-constrained paths, in *1998 IEEE International Conference on Communications, 1998. ICC 98. Conference Record*, vol. 2 (IEEE, Piscataway, 1998), pp. 874–879
69. T. Korkmaz, M. Krunz, Multi-constrained optimal path selection, in *IEEE INFOCOM*, vol. 2, no. April. Citeseer (2001), pp. 834–843
70. P. Khadivi, S. Samavi, T.D. Todd, Multi-constraint Qos routing using a new single mixed metrics. J. Netw. Comput. Appl. **31**(4), 656–676 (2008)
71. G. Cheng, N. Ansari, A new heuristics for finding the delay constrained least cost path, in *Global Telecommunications Conference, 2003. GLOBECOM '03.*, vol. 7 (IEEE, Piscataway, 2003), pp. 3711–3715
72. A. Jüttner, B. Szviatovski, I. Mécs, Z. Rajkó, Lagrange relaxation based method for the Qos routing problem, in *Proceedings IEEE INFOCOM 2001. Conference on Computer Communications. Twentieth Annual Joint Conference of the IEEE Computer and Communications Society (Cat. No.01CH37213)*, vol. 2 (2001), pp. 859–868
73. H. Yao, T. Mai, J. Wang, Z. Ji, C. Jiang, Y. Qian, Resource trading in blockchain-based industrial internet of things. IEEE Trans. Ind. Inform. **15**(6), 3602–3609 (2019)
74. F. Li, H. Yao, J. Du, C. Jiang, Y. Qian, Stackelberg game-based computation offloading in social and cognitive industrial internet of things. IEEE Trans. Ind. Inform. **16**(8), 5444–5455 (2020)
75. J. Wang, C. Jiang, K. Zhang, X. Hou, Y. Ren, Y. Qian, Distributed q-learning aided heterogeneous network association for energy-efficient IIoT. IEEE Trans. Ind. Inform. **16**(4), 2756–2764 (2020)
76. J.W. Guck, A. Van Bemten, M. Reisslein, W. Kellerer, Unicast QoS routing algorithms for SDN: a comprehensive survey and performance evaluation. IEEE Commun. Surv. Tutorials **20**(1), 388–415 (2018)
77. H. Yao, M. Li, J. Du, P. Zhang, C. Jiang, Z. Han, Artificial intelligence for information-centric networks. IEEE Commun. Mag. **57**(6), 47–53 (2019)
78. J. Wang, C. Jiang, H. Zhang, Y. Ren, K.-C. Chen, L. Hanzo, Thirty years of machine learning: the road to pareto-optimal wireless networks. IEEE Commun. Surv. Tutorials **22**(3), 1472–1514 (2020)
79. K. Arulkumaran, M.P. Deisenroth, M. Brundage, A.A. Bharath, Deep reinforcement learning: a brief survey. IEEE Signal Process. Mag. **34**(6), 26–38 (2017)
80. V. Mnih, K. Kavukcuoglu, D. Silver, A.A. Rusu, J. Veness, M.G. Bellemare, A. Graves, M. Riedmiller, A.K. Fidjeland, G. Ostrovski, et al., Human-level control through deep reinforcement learning. Nature **518**(7540), 529–533 (2015)
81. B.A.A. Nunes, M. Mendonca, X. Nguyen, K. Obraczka, T. Turletti, A survey of software-defined networking: past, present, and future of programmable networks. IEEE Commun. Surv. Tutorials **16**(3), 1617–1634 (2014)
82. C. Yang, J. Li, A. Anpalagan, M. Guizani, Joint power coordination for spectral-and-energy efficiency in heterogeneous small cell networks: a bargaining game-theoretic perspective. IEEE Trans. Wirel. Commun. **15**(2), 1364–1376 (2016)
83. R.W. Heath, M. Kountouris, T. Bai, Modeling heterogeneous network interference using poisson point processes. IEEE Trans. Signal Process. **61**(16), 4114–4126 (2013)
84. C. Yang, Y. Yao, Z. Chen, B. Xia, Analysis on cache-enabled wireless heterogeneous networks. IEEE Trans. Wirel. Commun. **15**(1), 131–145 (2016)

85. A. Bashandy, E. Chong, A. Ghafoor, Network modeling and jitter control for multimedia communication over broadband network, in *IEEE International Conference on Computer Communications*, New York, vol. 2 (1999), pp. 559–566
86. M. Abou-El-Ata, A. Hariri, The M/M/c/N queue with balking and reneging. Comput. Oper. Res. **19**(8), 713–716 (1992)
87. Z. Rongcai, Z. Shuo, Network traffic generation: a combination of stochastic and self-similar, in *International Conference on Advanced Computer Control*, Shenyang (2010), pp. 171–175
88. N. Sapountzis, T. Spyropoulos, N. Nikaein, U. Salim, Joint optimization of user association and dynamic TDD for ultra-dense networks, in *IEEE INFOCOM 2018 – IEEE Conference on Computer Communications*, Honolulu (2018), pp. 2681–2689
89. Y. Wang, H.H. Yang, Q. Zhu, T.Q.S. Quek, Analysis of packet throughput in spatiotemporal hetnets with scheduling and various traffic loads. IEEE Wirel. Commun. Lett. **9**(1), 95–98 (2020)
90. K. Wang, Optimal control of an M/E/k/1 queueing system with removable service station subject to breakdowns. J. Oper. Res. Soc. **48**(9), 936–942 (1997)
91. A. El-Naggar, A. Shalaby, Biomimicry to network on chip: router heart rate, in *2015 27th International Conference on Microelectronics (ICM)*, Casablanca (2015), pp. 162–165
92. M. Zhang, Model analysis of risk queuing and its application in port bottlenecks management, in *2017 3rd IEEE International Conference on Computer and Communications (ICCC)*, Chengdu (2017), pp. 1009–1013
93. A.R. Bashandy, E.K.P. Chong, A. Ghafoor, Generalized quality-of-service routing with resource allocation. IEEE J. Select. Areas Commun. **23**(2), 450–463 (2005)
94. C. Meister, R. Cotterell, T. Vieira, If beam search is the answer, what was the question? in *Proceedings of the 2020 Conference on Empirical Methods in Natural Language Processing (EMNLP)*. Online: Association for Computational Linguistics (2020), pp. 2173–2185
95. J.A. Boyan, M.L. Littman, Packet routing in dynamically changing networks: a reinforcement learning approach, in *Proceedings of the 6th International Conference on Neural Information Processing Systems* (Morgan Kaufmann Publishers Inc., San Francisco, 1993), pp. 671–678
96. Y. Li, Deep reinforcement learning: an overview. Computing Research Repository. abs/1701.07274 (2017)
97. T.P. Lillicrap, J.J. Hunt, A. Pritzel, N. Heess, T. Erez, Y. Tassa, D. Silver, D. Wierstra, Continuous control with deep reinforcement learning. Comput. Sci. **8**(6), A187 (2015)
98. I. Goodfellow, Y. Bengio, A. Courville, *Deep Learning* (MIT Press, Cambridge, 2016). http://www.deeplearningbook.org
99. V. Mnih, K. Kavukcuoglu, D. Silver, A.A. Rusu, J. Veness, M.G. Bellemare, A. Graves, M. Riedmiller, A.K. Fidjeland, G. Ostrovski, et al., Human-level control through deep reinforcement learning. Nature **518**(7540), 529–533 (2015)
100. Y. Zhong, M. Haenggi, T.Q.S. Quek, W. Zhang, On the stability of static poisson networks under random access. IEEE Trans. Commun. **64**(7), 2985–2998 (2016)
101. S. Henna, M. Sajeel, F. Hussain, M. Asfand-e yar, M. Tauqir, A fair contention access scheme for low-priority traffic in wireless body area networks. Sensors **17**(9), 1931 (2017)
102. Y.R.B. Al-Mayouf, N.F. Abdullah, O.A. Mahdi, S. Khan, M. Ismail, M. Guizani, S.H. Ahmed, Real-time intersection-based segment aware routing algorithm for urban vehicular networks. IEEE Trans. Intell. Transport. Syst. **19**(7), 1–1 (2018)
103. P.A. Humblet, Another adaptive distributed shortest path algorithm. IEEE Trans. Commun. **39**(6), 995–1003 (2002)
104. S. Vutukury, J.J. Garcia-Luna-Aceves, MDVA: a distance-vector multipath routing protocol, in *IEEE International Conference on Computer Communications (INFOCOM)*, Anchorage, vol. 1 (2001), pp. 557–564
105. E.W. Dijkstra, A note on two problems in connexion with graphs. Numer. Math. **1**(1), 269–271 (1959)
106. L. Zou, M. Lu, Z. Xiong, A distributed algorithm for the dead end problem of location based routing in sensor networks. IEEE Trans. Veh. Technol. **54**(4), 1509–1522 (2005)

107. M.H. Eiza, T. Owens, Q. Ni, Q. Shi, Situation-aware QoS routing algorithm for vehicular ad hoc networks. IEEE Trans. Veh. Technol. **64**(12), 5520–5535 (2015)
108. M.R. Garey, D.S. Johnson, *Computers and Intractability: A Guide to the Theory of NP-Completeness* (W. H. Freeman & Co., New York, 1979)
109. Z. Wang, J. Crowcroft, Quality-of-service routing for supporting multimedia applications. IEEE J, Select. Areas Commun. **14**(7), 1228–1234 (1996)
110. L. Hanzo, R. Tafazolli, A survey of QoS routing solutions for mobile ad hoc networks. IEEE Commun. Surv. Tutorials **9**(2), 50–70 (2007)
111. Y. Bejerano, Y. Breitbart, A. Orda, R. Rastogi, A. Sprintson, Algorithms for computing QoS paths with restoration. IEEE/ACM Trans. Netw. **13**(3), 648–661 (2005)
112. H. Chang, X. Yin, H. Yao, J. Wang, R. Gao, J. An, L. Hanzo, Low-complexity adaptive optics aided orbital angular momentum based wireless communications. IEEE Trans. Veh. Technol. **70**(8), 7812–7824 (2021)
113. R. Battiti, A. Bertossi, D. Cavallaro, A randomized saturation degree heuristic for channel assignment in cellular radio networks. IEEE Trans. Veh. Technol. **50**(2), 364–374 (1999).
114. D. Turkovic, F. Kuipers, N. van Adrichem, K. Langendoen, Fast network congestion detection and avoidance using p4, in *Proceedings of the 2018 Workshop on Networking for Emerging Applications and Technologies*, New York (2018), pp. 45–51
115. H.S. Kim, H. Kim, J. Paek, S. Bahk, Load balancing under heavy traffic in RPL routing protocol for low power and lossy networks. IEEE Trans. Mob. Comput. **16**(4), 964–979 (2017)
116. H.F. Salama, D.S. Reeves, Y. Viniotis, A distributed algorithm for delay-constrained unicast routing, in *IEEE International Conference on Computer Communications (INFOCOM)*, Kobe, vol. 1 (1997), pp. 84–91
117. D.P. Kingma, J. Ba, Adam: a method for stochastic optimization, in *International Conference on Learning Representations*. abs/1412.6980 (2014)
118. H.D. Neve, P.V. Mieghem, A multiple quality of service routing algorithm for PNNI, in *ATM Workshop Proceedings*, Fairfax (2002), pp. 324–328

Chapter 5
Intelligent Resource Scheduling

Abstract The continued growth in the number and applications of Internet of Things (IoT) connected devices makes it more challenging to meet multi-dimensional QoS within the same IoT network. In this chapter, we first design a network slicing architecture over the SDN-based long-range wide area network. The SDN controller can dynamically split the network into multiple virtual networks according to different business requirements. Then, a Continuous-Decision virtual network embedding scheme relying on Reinforcement Learning (CDRL) is proposed, two traditional heuristic embedding algorithms as well as the classic reinforcement learning aided embedding algorithm are used for benchmarking our proposed CDRL algorithm. Finally, we propose a hybrid intelligent control architecture, which adopts the centralized training and distributed execution paradigm. A centralized critic is introduced to ease the training process of the distributed network nodes. Besides, considering the competitive behavior of users, we formulate the resource allocation problem as a multi-user competition game model. Based on this, we proposed a multi-agent reinforcement learning-based SFCs deployment algorithm.

Keywords Network slicing · Long-range wide area network · Service function chain · Hybrid intelligent control

The continued growth in the number and applications of Internet of Things (IoT) connected devices makes it more challenging to meet multi-dimensional QoS within the same IoT network. In this chapter, we first design a network slicing architecture over the SDN-based long-range wide area network. The SDN controller can dynamically split the network into multiple virtual networks according to different business requirements [1]. Then, a Continuous-Decision virtual network embedding scheme relying on Reinforcement Learning (CDRL) is proposed, two traditional heuristic embedding algorithms as well as the classic reinforcement learning aided embedding algorithm are used for benchmarking our proposed CDRL algorithm [2]. Finally, we propose a hybrid intelligent control architecture, which adopts the centralized training and distributed execution paradigm. A centralized

© The Author(s), under exclusive license to Springer Nature Switzerland AG 2023   211
H. Yao, M. Guizani, *Intelligent Internet of Things Networks*, Wireless Networks,
https://doi.org/10.1007/978-3-031-26987-5_5

critic is introduced to ease the training process of the distributed network nodes [3]. Besides, considering the competitive behavior of users, we formulate the resource allocation problem as a multi-user competition game model. Based on this, we proposed a multi-agent reinforcement learning-based SFCs deployment algorithm.

5.1 Transfer Reinforcement Learning aided Network Slicing Optimization

Industrial IoT as the most powerful and exciting technology has quickly become a disruptive force reshaping how we live and work [4]. Massive industrial devices are now connected to the network, allowing for data collecting, exchanging, and analyzing. According to the Juniper's report, there will be 37 billion industrial IoT devices by 2025. Faced with such massive devices, IoT network technology is now confronted with unprecedented challenges.

Currently, the industrial IoT network market is dominated by the Long-Range Wide Area Network (LoRaWAN) technology [5]. It is a non-cellular wireless wide area network technology, which uses the LoRa radio modulation technology to provide low cost, low power, and long-range communication. Such appealing characteristics enable LoRaWAN suited for a wide range of industrial applications, particularly in distant areas where businesses like petroleum drilling, mining, and construction operate. However, with the emergence of new applications, providing the QoS to different IoT devices becomes a critical challenge in LoRaWAN [6].

Recently, network slicing technology has emerged as a viable solution to address this challenge. Network slicing is the process of dividing a single physical network into many logical (virtual) networks using network virtualization (i.e., NFV and SDN) [7]. Different slices have different logical topology, security rules, and performance characteristics for the sake of fulfilling different business purposes. In this paper, we design a network slicing architecture over SDN-based LoRaWAN, where the SDN controller can dynamically partition LoRa gateways' resources (e.g., physical channel) into several virtual networks on the fly [8].

Meanwhile, considering the limited resources on each gateway, designing an efficient slice resource optimization scheme is another crucial problem. The gateway should be able to configure slice parameters (e.g., Bandwidth (BW), Spreading Factor (SF), Transmission Power (TP)) to satisfy the distinct QoS [9]. To address this issue, we propose a DDPG based slice resource optimization algorithm [10]. The LoRa gateways using the DDPG are able to improve the performance by exploring the environment and learning directly from their experiences.

While such learning mechanism can converge to the optimal policy, in the end, it has to take a large number of training episodes. Especially in our scenario, each slice agent on each LoRa gateway has to learn from scratch (i.e., randomized policy), thereby resulting in a long learning time to reach the system's optimal performance. To accelerate the training process, we introduce the transfer learning framework

[11]. Transfer learning is a machine learning technique where experience gained acquired from one task can be transferred to other related tasks. Therefore, to accelerate the learning process across multiple LoRa gateways, we propose a Transfer learning-based Multi-agent DDPG (TMDDPG) algorithm.

5.1.1   System Model

In this section, we design a SDN-based network slicing architecture in LoRaWAN. Then, we present the system model and problem formulation of slice optimization.

5.1.1.1   Network Slicing Architecture

A typical LoRaWAN architecture consists of end devices, LoRa gateways, network servers, and application servers. The gateways are responsible for forwarding messages from end devices to network servers. The network servers are responsible for network management functions, including Over-The-Air-Activation, message routing, acknowledgment of messages, and adaptive data rate control. The application servers are used to process application-specific data messages received from end devices.

In this paper, we leverage SDN and NFV technologies to enhance the flexibility and programmability of the LoRaWAN. As shown in Fig. 5.1, we present a network slicing architecture over SDN-based LoRaWAN. The architecture runs on open-spec commodity compute and networking hardware and connects with the LoRa gateways. All the network functions are built on micro-services, hosted into containers, and orchestrated with Kubernetes. In our architecture, it contains three layers, named Application Plane, Control Plane, and Infrastructure Plane.

Application Plane In this plane, the applications can issue their QoS requirements and desired network behavior to the control plane via northbound application program interfaces. Meanwhile, it can leverage network information (e.g., network topology, network state) for its internal decision-making purposes. According to different QoS, the applications can be classified into three types, including best effort (BE), reliability aware (RA), and urgency and reliability aware (UR). The URA application requires the highest priority. Examples include emergency alerting and robot arm control. The RA applications require lower priority (e.g., security systems), while the BE requires the lowest priority (e.g., smart metering applications) [12].

Control Plane The control plane is composed of multiple SDN controllers. The controller can translate the application requirements to the infrastructure plane and provide global network abstractions of the LoRaWAN to the application plane [13]. According to the complete knowledge of network state and applications QoS, the

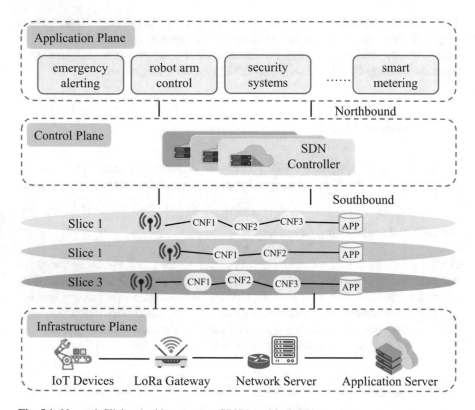

Fig. 5.1 Network Slicing Architecture over SDN-based LoRaWAN

controller can split physical network resources (e.g., channels on LoRa gateways) for different slices and assign IoT devices to the corresponding slice.

Infrastructure Plane The infrastructure plane consists of end devices, LoRa gateways, network servers, and applications servers. The LoRa gateways are connected in a star of star network topology. The network servers and application servers are hosted into containers. Each gateway will reserve channels resources for each slice according to the configuration issued by the SDN controller. And the container-based network server will be dynamically orchestrated according to controller commands [14].

5.1.1.2   System Model

Consider a network consisting of a set of LoRa gateways $G = \{1, \ldots, G\}$, and a set of industrial IoT devices $\mathcal{D} = \{1, \ldots, D\}$. We assume that each device will be assigned to a specified network slice $j \in \mathcal{J} = \{1, \ldots, J\}$ of the closest gateway according to its QoS. Each gateway will reserve its channel resource $C = \{1, \ldots, C\}$

for each slice, where $c_{j,g}$ denote the channel associated with the slice j on the gateway g. Each channel has its corresponding bandwidth $b \in \mathcal{B} = \{1, \ldots, B\}$. To evaluate the slice performance, we design three evaluation metrics, named Throughput&Delay Metric, Energy Efficiency Metric, and Reliability Metric. We will detail the design in the following.

Throughput and Delay Metric

The first evaluation indicator is throughput and delay. In LoRaWAN, each device adopts a specific spreading factor (varying between 7 and 12) for message transmission, which means each symbol will be encoded into 2^{SF} signals (chips). Thus, the LoRa modulation data rate can be formulated as:

$$r_d = SF \cdot \frac{b_{j,g}}{2^{SF}} \cdot CR \quad bits/s, \forall d \in \mathcal{D}_{j,g}, \tag{5.1}$$

where r_d denotes the LoRa modulation bit rate of device d, CR denotes the code rate, and $b_{j,g}$ denotes the bandwidth assigned for slice j on the gateway g. Then, the transmission delay can be expressed as:

$$t_d = \frac{L}{r_d} \quad seconds, \forall d \in \mathcal{D}_{j,g}, \tag{5.2}$$

where L is the packet length. Based on this, we define a QoS metric $u_{QoS}^{\mathcal{D}_{j,g}}$ to measure the delay and throughput of slice j, which can be formulated as:

$$u_{QoS}^{\mathcal{D}_{j,g}} = \sum_{d \in \mathcal{D}_{j,g}} (\bar{r}_d + (1 - \bar{t}_d)), \tag{5.3}$$

where $\bar{r}_d$ and $\bar{t}_d$ indicate the normalized value of throughput and transmission delay, respectively.

Energy Efficiency Metric

Another evaluation metric of slice performance is energy efficiency. The energy consumption consists of two states, including active mode consumption $P_{j,g}^{ac}$ and sleep mode $P_{j,g}^{sleep}$. Hence, the energy consumption of slice j during a slicing interval time T can be described as:

$$P_{j,g}^{tt} = P_{j,g}^{ac} T_{active} + P_{j,g}^{sleep} T_{sleep}, \tag{5.4}$$

where T_{active} is related to the r_d and $P_{j,g}^{ac}$ is related to the transmission power. According to pervious works [15], we formulate the unit power consumption as:

$$u_{EE}^{\mathcal{D}_{j,g}} = \sum_{d \in \mathcal{D}_{j,g}} (1 - P_{j,g}^{\overline{tt}}), \tag{5.5}$$

where $P_{j,g}^{\overline{tt}}$ is the normalized energy consumption.

Reliability Metric

The third evaluation indicator is reliability. Configuring low SF and TP values may cause packet loss due to sensitivity, inter-SF, and intra-SF interference. To evaluate the reliability of transmission, we design the packet success rate metric $u_{REL}^{\mathcal{D}_{j,g}}$, which can be formulated as:

$$u_{REL}^{\mathcal{D}_{j,g}} = \sum_{d \in \mathcal{D}_{j,g}} PSR_{d,j}, \tag{5.6}$$

$$with \quad PSR_{d,j} = PR_{d,j}^{tra} + PR_{d,j}^{tre} + PR_{d,j}^{sen},$$

where $PSR_{d,j}$, $PR_{d,j}^{tra}$, $PR_{d,j}^{tre}$, and $PR_{d,j}^{sen}$ are binary variable. The $PR_{d,j}^{tra}$ indicates the packets lost caused by collisions that occur between two end devices configured with the same SF. According to the random access formula, the $PR_{d,j}^{tra}$ can be described as:

$$PR_{d,j}^{tra} = 1 - e^{-2G_{SF}}, \tag{5.7}$$

where G_{SF} is the number of packets generated when one packet is transmitted.

The $PR_{d,j}^{tre}$ indicates the packets lost due to the inter-SF interference. The devices experience a signal-to-interference-plus-noise ratio (SINR), which can be described as:

$$SINR_{i,j} = \frac{P_i^{rx}}{\sigma^2 + \sum_{n \in \partial_j} P_{n,j}^{rx}}, \tag{5.8}$$

where P_i^{ac} is the transmission power with $SF = i$, and σ^2 is the white Gaussian Noise. The below matrix denotes the minimum signal power margin threshold with other SF configuration [16].

$$
\begin{array}{c c c c c c c}
 & SF_7 & SF_8 & SF_9 & SF_{10} & SF_{11} & SF_{12} \\
SF_7 & -6 & 16 & 18 & 19 & 19 & 20 \\
SF_8 & 24 & -6 & 20 & 22 & 22 & 22 \\
SF_9 & 27 & 27 & -6 & 23 & 25 & 25 \\
SF_{10} & 30 & 30 & 30 & -6 & 26 & 28 \\
SF_{11} & 33 & 33 & 33 & 33 & -6 & 29 \\
SF_{12} & 36 & 36 & 36 & 36 & 36 & -6
\end{array}.
$$

Hence, the $PR_{d,j}^{tre}$ can be formulated as:

$$
PR_{d,g}^{tre} = \begin{cases} 0, & \text{if packet survives interference} \\ 1, & \text{Otherwise} \end{cases}.
$$

In addition, the $PR_{d,j}^{sen}$ indicates the packets lost when a packet is transmitted to the gateways below sensitivity, which can be expressed as:

$$
PR_{d,g}^{sen} = \begin{cases} 0, & \text{if packet successfully reaches } j \in \mathcal{J} \\ 1, & \text{Otherwise} \end{cases}.
$$

5.1.1.3   Problem Formulation

At this stage, we can formulate a multi-objective optimization problem. We search for the optimum SF and TP configuration that can simultaneously enhance the slice's QoS, energy efficiency, and reliability. This optimization problem can be formulated as:

$$
\max u^{\mathcal{D}_{j,g}} = u_{QoS}^{\mathcal{D}_{j,g}} + u_{EE}^{\mathcal{D}_{j,g}} + u_{REL}^{\mathcal{D}_{j,g}}, \tag{5.9}
$$

subject to the following constraints:

$$
C1 : d_{j,g} \bigcap d_{j',g} = \varnothing, \forall j, j' \in \mathcal{J}, \forall g \in \mathcal{G} \tag{5.10a}
$$

$$
C2 : d_{j,g} \bigcap d_{j,g'} = \varnothing, \forall j \in \mathcal{J}, \forall g, g' \in \mathcal{G} \tag{5.10b}
$$

$$
C3 : 0 \leq P_{j,g}^{ac} \leq P^{max}, \tag{5.10c}
$$

$$
C4 : \sum_{d \in \mathcal{D}_{j,g}} r_d \leq R_{j,g}^{max}. \tag{5.10d}
$$

The constraint $C1$ ensures that one device can only be allocated to one slice. $C2$ ensures that one device can only be allocated to one gateway. $C3$ ensures that the transmission power is limited to the maximum power. And $C4$ ensures that the total

Table 5.1 List of main notations

Parameter	Definition
SF	Spreading factor
TP	Transmission power
G	Number of gateways
J	Number of slices
C	Number of channel
D	Number of industrial IoT devices
r_d	The LoRa modulation data rate of device d
$b_{j,g}$	The bandwidth assigned for slice j on the gateway g
$P_{j,g}^{tt}$	The energy consumption of slice j during a slicing interval time T
$PR_{d,j}^{tra}$	The packets lost caused by intra-SF collisions
$PR_{d,j}^{tre}$	The packets lost caused by inter-SF collisions
$PR_{d,j}^{sen}$	The packets lost when a packet is transmitted to the gateways below sensitivity
$u_{QoS}^{\mathcal{D}_{j,g}}$	The satisfaction rate of slice j in terms of delay and throughput
$u_{EE}^{\mathcal{D}_{j,g}}$	The satisfaction rate of slice j in terms of energy efficiency
$u_{REL}^{\mathcal{D}_{j,g}}$	The satisfaction rate of slice j in terms of reliability

transmission power is limited to the maximum rate [17]. As shown in Table 5.1, we list the notations of this paper.

5.1.2 Transfer Multi-agent Reinforcement Learning

In this paper, we introduce the DDPG algorithm to search the optimal SF and TP parameters of each slice. Then, to accelerate the learning rate, the transfer learning framework is introduced.

5.1.2.1 Markov Decision Process

The slice optimization problem can be described as an independent Markov Decision Process (MDP) [18]. Formally, an MDP can be formalized as a 4-tuple $< S, A, \pi, R >$, where S is the state space, A is the action space, Π is the policy space, and R is the immediate rewards. At each step, the slice agent takes an action $a \in A$ according to current policy $\pi(a|s)$ and its observation $s \in S$. Then, the underlying environment will generate an immediate reward R, and the state s will transit to a new state $s' \in S$. Specifically, in our scenario, we will define the three components of MDP in the following.

State Definition

In this paper, we define the environment state as the device information and packet information. The device information includes spreading factors, transmission power, bandwidth, and energy consumed (ENY). And the packet information includes signal-to-noise ratio (SNR), received signal strength indicator (RSSI), and the total number of packets during a unit time (Num). These values are encoded in an one-hot format, which can be represented as:

$$S = (\bar{SF}, \bar{TP}, \bar{BW}, \bar{ENY}, \bar{SNR}, \bar{RSSI}, \bar{Num}). \tag{5.11}$$

All the values are average values over an interval period.

Action Definition

In this paper, we define the action as:

$$A = (SF, TP). \tag{5.12}$$

In LoRa, the SF's value range is $\{7, 8, 9, 10, 11, 12\}$, and TP's value range is $\{2, 5, 8, 11, 14\,\text{dBm}\}$. Therefore, there exist 30 actions in the action space.

Reward Function

In this paper, the optimization goal needs to consider throughput, energy efficiency, and reliability simultaneously. Thus, we define the reward function as:

$$R(a, s) = u^{\mathcal{D}_{j,g}} = \alpha u_{QoS}^{\mathcal{D}_{j,g}} + \beta u_{EE}^{\mathcal{D}_{j,g}} + \gamma u_{REL}^{\mathcal{D}_{j,g}}, \tag{5.13}$$

where α, β, and γ are system weight parameters. Different values indicate different preferences of the slice QoS requirement. For example, a larger γ should be adopted in the URA slice for the sake of better transmission reliability.

5.1.2.2 Deep Deterministic Policy Gradient

In this paper, we introduce the DDPG to learn the optimal slice configuration policy. DDPG is a model-free off-policy RL algorithm that combines the deep Q-Network (DQN) algorithm and deterministic policy gradient (DPG) algorithm [19]. As shown in Fig. 5.2, it is composed of two neural network components, termed as critic and actor. The actor function $\mu(s|\theta^{\mu})$ specifies action a given the current state s of

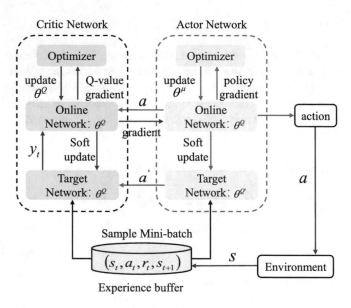

Fig. 5.2 The DDPG algorithm

the environments. Critic value function $Q(s, a|\theta^Q)$ specifies a signal (TD Error) to criticize the action made by the actor.

For the actor network, the objective function can be described as:

$$J(\theta^\mu) = E_{\theta^\mu}[r_1 + \gamma r_2 + \gamma^2 r_3 + \ldots]. \qquad (5.14)$$

The actor network will update the parameters θ^μ toward the direction of increasing $J(\theta)$. The gradient of the objective $J(\theta^\mu)$ can be expressed as:

$$\frac{\partial J(\theta^\mu)}{\partial \theta^\mu} = E_s \left[\frac{\partial Q(s, a|\theta^Q)}{\partial a} \frac{\partial \mu(s|\theta^\mu)}{\partial \theta^\mu} \right]. \qquad (5.15)$$

For the critic network, it calculates the Q-value of the observation-action pair (s, a) to measure the profit of action a under a specific state s. The update of the critic network θ^Q can be described as:

$$\frac{\partial L(\theta^Q)}{\partial \theta^Q} = E_{s,a,r,s'} \left[(TargetQ - Q(s, a|\theta^Q)) \frac{\partial Q(s, a|\theta^Q)}{\partial \theta^Q} \right], \qquad (5.16)$$

where

$$TargetQ = r + \gamma Q'(s', \mu(s'|\theta^{\mu'})|\theta^{Q'}). \qquad (5.17)$$

Besides, to stabilize the training process, DDPG adopts the Experience Replay and Target Networks scheme.

Experience Reply Experience Replay is a replay memory technique, where agent's experiences are stored as a tuple of $[s_t, a_t, r_t, s_{t+1}]$ in a replay buffer D. During the training, the RL agent randomly drew a minibatch of experience from D to train the network. Such storage-sampling act effectively addresses the unstable training problem caused by the autocorrelation among training data [20].

Target Network During the training, since the learning object constantly changes, the value TD estimations can easily spiral out of control. To mitigate that risk, the target network is introduced. The target network's weights are fixed during the learning process and periodically reset to the original network's values.

In DDPG, we define the target critic network and target actor network to calculate the Q-value for the next state in TD-error computations:

$$policy\ network \begin{cases} online: \mu(s|\theta^{\mu}): gradient\ update\ \theta^{\mu} \\ target: \mu(s|\theta^{\mu'}): soft\ update\ \theta^{\mu'} \end{cases} \tag{5.18}$$

$$Q\ network \begin{cases} online: Q(s,a|\theta^{Q}): gradient\ update\ \theta^{Q} \\ target: Q(s,a|\theta^{Q'}): soft\ update\ \theta^{Q'} \end{cases}. \tag{5.19}$$

In DDPG, the weights of targets are updated based on the main networks periodically, which can be formulated as:

$$soft\ update: \begin{cases} \theta^{Q'} \leftarrow \tau\theta^{Q} + (1-\tau)\theta^{Q'} \\ \theta^{\mu'} \leftarrow \tau\theta^{\mu} + (1-\tau)\theta^{\mu'} \end{cases} \tag{5.20}$$

5.1.2.3 Transfer Reinforcement Learning

As discussed above, the DDPG agent can learn from its experience and improve its performance by exploring the environment. However, such learning mechanism has to take a large number of training episodes to converge the optimal value. Especially in our scenario, multiple slice agents are co-existed. Each RL agent has to learn from scratch and therefore reduce the system utility.

To accelerate the learning process across multiple agents, the transfer learning techniques are integrated into our method to facilitate the learning process. Transfer learning is a machine learning method that the learned models trained from a task are reused as the starting point for new models on another task [21]. Different from the isolated learning paradigm, transfer learning exploits the knowledge acquired from previous tasks to improve generalization and learning rate about related ones.

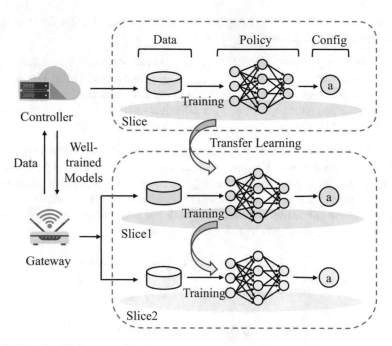

Fig. 5.3 Transfer reinforcement learning

Combining with the transfer learning methods, as shown in Fig. 5.3, we propose a transfer multi-agent Deep Deterministic Policy Gradient (TMDDPG) scheme. At the first stage, the centralized controller gathers the data experience $[s_t, a_t, r_t, s_{t+1}]$ from the distributed gateways to construct a replay buffer and train the model. After convergence, the well-trained network models will be issued to each gateway and used as the starting point on local slice optimization tasks. Moreover, the experience memory pool of each slice agent is imported by well-trained model experience in the SDN controller.

The slice optimization policy of the TMDDPG algorithm is shown in detail in Algorithm 5.1.

5.1.3   Experiments and Simulation Results

In this section, we present the simulation results to evaluate the validity of the proposed algorithm. Our experiments simulate Ubuntu 16.04 with 32g RAM, Nvidia RTX 2060, and intel i7-10875H. We use OMNET++ for building the LoRaWAN simulation environment, and PyTorch 1.4.0 for implementing neural networks.

Algorithm 5.1 The TMDDPG based slicing optimization algorithm

---

Import pre-trained model and replay buffer data from SDN controller

Initialized online network weights θ^Q, θ^μ

Initialized target network weights $\theta^{Q'} \leftarrow \theta^Q, \theta^{\mu'} \leftarrow \theta^\mu$

for step = 1 to maximum episode length do **do**

    Observe initial state s_0

    for step = 1 to maximum episode duration **do**

        Takes an action according to the current policy:

        $a_t = \mu(s_t|\theta^\mu) + \mathcal{N}_t$

        Execute action a_t and receive a new state s_{t+1}

        Storage the transition in R

        Sample a batch of experiences

        $R \times (s, a, r, s')$ from replay buffer

        Calculate TD target of Q-network:

        $y_i = r_i + \gamma Q'(s_{i+1}, \mu'(s_{i+1}|\theta^{\mu'})|\theta^{Q'})$

        Update online Q-network by minimizing the loss:

        $L = \frac{1}{N} \sum_i (y_i - Q(s_i, a_i|\theta^Q))^2$

        Update online policy network with:

        $\nabla_{\theta^\mu} J \approx \frac{1}{N} \sum_i \nabla_a Q(s, a|\theta^Q) \nabla_{\theta^\mu}|_{s=s_i, a=\mu s_i} \mu(s|\theta^\mu)|_{s_i}$

        Update the target networks by:

        $\theta^{Q'} \leftarrow \tau\theta^Q + (1-\tau)\theta^{Q'}$

        $\theta^{\mu'} \leftarrow \tau\theta^\mu + (1-\tau)\theta^{\mu'}$

    end for

end for

5.1.3.1 Simulation Settings

In our experiment, we simulate a LoRa-based network with 4 gateways (i.e., $G = 4$). The number of slices on each gateway is 3, and the number of industrial IoT device assigned to each slice range from 100 to 1000. We assume that gateways and devices are uniformly distributed in a cell of a 10 km radius. In addition, the number of channels per slice varies from 1–3 and each channel Bandwidth is 125 kHz. The SF's value range is $\{7, 8, 9, 10, 11, 12\}$, and TP's value range is $\{2, 5, 8, 11, 14 \text{dBm}\}$. The detailed parameters setting can be found in Table 5.2.

5.1.3.2 Convergence Analysis

First, we evaluate the convergence of our proposed algorithm. We adopt two other reinforcement learning algorithms, DDPG and DQN, as the baseline algorithms. The configuration of TMDDPG, DDPG, and DQN are shown in Table 5.3.

Table 5.2 Simulation parameters

Parameter	Value
Number of gateways	4
Number of slice	3
Number of devices per slice	100–1000
Number of channels per slice	1–3
Channel bandwidth	125 kHz
Gateways and devices distribution	Uniform distribution
Training batch size	25000 steps
Spreading factor	7–12
Transmit power	2–14 dBm

Table 5.3 TMDDPG & DDPG & DQN configuration

Parameter	Value
Episode maximum duration	100 s (100 steps)
Maximum episode	20,000 episodes
Discount factor (γ)	0.95
Learning rate (α)	0.01
Training batch size	25,000 steps
Replay buffer size	100
Minibatch size	10
Hidden layer size	64
Step duration (τ)	1 s

As shown in Fig. 5.4a, the learning process of three algorithms is demonstrated. We notice that three algorithms can all converge to a near reward. This demonstrates that reinforcement learning is capable of improving policy performance through interacting with the environment. Also, we notice that TMDDPG exhibits a better convergence compared to the DDPG and the DQN. The TMDDPG can obtain stable reward around 500 episodes, while DQN around 1300 episodes and DDPG around 1700 episodes. This is because TMDDPG acquires knowledge from previous tasks, without having to restart training from the scratch.

5.1.3.3   Performance Analysis

In this part, we evaluate the algorithm performance in terms of network reliability. Two classical SF-TP adjust algorithms, termed dynamic random (DR) algorithm and dynamic adaptive (DA), are set as the baseline, where the DR configure the SF and TP values randomly, and DA configure the SF and TP values based on link quality with specific SF-TP pair, including (7 dbm, 2 dbm), (8 dbm, 5 dbm), (9 dbm, 8 dbm), (10 dbm, 11 dbm), (11 dbm, 14 dbm) and (12 dbm, 14 dbm).

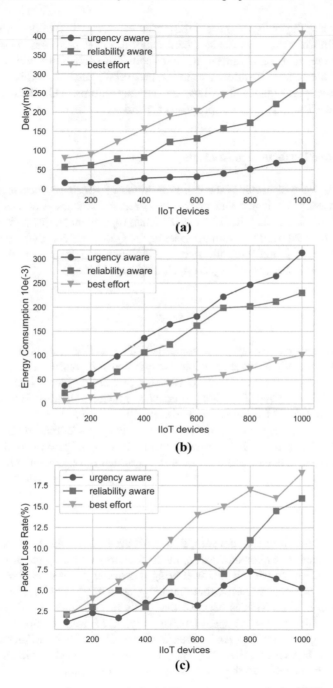

Fig. 5.4 Convergence performance analysis. (**a**) The convergence analysis. (**b**) The performance evaluation

As shown in Fig. 5.4b, we present the packet loss rate (PLR) of the different algorithms. With the number of devices increasing, the PLR increases. This is mainly caused by the increase of inter- and intra-collision probability with the number of devices increasing. Besides, we notice that reinforcement learning algorithms exhibit better performance than DA and DR algorithms. This demonstrates the validity and effectiveness of our proposed algorithm.

5.1.3.4   Slice Performance Analysis

In this part, we will evaluate the different kinds of slice's performance in terms of delay, energy efficiency, and packet loss rate. As discussed above, we define three types of slices, i.e., UR, RA, and BE. Each type agent adopts different weight parameters in the reward function for meeting the QoS preference, where α indicates the QoS weight, β indicates the energy efficiency weight, and γ indicates the reliability weight. We detail the parameter's value in Table 5.4.

Delay Evaluation

Firstly, we evaluate the delay performance of different slices. As shown in Fig. 5.5a, with the number of devices increasing, the delay also increases. This is caused by inter- and intra-interference. With the collision probability increasing, the gateway has to set higher SF and TP values to enhance system reliability. Hence, the data rate will decrease and the transmission delay will increase.

Also, we notice that the UR and RA's delay is always lower than the BE slice. The reason is that UR and RA slices' reward functions are configured with the higher α and γ. The agent will gain more profit with the higher throughput and lower delay.

Energy Evaluation

Then, we present the energy consumption of different slices. As shown in Fig. 5.5b, with the number of devices increasing, the total energy consumption also increases. That is because more participants will bring more energy consumption to the total consumption. Besides, we notice that the energy consumption of UR is always lower than the RA and BE. This is because that UR's reward function adopts a higher value of γ, which forces the agent to take a higher TP value compared to the RA and BE for higher reliability. And increasing SF and TP values will increase energy consumption. In addition, the BE's reward adopts the highest β, which drives the agent to adjust its policy to the lower consumptions direction.

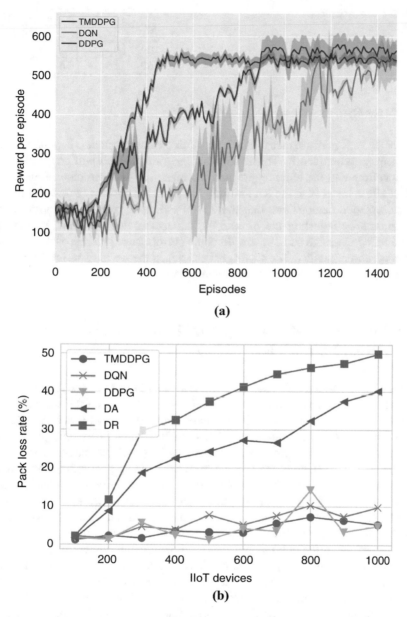

Fig. 5.5 Slice performance analysis. (**a**) The delay in different slices. (**b**) The Energy consumption in different slices. (**c**) Packet loss rate in different slices

Table 5.4 Slice parameters
configuration

Slice	α value	β value	γ value
UR	0.4	0.1	0.5
RA	0.4	0.2	0.4
BE	0.3	0.4	0.3

Packet Loss Rate Evaluation

We plot the PLR performance in Fig. 5.5c. With the devices increasing, the PLR will
also increase subsequently. This is because that more packets will be transferred at
the same time and therefore aggravated the inter- and intra-problem. Besides, the
PRL of UR is lower than the RA and BE, and there are obvious reasons for that.

These results demonstrate that the TMDDPG can dynamically optimize slice
performance by adjusting the SF and TP parameters configuration. Besides, the
reward function design can directly affect the performance in terms of delay, energy
efficiency, and reliability. The slice agent can adjust the reward's weights (i.e., α, β,
γ) for acquiring different performance preferences.

5.2   Reinforcement Learning-Based Continuous-Decision Virtual Network Embedding

Network Virtualization (NV) has attracted more and more attention [22–24], which
allows multiple virtual networks to share limited resources on the same substrate
network. Specifically, a new business model Infrastructure as a Service (IaaS) is
enabled by the Internet Service Providers (ISPs), which hosts multiple network
services on its infrastructure. Therefore, embedding decision-making has become
one of the important tasks for ISP, because an optimal embedding decision can
provide services for more virtual networks and make resource utilization more
effective [25]. For the sake of simplification, a virtual network can be modeled
as multiple virtual nodes, which are connected by virtual links. The purpose of
virtual network embedding is to map virtual networks to shared physical networks
while providing sufficient computation and bandwidth resources for requests.
Virtual Network Embedding (VNE) has played a critical role in network resource
allocation problems, which is considered as the main implementation method of
NV [26, 27]. To elaborate a little further, SPs embed virtual network requests into
the substrate network and offer the corresponding services for the sake of obtaining
revenue. Additionally, the substrate network is comprised with substrate nodes with
computational capability as well as substrate links with certain bandwidth. By
contrast, the virtual network contains nodes having requirement of computational
resources and links with requirement of transmission bandwidth.

Specifically, VNE process can be divided into two phases, i.e., node embedding
and link embedding. Moreover, different optimization objectives and constraints in

different contexts may formulate different VNE problems, which have been proved to be NP-Hard [27]. Liu et al. [28] concluded that solving VNE problems mainly relied on direct solutions and heuristic solutions. However, direct solutions such as Integer Linear Programming (ILP) aim at finding the optimal embedding results yet with high computational complexity, which are not suitable for large-scale networks. Additionally, most of the heuristic algorithms [29–32] are based on static decision-making process, where network parameters cannot be adaptively optimized corresponding to dynamic network states.

With the development of artificial intelligence and machine learning [33], more and more intelligent algorithms have been taken into account for solving the VNE problems. In comparison to the heuristic algorithms, machine learning [34] algorithms can intelligently adjust network parameters relying on available data and make predictions for the future. Moreover, as one famous member of machine learning family, reinforcement learning [35, 36] methods are characterized with self-adaptivity in which the agent interacts with the environment to perform specific actions. Combining with the deep neural network [37], Deep Reinforcement Learning (DRL) algorithms are proposed for extracting and analyzing information of the substrate network relying on the perception of the neural network. Inspired by these, in this paper, the seq2seq model[1] is applied to formulate the node embedding process. Furthermore, a Continuous-Decision VNE scheme relying on Reinforcement Learning (CDRL) algorithm is proposed for solving the VNE problem and for optimizing their performance.

5.2.1   System Model and Evaluation Metrics

5.2.1.1   System Model

As mentioned above, VNE is the process of assigning virtual network requests to the substrate network according to their resource requirements. For the sake of analysis, the substrate network can be formulated by an undirected graph $G^S = (N^S, L^S, A_N^S, A_L^S)$, where N^S and L^S denote substrate nodes and links, while A_N^S and A_L^S represent the computational capability of nodes and the bandwidth of links, respectively. Moreover, let P^S represent a set of acyclic paths in the substrate network. Similarly, we also utilize an undirected graph $G^V = (N^V, L^V, C_N^V, C_L^V)$ to formulate a virtual network, where N^V and L^V are nodes and links in G^V, while C_N^V and C_L^V are computational resource requirements of nodes and bandwidth requirements of links, respectively. Relying on the symbols defined above, the VNE process can be easily expressed by $M : G^V(N^V, L^V) \rightarrow G^S(N', L')$, where

---

[1] The seq2seq is a common model in the field of Natural Language processing (NLP), which is a widely used architecture for machine translation and summarization relying on a recurrent neural network as one of its building blocks [38, 39].

$N' \subset N^S$ and $L' \subset L^S$. In our resource allocation problem, one objective is to find the node embedding relation $X = \{x_{ij} \mid n_i \in N^V, n_j \in N^S\}$, which can be formulated by:

$$\sum_{j=1}^{|N^S|} x_{ij} = 1, \tag{5.21}$$

where $x_{ij} = 1$ denotes that the virtual node n_i is embedded to the substrate node n_j, and we have $n_i \in N^V$, $n_j \in N^S$. Hence, another objective is to find the link embedding relation $Y = \{y_{ij} \mid l_i \in L^V, l_j \in L^S\}$, which can be given by:

$$\sum_{j=1}^{|L^S|} y_{ij} \geq 1, \tag{5.22}$$

where $y_{ij} \geq 1$ means that link l_i is embedded to one or more substrate links, and we have $l_i \in L^V$, $l_j \in L^S$. If a virtual network request G^V is eligible to be successfully embedded to the substrate network, it needs to satisfy the constrain of computational capability, which can be expressed by:

$$x_{ij} \cdot C_{n_i}^V \leq x_{ij} \cdot A_{n_j}^S, \tag{5.23}$$

where $A_{n_j}^S$ denotes the actual computational capability of n_j, while $C_{n_i}^V$ denotes the node resource requirement of n_i, and we have $n_i \in N^V, n_j \in N^S$. Furthermore, the bandwidth constrain also needs to be satisfied, which can be given by:

$$y_{ij} \cdot C_{l_i}^V \leq y_{ij} \cdot A_{l_j}^S, \tag{5.24}$$

where $A_{l_j}^S$ denotes the actual bandwidth of l_j, while $C_{l_i}^V$ denotes the bandwidth requirement of l_i, and we have $l_i \in L^V$, $l_j \in L^S$.

For the sake of simplification, figures in Fig. 5.6 are used to clarify our VNE process. Figure 5.6 is consisting of request 1 and request 2 as well as a substrate network, where request 1 has 2 virtual nodes and 1 virtual links, while request 2 has 4 virtual nodes and 2 links. Moreover, the value on the line represents the bandwidth requirement of link, while the value next to the node denotes the computational resource requirement of node. Furthermore, for the substrate network in Fig. 5.6, the value on the line represents the actual bandwidth of substrate link, while the value next to the node denotes the actual computational capability of substrate node. Whatever the case is, different embedding algorithms all aim at looking for an optimal embedding strategy, in which less resource is consumed in the case of successful embedding and ISPs gain much more revenue.

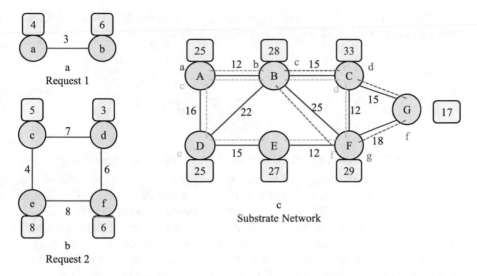

Fig. 5.6 Embedding results from the virtual network requests to the substrate network. Virtual nodes a, b are embedded to the substrate nodes A, B while the virtual link ab is embedded to the substrate link AB, and virtual nodes c, d, e, f are embedded to the substrate nodes A, C, D, F while the virtual links cd, df, ef, ce are embedded to the substrate paths AC, CF, DF, AD. With the aid of criterion of consuming less resource to achieve embedding result, it can be easily found that the bandwidth consumed by AC and DF are not optimal. Furthermore, After virtual nodes c, d, e, f are transferred to the substrate nodes B, C, F, G some bandwidth resources are saved for accepting other virtual network requests

5.2.1.2 Evaluation Metrics

The revenue of ISP, namely $R(G^V, t, t_d)$, is computed in line with the duration time t_d of virtual network request, consumed node computational resources $CPU(n^V)$ as well as the link bandwidth $BW(l^V)$. Hence, the revenue can be formulated by:

$$R(G^V, t, t_d) =$$

$$t_d \left[w_c \sum_{n^V \in N^V} CPU(n^V) + w_b \sum_{l^V \in L^V} BW(l^V) \right], \tag{5.25}$$

where $CPU(n^V)$ is the node computational resources consumed by the virtual node n^V, while $BW(l^V)$ is the bandwidth consumed by the virtual link l^V. Furthermore, w_c and w_b are revenue weights of nodes and links in the substrate network, which are only determined by the ISP. Specifically, some ISPs pay more attention to the benefits from consuming computational resources of nodes, then w_c is correspondingly a little bit higher. By contrast, if ISPs care much more about benefits from consuming bandwidth of links, w_b is correspondingly a little bit higher.

After a virtual network request is successfully embedded to the substrate network, the resource consumption can be calculated by:

$$C(G^V, t, t_d) =$$

$$t_d \left[\sum_{n^V \in N^V} CPU(n^V) + \sum_{l^V \in L^V} \sum_{l^S \in L^S} BW(f_{l^S}^{l^V}) \right], \tag{5.26}$$

where $BW(f_{l^S}^{l^V})$ is the bandwidth consumed by l^V in the substrate link l^S, while $CPU(n^V)$ denotes the computational resources consumed by n^V. Moreover, during the link embedding stage, a virtual link l^V may be assigned to multiple substrate links, so the total bandwidth consumption of G^V needs to be calculated.

For the sake of easy comparison, there are generally three metrics to evaluate the performance of different VNE algorithms. The first metric is the *long − term average revenue*, which calculates the average revenue over an infinite period of time to evaluate the overall impact of a VNE algorithm. We have:

$$Rev = \lim_{T \to \infty} \frac{\sum_{t=0}^{T} R(G^V, t, t_d)}{T}, \tag{5.27}$$

where T denotes the elapsed time. As mentioned above, VNE aims at getting as much revenue as possible with limited resource consumption, which gives us a second evaluation metric, namely *long − term average revenue to cost ratio*. It can be calculated by:

$$Rev2Cos = \lim_{T \to \infty} \frac{\sum_{t=0}^{T} R(G^V, t, t_d)}{\sum_{t=0}^{T} C(G^V, t, t_d)}, \tag{5.28}$$

where a higher *long − term average revenue to cost ratio* indicates a higher resource utilization efficiency to meet the resource requirements of more virtual network requests. The last evaluation metric is the *long − term acceptance ratio*, which calculates the percentage of all virtual network requests that are successfully embedded. We have:

$$Acp = \lim_{T \to \infty} \frac{\sum_{t=0}^{T} Acp(G^V, t, t_d)}{\sum_{t=0}^{T} All(G^V, t, t_d)}, \tag{5.29}$$

where $Acp(G^V, t, t_d)$ denotes the number of accepted virtual network requests over time horizon T. Furthermore, we utilize these three metrics to evaluate performance of our proposed CDRL algorithm with other algorithms in following sections.

5.2.2   Embedding Algorithm

In this section, our CDRL algorithm and seq2seq model are described in detail. The reinforcement learning agent is applied to extract the information of nodes in the substrate network and form a feature matrix as the input of the seq2seq model. Then the model outputs the node embedding results of current virtual network request. The policy gradient algorithm is utilized to update the network parameters, and finally a model with a good embedding mechanism is produced.

5.2.2.1   Seq2seq Model

The seq2seq model is commonly utilized to solve some sequence-to-sequence problems such as machine translation, QA system, etc. It is divided into two parts, as shown in Fig. 5.7, the encoder and decoder part, respectively. Each cell in two parts has a hidden vector h_t, which is related to the input vector x_t and the hidden vector of the previous cell h_{t-1}. The h_t can be calculated by:

$$h_t = f(x_t, h_{t-1}), \tag{5.30}$$

where f is a nonlinear activation function tanh or sigmoid. For the sake of simplification, we assume that a pair of source-target vectors is represented by $(A, B, C) \rightarrow (W, X, Y, Z)$. Specifically, in encoder part, the source vectors (A, B, C) are integrated into a encode vector E, which is actually the last hidden vector h_t in encoder part and represents the semantics of all input vectors. Then vector E and the target vectors (SOS, W, X, Y, Z) are used as the input of the decoder part, where SOS is the start signal of the decoder part. Furthermore, the training process of seq2seq model aims at making the output of the decoder part get close to the target vectors (W, X, Y, Z, EOS), where EOS is the end signal of the decoder part. In comparison to the VNE, we found the VNE problem can also be considered as a sequence-to-sequence problem in which information of the substrate network is encoded in the encoder part and be sent to the decoder part

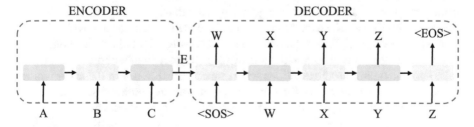

Fig. 5.7 The seq2seq model

for decoding. As a result, we can get continuous node embedding results of current virtual network request.

In order to explain this process more clearly, here we define an input sequence $x_1, x_2, \ldots, x_t$ and an output sequence $y_1, y_2, \ldots, y_t$. Hence, for the seq2seq model, the purpose is to maximize the probability:

$$\prod_{t=1}^{T} P(y_t \mid v, y_1, \ldots, y_{t-1}), \qquad (5.31)$$

where v denotes the input vector $x_1, x_2, \ldots, x_t$. Different from the encoder part, the hidden vector h_t in decoder part is calculated by:

$$h_t = g(v, y_{t-1}, h_{t-1}), \qquad (5.32)$$

which is related to the output y_{t-1} of the last cell, where g denotes a nonlinear activation function. In order to prevent the numerical underflow problem of Eq. (5.31), the objective function of the seq2seq model can be rewritten as a log likelihood conditional probability function, which can be formulated by:

$$\max_{\theta} \frac{1}{T} \sum_{t=1}^{T} \log P_{\theta}(y_t \mid x_t), \qquad (5.33)$$

where θ denotes the parameters of the seq2seq model.

5.2.2.2   Information Extraction

In order to vectorize the global state information of the substrate network, we need to extract the state information of each substrate node and link. Nodes in the substrate network have a range of characteristics such as remaining computational capability, bandwidth of adjacent links as well as the connection degree. Generally speaking, the more that features are extracted by the reinforcement learning agent, the more that the feature matrix can represent the entire substrate network. However, if the dimension of the feature matrix is large enough, it will lead to a high complexity, and hence will be likely to be over-fitting. As for the benchmark algorithm RLVNE [40], Yao et al. extracted four types of state information, including the computing capacity, degree, sum of bandwidth and average distance to other host nodes, which take into consideration the positions where other virtual nodes in the same request are embedded. However, as for our seq2seq model, it continuously outputs the embedding results of all virtual nodes in current virtual network each time, so we cannot take the average distance to other host nodes into account. Hence, three pieces of general information of the substrate network can be extracted in order to formulate the feature matrix, i.e.:

1. Computational capability (CPU): The remaining computational capability has a great impact on embedding results during the node embedding process, so this dimension needs to be taken into account in our feature matrix. The CPU can be given by:

$$CPU(n^s) = CPU(n^s)' - \sum_{n^v \to n^s} CPU(n^v), \qquad (5.34)$$

where $CPU(n^s)$ denotes the initialization CPU value of n^s, while $\sum_{n^v \to n^s} CPU(n^v)$ represents the sum of CPU value of all virtual nodes embedded to n^s.

2. Degree (DEG): The degree is the number of links directly connected to the substrate node, which reflects the connectivity of the substrate network. Furthermore, the higher the degree, the easier it is to find the path between the other nodes. The DEG can be calculated as:

$$DEG(n^s) = \sum_{n \in N^s} L(n^s, n), \qquad (5.35)$$

where $L(n^s, n)$ is equal to 1 when n^s and n are connected, 0 when not connected.

3. Sum of Bandwidth (SUM): Each substrate node may be connected to more than one link, we calculate the sum of bandwidth of all links connected to it, which can be formulated as:

$$SUM = \sum_{l^s \in L^{n^s}} BW(l^s), \qquad (5.36)$$

where L^{n^s} denotes the substrate links connected to node n^s, while l^s denotes one of the L^{n^s}. Moreover, a higher SUM means more likely to complete the link embedding in this substrate node.

After these three features are extracted, their normalized values are connected into a feature matrix A, which can be given by:

$$A = [CPU, DEG, SUM]. \qquad (5.37)$$

The matrix A is used as the input of the seq2seq model and is updated as the state of the substrate network changes.

5.2.2.3 Markov Decision Process

For our VNE problem, we have no way to obtain sufficient embedding data with labels, so we cannot solve this problem by using supervised learning algorithms. Hence, we introduce the reinforcement learning algorithm [42] to solve this problem. As is well known, the reinforcement learning algorithm is a method which is learned in practice and judges the performance of the current operation based on the value of a reward signal. For the sake of simplification, the process of reinforcement learning is usually assumed that there is Markov property between state probability transitions, and hence we use a Markov Decision Process (MDP) to model the reinforcement learning. In our model, the node embedding is considered as a continuous decision process, where the decision-making agent aims at collecting information of the substrate network to make node embedding decisions and the revenue can be obtained after the embedding. Therefore, the node embedding process of all virtual network requests can be simulated as (S,A,P,R,γ), where S and A represent two finite sets of states and actions, while P represents the state transition probability. It can be given by:

$$P_{ss'}^a = P\,[S_{t+1} = s' \mid S_t = s, A_t = a], \tag{5.38}$$

where S_t and A_t are state and action at time t. Moreover, R is the reward function, which can be expressed by:

$$R_s^a = E\,[R_{t+1} \mid S_t = s, A_t = a], \tag{5.39}$$

where R_s^a denotes the reward after action a is performed at state s, while $\gamma \in [0, 1]$ is a discount factor when the total discounted reward G_t is calculated from time t. The G_t can be formulated by:

$$G_t = R_{t+1} + \gamma R_{t+2} + \ldots = \sum_{k=0}^{\infty} \gamma^k R_{t+k+1}, \tag{5.40}$$

where G_t reflects the impact on all subsequent states after action a is performed, which is reduced over time. Furthermore, action policy π is a distribution of actions at state s, which can be expressed by:

$$\pi(a|s) = P[A_t = a \mid S_t = s]. \tag{5.41}$$

Additionally, the task of the decision-making agent is to find the best policy π which can be achieved by maximizing the reward. For MDP, there are two value functions including the state-value as well as the action-value. Moreover, the state-value function $V_\pi(s)$ is only related to the current state s, which can be formulated by:

$$V_\pi(s) = E_\pi[G_t \mid S_t = s]. \tag{5.42}$$

Similarly, the action-value function $q_\pi(s, a)$ is related to current action a as well as the state s, which can be expressed by:

$$q_\pi(s, a) = E_\pi[G_t \mid S_t = s, A_t = a]. \tag{5.43}$$

According to the expression of the value function, we can derive the recursive relationship of the value function. For example, for the state-value function $v_\pi(s)$, we have:

$$v_\pi(s) = E_\pi[R_{t+1} + \gamma R_{t+2} + \gamma^2 R_{t+3} + \ldots \mid S_t = s] \tag{5.44}$$

$$= E_\pi[R_{t+1} + \gamma(R_{t+2} + \gamma R_{t+3} + \ldots \mid S_t = s] \tag{5.45}$$

$$= E_\pi[R_{t+1} + \gamma G_{t+1} \mid S_t = s] \tag{5.46}$$

$$- E_\pi[R_{t+1} + \gamma v_\pi(S_{t+1}) \mid S_t - s]. \tag{5.47}$$

Obviously, there is a recursive relationship between the state S_t and S_{t+1}, which can be given by:

$$v_\pi(s) = E_\pi[R_{t+1} + \gamma v_\pi(S_{t+1}) \mid S_t = s]. \tag{5.48}$$

This recursive equation is generally called the Bellman equation, which indicates that the state-value of a state is composed of the reward of the state and the subsequent state-value combined in a certain ratio. In the same way, we can get the Bellman equation of the action-value function $q_\pi(s, a)$:

$$q_\pi(s, a) = E_\pi[R_{t+1} + \gamma q_\pi(S_{t+1}, A_{t+1}) \mid S_t = s, A_t = a]. \tag{5.49}$$

According to the above definition, we can easily conclude the translation relationship between $V_\pi(s)$ and $q_\pi(s, a)$, which can be given by:

$$V_\pi(s) = \sum_{a \in A} \pi(a|s) q_\pi(s|a), \tag{5.50}$$

$$q_\pi(s, a) = R_s^a + \gamma \sum_{s' \in s} P_{ss'}^a V_\pi(s'). \tag{5.51}$$

Suppose now the decision-making agent receives a virtual request with p numbers of nodes, it has to make p numbers of embedding decisions at time t to embed these nodes to the substrate network. Obviously, S_t can be expressed by the feature matrix which is inputted to the model at time t. Moreover, the action A_t at time t is:

$$A_t = \{(n_t^v, n_t^s) \mid n_t^s \in N_t^S \cap N^S(n_t^s)\}, \tag{5.52}$$

where $N^S(n_t^s)$ denotes the substrate nodes which meet the CPU requirement of n_t^s. After action a_t is performed, the decision-making agent receives a reward signal R_t.

However, if we use the value-based reinforcement learning algorithms such as Q-learning algorithm [43] to make embedding decisions, we need to calculate the reward obtained by executing different actions under each state and choose the action with the largest reward. Obviously, the value-based algorithm does not fit our problem well. Because our state space is made up of continuous values and we cannot get the transition probability distribution between different states. Hence, we introduce a policy-based algorithm to address with the MDP, which directly optimizes the policy of actions. Furthermore, details of the policy gradient algorithm are introduced in the next section.

5.2.2.4   Policy Gradient

Policy gradient [44] algorithm is a reinforcement learning algorithm which optimizes the policy of actions directly. At the beginning, the policy π can be described as a function containing the parameter θ:

$$\pi_\theta(s, a) = P(a|s, \theta) \approx \pi(a|s). \tag{5.53}$$

After the policy function is represented as a continuous function, we can use continuous function optimization methods such as gradient descent algorithm to optimize the strategy. The optimization function can be expressed by:

$$J(\theta) = \sum_s d^{\pi_\theta}(s) \sum_a \pi_\theta(s, a) R_s^a, \tag{5.54}$$

where θ denotes the parameters of policy gradient algorithm, while $d^{\pi_\theta}(s)$ is the probability distribution of states. Moreover, when gradient of the optimization function $J(\theta)$ is calculated, there is a little trick named likelihood ratio, which is:

$$\begin{aligned}
\nabla_\theta \pi_\theta(s, a) &= \pi_\theta(s, a) \frac{\nabla_\theta \pi_\theta(s, a)}{\pi_\theta(s, a)} \\
&= \pi_\theta(s, a) \nabla_\theta \log \pi_\theta(s, a).
\end{aligned} \tag{5.55}$$

Relying on the likelihood ratio, gradient of the objective function $J(\theta)$ can be formulated by:

$$\begin{aligned}
\nabla_\theta J(\theta) &= \sum_{s \in S} d(s) \sum_{a \in A} \pi_\theta(s, a) \nabla_\theta \log \pi_\theta(s, a) R_s^a \\
&= E_{\pi_\theta}[\nabla_\theta \log \pi_\theta(s, a) r],
\end{aligned} \tag{5.56}$$

where r is the total reward over the entire process. Furthermore, according to the policy gradient theorem [45], under the multi-step MDP, we have:

$$\nabla_\theta J(\theta) = E_{\pi_\theta}[\nabla_\theta \log \pi_\theta(s, a) Q^{\pi_\theta}(s, a)], \tag{5.57}$$

where $Q^{\pi_\theta}(s, a)$ denotes the sum of multi-step reward. During the actual optimization process, unbiased sampling is performed on $Q^{\pi_\theta}(s_t, a_t)$. Then we can get v_t, and the parameters can be updated by the form of:

$$\theta = \theta + \alpha \nabla_\theta \log \pi_\theta(s_t, a_t) v_t. \tag{5.58}$$

5.2.2.5   Training and Testing

The training process of our model is shown in Fig. 5.8, and network parameters of the seq2seq model need to be initialized at the beginning. In the training process, we input the feature matrix of the substrate network extracted by the reinforcement learning agent and get the encode vector E. Furthermore, the encode vector E and the start signal SOS are inputted into the decoder part. Let the output N_1 represent the location where the first virtual node is embedded to the substrate network. Then, it is calculated by a softmax function, which can be expressed by:

$$p_i = \frac{e^{h_i}}{\sum_j e^{h_j}}, \tag{5.59}$$

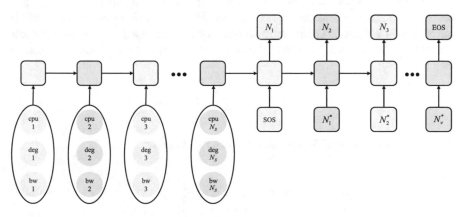

Fig. 5.8 The seq2seq training model. During the training process, the start decoding signal SOS and hand-crafted labels N_1^*, N_2^*,..., N_v^* are inputted to the decoder part. The output is the decoding result N_1, N_2,.., N_v and the end decoding signal EOS. Each output vector N corresponds to the location of the substrate network node to which the current virtual node is embedded

where h denotes the output vector of each cell in decoder part. Actually, parameters in our model are randomly initialized, and the node with the highest probability may not be the best result. Therefore, according to the probability of the output in the softmax layer, we select one node from the set of nodes with sufficient available resources as the host node. Moreover, after all the virtual nodes of the current virtual network are embedded, the Breadth-First-Search (BFS) algorithm is applied to find the path with the optimal bandwidth consumption between each pair of virtual nodes [46].

Furthermore, as for the supervised learning, each piece of data corresponds to a specific label. And the label is compared with the output of training process, where the cross-entropy or Mean Squared Error (MSE) formulation is used to calculate the error between them. Then some algorithms are used to minimize this error and update the parameters of the model. However, the reinforcement learning algorithm is a completely different algorithm from the supervised learning. Specifically, the reinforcement learning agent determines the effectiveness of each action based on the reward signal. If current action brings a large reward signal, it means the action is valid and will be encouraged to happen in next epoch. By contrast, if the current action brings a small or even negative reward signal, this action will be tried to avoid to happen in next epoch. Therefore, the reward signal is significant for agent to train the model and make embedding decisions. So in our experimental settings, we take the long-term average revenue to cost ratio as the reward signal, which is because this metric reflects the utilization efficiency of the substrate resources, and it also has a significant impact on the other two metrics.

To implement our algorithm, we need to assign a label to each output of the decoder part, and then update the parameters based on the error between the label and the specific output. Therefore, we use the hand-crafted label which is determined by an $|n_s|$ dimensions vector y_i to denote the label of virtual node, where is the number of substrate nodes. Specifically, if the label of a virtual node is set as the i-th substrate node, the i-th dimension of y_i is set to 1, while the other dimensions are set to 0. Moreover, the error between y_i and output vector p_i can be calculated by:

$$L(y, p) = -\sum_i y_i log(p_i),$$

(5.60)

and the cross-entropy of the decoder part can be given by:

$$loss = \sum_{i=1}^{|N^V|} L_i(y, p).$$

(5.61)

Then we use the gradient descent algorithm to calculate the gradient g_f of loss, then multiply the reward signal r and the learning rate α to get the parameter update formula of our model, which can be given by:

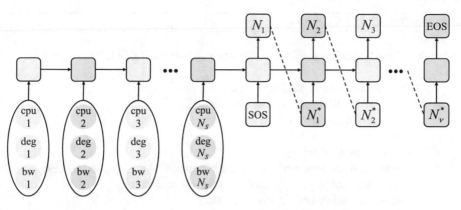

Fig. 5.9 The seq2seq testing model. During the testing process, we can find that when the start decoding signal SOS is inputted to the decoder part, the decoding process begins. Obviously, the output of the previous cell is used as the input to the next cell, and the decoding process ends when the output is EOS

$$g = \alpha \cdot r \cdot g_f. \tag{5.62}$$

The learning rate α is a significant parameter in the process of updating the model parameters, which determines the speed of network updates. If α is set incorrectly, it may cause over-fitting or under-fitting of the network, affecting the final performance of the model. Hence, we need to find a moderate α to keep a stable update process and converge speed. The batch gradient descent algorithm is applied to update the network, which is beneficial in terms of improving the converge speed as well as making the network stable. Specifically, during the training process, we created a gradient stack to temporarily store these calculated gradients instead of using them immediately. When the number of storage reaches the batch amount, the model parameters are updated. However, if the link embedding process fails because of insufficient bandwidth, g_f will be deleted as we are unable to identify the reward signal. Our training algorithm is represented in Algorithm 5.2.

In testing process, the performance of our network is tested with the online virtual network requests which are slightly different from training process. The testing model is shown in Fig. 5.9, which shows that the output of each cell in decoder part is applied as the input of the next cell. The embedding result adopts a beam search strategy [47], where the nodes with the highest product of probabilities are chosen as host nodes. Algorithm 5.3 is our testing algorithm.

5.2.2.6 Computational Complexity

For our CDRL algorithm, the computational complexity is $O(v(mr + p + q))$, where v and m are the number of virtual network requests and the substrate nodes,

---

Algorithm 5.2 Training process

Input:

   Number of epochs $epochNum$; Training data $trainingSet$; Batch size;

Output:

   Trained parameters in seq2seq model;

 1: Initialize all the parameters in seq2seq model;
 2: **while** $epoch < epochNum$ **do**
 3:     count=0
 4:     **for** $request \in trainingSet$ **do**
 5:         input feature matrix of the substrate network to the seq2seq model;
 6:         get output of seq2seq model and select host nodes;
 7:         **if** embedded ($\forall$ node $\in request$) **then**
 8:             compute gradient g_f of $request$ and add to stack;
 9:             link embedding process;
10:         **end if**
11:         **if** embedded ($\forall$ node $\in request$, $\forall$ link $\in request$) **then**
12:             compute the revenue to cost ratio as the reward signal;
13:             multiply reward signal and g_f to compute the final gradient;
14:         **else**
15:             delete g_f;
16:         **end if**
17:         ++count
18:         **if** count equals to the batch size **then**
19:             update network parameters;
20:             count=0;
21:         **end if**
22:     **end for**
23:     $++epoch$;
24: **end while**

---

p and q are the maximum number of nodes and links in virtual network requests, respectively. Moreover, r is the dimension of the feature matrix.

Proof: for every virtual network request, the computational complexity of computing the feature matrix is $O(mr)$. After the feature matrix of the substrate network is inputted to our seq2seq model, the computational complexity is $O(p)$. Furthermore, when virtual nodes are successfully embedded, the computational complexity of the link embedding process is $O(q)$. Hence, for all v numbers of virtual network requests, the computational complexity is $O(v(mr + p + q))$.

Algorithm 5.3 Testing process

Input: : Testing data $testingSet$; Trained parameters in seq2seq model;
Output: : Long-term average revenue; Long-term average acceptance ratio; Long-term average revenue to cost ratio;
 1: **for** $request \in testingSet$ **do**
 2:     input feature matrix of the substrate network to the trained seq2seq model;
 3:     get output of the seq2seq model;
 4:     select host nodes using beam search strategy;
 5:     **if** embedded ($\forall$ node $\in request$) **then**
 6:        link embedding process;
 7:     **else**
 8:        return FALSE;
 9:     **end if**
10:     **if** embedded ($\forall$ node $\in request$, $\forall$ link $\in request$) **then**
11:        return embedding result;
12:     **else**
13:        return FALSE;
14:     **end if**
15: **end for**

5.2.3   Experiments and Simulation Results

5.2.3.1   Experiment Setup

We used the GT-ITM tool [48] to generate a substrate network of 100 nodes and 550 links, which belongs to a medium ISP scale. The computational capability of nodes satisfies the uniform distribution between (50, 100), while the bandwidth of links satisfies the uniform distribution between (20, 50).

After the substrate network is generated, 2000 numbers of virtual networks will be produced. Furthermore, in these 2000 numbers of virtual networks, the training set consists of the first 1000 virtual networks, while the remaining 1000 virtual networks form the testing set. Specifically, every virtual network request has 2-10 nodes and the computational capability satisfies a uniform distribution between (0, 50), while the bandwidth of each link satisfies a uniform distribution between (0,50). The connection probability between virtual nodes is 0.5, which means the average number of links is $\frac{(n(n-1))}{4}$, where n represents the number of virtual nodes. The arrival speed of virtual network requests obeys a Poisson distribution, where 4 requests will be reached within 100 time units, while the duration time of virtual network requests follows an exponential distribution with an average duration of 1000 time units. Hence, we have built a timeline with a length of 50,000 time units.

We apply the TensorFlow [49] frame to build the seq2seq model. Specifically, the encoder part has 100 LSTM cells which represent 100 substrate nodes, while the number of cells in the decoder part is determined by the number of virtual nodes in the current virtual network request. In general, the number of nodes in the encoder part is fixed, while the number of nodes in the decoder part varies with different virtual requests. Network parameters are initialized according to the normal distribution.

5.2.3.2 Training Performance

In comparison to the supervised learning, the converge speed of the reinforcement learning is slower, which is because it has no accurate label and network parameters need to be updated based on the hand-crafted labels and reward signal. During the training process, we utilized the training set with 1000 virtual requests to train the seq2seq model with random parameters initialization for 100 epochs. From the loss changing process in Fig. 5.10, we can conclude that the model converges well. Furthermore, Fig. 5.11 shows these three metrics changing processes within 100 epochs. Furthermore, we can find at the beginning of the training process, every metric is relatively low because the parameters are randomly initialized. As the training process continues, the reinforcement learning agent applies the policy gradient algorithm to update the network according to the reward signal, where the embedding policies are adjusted according to the magnitude of the reward signal. With the constant update of network parameters, the model performance is getting better. Furthermore, after 100 epochs, because of the limitation of the resources of the substrate network, these three metrics have reached a certain point.

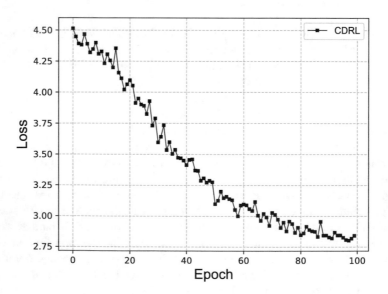

Fig. 5.10 Loss on the training set

Fig. 5.11 The algorithm performance on the training set. (**a**) The changing process of the long-term average revenue. (**b**) The changing process of the long-term average Rev2Cost ratio. (**c**) The changing process of the long-term average acceptance ratio

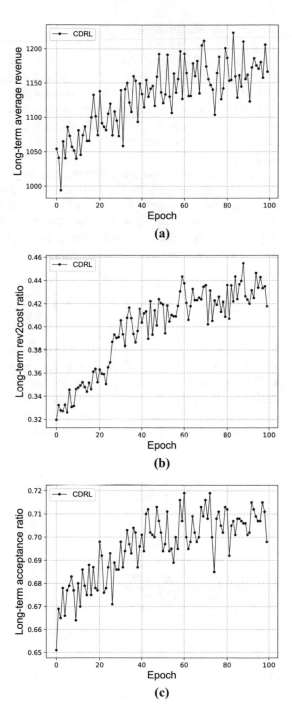

(a)

(b)

(c)

### 5.2.3.3	Simulation Result

From figures of the training performance, we can conclude that our CDRL algorithm can be applied to the training set well. However, our optimization goal is not only perform well on the training set but also on the online virtual network requests. This requires us to test our algorithm on the testing set which is different from the training set. Moreover, if our CDRL algorithm performs well in both training and testing sets, it proves that our proposed algorithm is robust and our research topic is meaningful.

In testing process, the node with the highest probability is selected. In order to test our algorithm performance convincingly, the contrast experiments with other three algorithms are conducted, which includes two heuristic algorithms as well as a reinforcement learning aided algorithm. The first is the baseline algorithm proposed in [29], which introduced the path splitting and migration to the link embedding part. The second algorithm is the NodeRank algorithm proposed in [31], which calculated $H(n^s)$:

$$H(n^s) = CPU(n^s) \sum_{l^s \in L(n^s)} BW(l^s), \tag{5.63}$$

to rank substrate nodes, where the $H(n^s)$ represents the resource availability of node n^s. The last algorithm is RLVNE algorithm proposed in [40], which introduced a policy network based on the reinforcement learning to make node embedding decisions. The policy network architecture is shown in Fig. 5.12.

In order to avoid interference caused by random initialization of neural network parameters, the CDRL and RLVNE algorithms need to run for 30 different

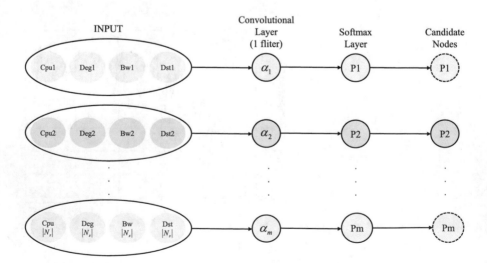

Fig. 5.12 The RLVNE model

initializations and to apply the testing set for the sake of evaluating the algorithm performance. The evaluation metrics of four different algorithms are shown in Figs. 5.8, 5.9, and 5.10. Furthermore, we added the error bar for each time slot. As we can see from Fig. 5.13a–c, the long-term average revenue and acceptance ratio are relatively high in previous epochs, because the substrate network has more available resources at the beginning. As the testing process continues, available substrate resources are gradually consumed, which results in the decrease of the long-term average revenue and acceptance ratio. By contrast, the long-term revenue to cost ratio has no particularly substantial fluctuation from the beginning to the end, because it has no relationship with the available resources. Moreover, as we can see from the average metric figures and error bars, we can conclude that our algorithm has significantly improved these three evaluation metrics.

For the sake of determining whether our algorithm is better than RLVNE, a statistical hypothesis test is performed based on 30 testing results. The Wilcoxon [50] testing method is chosen and the average value over all time slots is regarded as the testing data. Furthermore, the original hypothesis H_0 is set as: RLVNE and CDRL algorithms have no significant difference on performance, while the alternative hypothesis H_1 is set as: CDRL algorithm performs better than RLVNE. According to the calculation results, the one-sided significance value $P < 0.05$, so H_0 is rejected and H_1 is accepted. Therefore, we can conclude that our algorithm is better than RLVNE.

5.3 Multi-agent Reinforcement Learning Aided Service Function Chain

The past decade has witnessed an exponential growth of IoT applications ranging from daily consumption to industrial production [18]. Billions of IoT devices (e.g., smart cameras, smart speakers) around the world are connected to the Internet. According to Gartner's prediction, the number of IoT devices will expand to more than 25 billion by 2025. Yet, at the same, such growth of IoT devices also poses great challenges to network service providers [51]. To meet different applications' requirements, diverse network proprietary hardware (i.e., firewall, code conversion, and network address translation) have to be implemented in the network. This rigid paradigm greatly reduces the scalability and flexibility of network system.

Recently, another breakthrough technology, network functions virtualization (NFV), offers significant opportunities to address these challenges [52]. NFV is a way to replace the dedicated appliance hardware with virtual machines [53]. These functions can be instantiated in virtual machines (VMs) running on standardized compute nodes (i.e., X86 servers), allowing them to be updated, patched, or replaced with ease [54]. Such a way can effectively reduce cost and accelerate service deployment with no more wholesale replacement of network hardware [55].

Fig. 5.13 Comparison of
four algorithms. (**a**) Average
revenue. (**b**) Average
Rev2Cost ratio. (**c**) Average
acceptance ratio

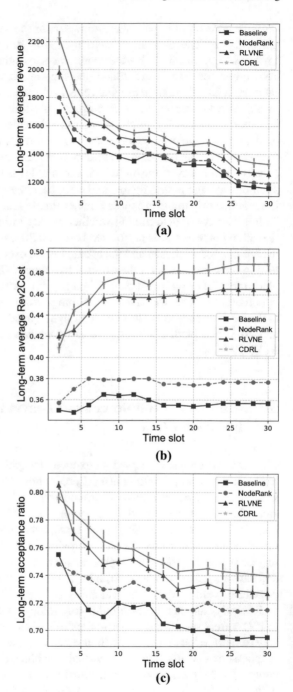

In a NFV-enabled IoT network, service providers can deploy new services in a fast, reliable, and cost-effective way to draw support from the service function chain (SFC) [56]. SFC is a mechanism that allows a sequence of heterogeneous VNFs to be connected to form a service enabled carrier for satisfying the different Quality of Service (QoS) [57]. It can flexibly meet the special demands of IoT business [58]. However, how to embed multiple SFCs requirements into a shared NFV-enabled network becomes a pivotal challenge [59].

Recently, an increasing amount of literature has investigated SFC deployment problem [60]. These methods can be classified into centralized solutions and distributed solutions. For centralized solutions, a controller is used to collect users' requirements and network state, calculate the optimal deployment policy, and then configure SFC in the network. For example, in [61], Ren et al. adopted a multicriteria-based arrangement scheme to orchestrate SFC in a software-defined network (SDN)-IoT network systems and designed a genetic algorithm (GA)-based method to optimize the service performance and resource consumption. However, the centralized solutions require the global information to calculate the deployment policy, which may incur privacy and scalability issues [62]. Recently, distributed solutions also have attracted lots of attention from academia [63]. For example, in [54], D'Oro et al. leveraged the non-cooperative game theory to implement a distributed scheme that can deal with scalability and privacy issues. However, distributed algorithms may suffer serious non-convergence problems. How the distributed network nodes can learn globally optimal deployment policy becomes a critical problem to distributed algorithms design.

Inspired by the success of reinforcement learning (RL), we try to use distributed RL to optimize the SFC deployment problem [64]. RL allows distributed agents to explore the environment and learn from their experiences without human heuristics [65, 66]. In this paper, we propose a distributed RL-based SFC deployment algorithm. To enhance the convergence efficiency, we design a hybrid control architecture and adopt a centralized learning and distributed execution framework. The centralized platform can simplify the learning process with global network information, whilst agents can make decisions based on their local observations in a distributed manner. Moreover, we propose an actor-critic-based multi-agent deep deterministic policy gradient (MADDPG) reinforcement learning algorithm to optimize the SFC deployment problem.

5.3.1 System Model and Problem Formulation

In this section, we first discuss the system model and problem formulation. Then, we model the SFC deployment problem as a multi-user competition game.

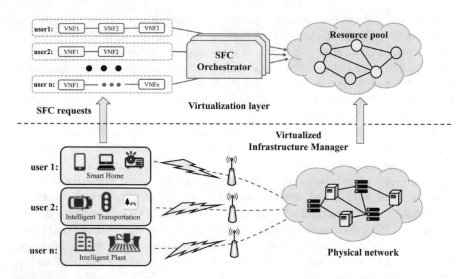

Fig. 5.14 The general architecture of IoT network

5.3.1.1 System Model

As shown in Fig. 5.14, we present a NFV-enabled IoT network. We model the network as a connected undirected graph $G = \{\mathcal{V}, \mathcal{E}\}$ along with $n = |\mathcal{V}|$ server nodes and $m = |\mathcal{E}|$ links, where server node $v \in \mathcal{V}$ has a resource capacity (i.e., CPU, memory), and link $e \in \mathcal{E}$ has a transmission capacity (i.e., bandwidth). We set the resource capacity of server node v as $C(v) = (C_v^{CPU}, C_v^{mem})$. For each link $e \in \mathcal{E}$, $C(e)$ denotes the transmission resources between server node v_i and v_j, where $i, j \in n$.

As discussed above, SFC is a set of VNFs that are arranged in a given order [67]. In this paper, we use $\mathcal{F} = \{F_1, F_2, \ldots, F_f\}$ to define the SFC, where $f = |\mathcal{F}|$, VNFs are deployed in the ordered sequence. Each SFC consists of a subset VNFs: $V_{\mathcal{F}} = \{v_{F_1}, v_{F_2}, \ldots, v_{F_f}\}$, where v_{F_f} denotes that the server node deploys the f-th VNF. Note that when the j-th VNF is executed on the server node, the server node's resource capacity $c(F_j)$ (i.e., computing resource, memory resource) will be consumed.

5.3.1.2 Multi-user Competition Game Model

As shown in Fig. 5.14, when a new request is arriving, the SFC orchestrator will dynamically deploy the SFC according to the user's demands. However, due to the privacy issue, the global orchestrator is not always available. Multiple users have to compete with the network resource to satisfy their demands. In this part, we model the SFC deployment problem among multiple users as a multi-user competition game. A network service (NS) composes of multiple VNFs. NS provider determines how to deploy these VNF instances in the network. We consider a group of users $\mathcal{N}$ that requires $N = |\mathcal{N}|$ different network services. The business traffic is generated by different network user $i \in \mathcal{N}$. We assume that the traffic is injected into the network through server node v_i and leaves from server node v_j. Each network user i requests a specific set of VNFs, which can be described as an ordered VNF chain $\mathcal{F} = \{F_1, F_2, \ldots, F_f\}$. We use $\mathbf{w}_i = (w_i^1, w_i^2, \ldots, w_i^f)$ to denote the network service configuration of user i, where $w_i^f \simeq v_j$ represents that user i's f-th VNF will be deployed in the server node v_j.

Then, we formulate the SFC deployment problem from the game-theoretic perspective. We model it as a non-cooperative game among a set of users $\mathcal{N}$. When the user i generates traffic and requests a network service, the SFC deployment decision $\mathbf{w}_i$ will be chosen from a finite action space $\mathcal{W}_i$ which includes all possible SFC deployment schemes of user i. The game model Ω can be defined by the following triple

$$\Omega = (\mathcal{N}, \mathcal{G}, \mathcal{W}^N, (C_v, C_e)_{v \in \mathcal{V}, e \in \mathcal{E}}), \tag{5.64}$$

where $\mathcal{W}^N$ represents the users' action space. We assume that the user gains $\psi(\mathbf{w}_i)$ by performing action $\mathbf{w}_i$. And we can define the optimization objective as

$$\max_{\mathbf{w}_i} \psi(\mathbf{w}_i), \forall \mathbf{w}_i \in \mathcal{W}_i, \forall i \in \mathcal{N}. \tag{5.65}$$

In the Eq. (5.65), if $\forall i \in \mathcal{N}$ and $\forall \mathbf{w}_i \in \mathcal{W}_i$, there exists an action $\mathbf{w}_i^*$ that satisfies

$$\psi(\mathbf{w}_i^*) \geqslant \psi(\mathbf{w}_i). \tag{5.66}$$

Then, we define the strategy set $(\mathbf{w}_1^*, \mathbf{w}_2^*, \ldots, \mathbf{w}_N^*)$ as the Nash Equilibrium. As shown in Fig. 5.15, multiple users need the network resources to provide services at the same time. Three users need to deploy the service chain on the service provider's infrastructure. For convenience, we denote user 1 deploys VNF F_1 on the server as "U1-V1." Each user will compete for resources to get more network resources. Service providers not only need to provide users with the best deployment strategy but also make rational use of network resources.

5.3.1.3   Problem Formulation

The total latency consists of two parts: processing delay and transmission delay. We define the processing delay of VNF as $d(v_{F_f})$, which includes calculation delay and caching delay. We define the transmission latency between server node v_i and server node v_j as $d(e_{v_i, v_j})$. Then, the total latency of the SFC in the network can be formulated as

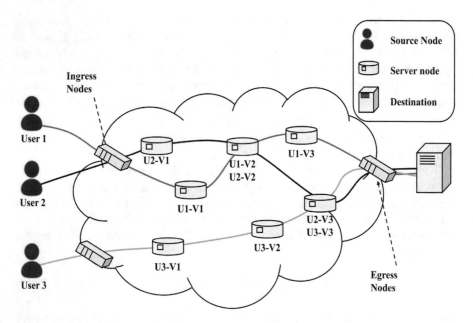

Fig. 5.15 An illustrative example of the competition between users for network resources

$$d(s, a) = d_\alpha \cdot \sum d(e_{v_i, v_j}) + d_\beta \cdot \sum d(v_{F_f}), \qquad (5.67)$$

where d_α and $d_\beta (d_\alpha, d_\beta > 0)$ are the constant reward coefficient. s is the remaining resource of network and $d(s, a)$ is the total delay obtained by executing action a.

As discussed above, network resources will be consumed when the SFC is deployed. Therefore, the capacity resources constraints must be satisfied

$$\sum_{\mathcal{F}} c(F_j) \leq C(v_i), \forall i = 1 \ldots n, \forall j = 1 \ldots f. \qquad (5.68)$$

Similarly, the bandwidth resources constraints can formulate as

$$\sum_{\mathcal{F}} c(e) \leq C(e), \forall e \in \mathcal{E}. \qquad (5.69)$$

When multiple SFC requirements arrive, how to distribute the load across the whole network is another problem. In this paper, the availability utilization of network server node v_i can be defined as

$$u(v_i) = \frac{C(v_i) - \sum_{\mathcal{F}} c(F_j)}{C(v_i)}, \forall i = 1 \ldots n, \forall j = 1 \ldots f. \qquad (5.70)$$

Similarly, we can define the availability utilization of link e as

$$u(e) = \frac{C(e) - \sum_{\mathcal{F}} c(e)}{C(e)}, \forall e \in \mathcal{E}. \tag{5.71}$$

For balancing network load, we design a general weight objective function:

$$\min \sum_{v_i \in \mathcal{V}} \|u(v_i)\|_2 + \sum_{e \in \mathcal{E}} \|u(e)\|_2. \tag{5.72}$$

We assume $u_1 + u_2 + \ldots + u_n$ as a constant. According to Cauchy–Bunyakovsky–Schwarz Inequality, when $u_1 = u_2 = \ldots = u_n$, the minimum value of $u_1^2 + u_2^2 + \ldots + u_n^2$ can be obtained. The smaller the value of Eq. (5.72), the more balanced the utilization of network resources. Service providers are more inclined to select server nodes and links with more resources to execute the action. When the SFC is deployed successfully, the revenue function can be described as

$$revenue(s, a) = \sum_{F \in \mathcal{F}}^{f} \frac{C_{v_F}}{C_F} + \sum_{L \in \mathcal{L}} \frac{B_{e_L}}{B_L}, \tag{5.73}$$

where C_F represents the resources needed to deploy virtual network function F and B_L represents the transmission resources needed to deploy in link L. We denote the SFC link deployed in the underlying network link as e_L. Let C_{v_F} and B_{e_L} represent the available resources of the mapping server node and link, respectively. The service provider inclines to select server nodes and links with more resources. It can be inferred from Eq. (5.73) that the network needs to consider load balancing issues when performing task processing.

In this paper, we consider the load-balanced problem and the service delay minimization problem at the same time. Therefore, the objective function of the SCF placement problem can be formalized as

$$\max \xi_1 \cdot revenue(s, a) - \xi_2 \cdot d(s, a), \tag{5.74}$$

$$\begin{aligned}
s.t. \quad & C_F \geq 0, \forall F \in \mathcal{F} \\
& B_L \geq 0, \forall L \in \mathcal{L} \\
& u(v_i) \geq 0, \forall v_i \in \mathcal{V} \\
& u(e) \geq 0, \forall e \in \mathcal{E}
\end{aligned} \tag{5.74a}$$

where ξ_1 and ξ_2 ($\xi_1, \xi_2 > 0$) are tunable parameters. As shown in Table 5.5, we list the important notations of this paper.

Table 5.5 List of main notations

Parameter	Description
$\mathcal{G}$	The network model as a connected graph with n server nodes and m links
$\mathcal{V}$	The set of server nodes in network graph, $\forall v \in \mathcal{V}$
$\mathcal{E}$	The set of links in network graph, $\forall e \in \mathcal{E}$
v	The server node in network graph
e	The link in network graph
$\mathcal{F}$	The service function chain model
F_i	The i-th VNF in SFC model $\mathcal{F}$
$V_{\mathcal{F}}$	The server node which the VNF deployed
$v_{\mathcal{F}_i}$	The server node which the i-th VNF deployed in
$\mathcal{N}$	The set of network users
$\mathbf{w}_i$	Action selected by user i
$\psi(\mathbf{w}_i)$	Revenue from select action $\mathbf{w}_i$
$\mathcal{W}^N$	The set of users $\mathcal{N}$'s action space
$C(*)$	Resources capacity of network
$d(s, a)$	The total latency of a SFC in the network
$c(*)$	Resources capacity required to deploy network functions
$u(*)$	The utilization of network resources

5.3.2   Multi-Agent Reinforcement Learning

In this section, we propose a centralized training and distributed executing reinforcement learning approach to solve the SFC deployment.

5.3.2.1   The Hybrid Control Framework

As discussed above, centralized optimization algorithms incur too much overhead in both communication and computation. The controller must continually monitor the global information of the network and reconfigure the network hardware to deploy new service functions. In contrast, as a distributed system, distributed algorithms suffer serious non-convergence problems. Therefore, learning the optimal policy is a very difficult problem, especially with the localized observation of the environment. How to learn the global optimal scheduling strategy of distributed schemes becomes a critical problem.

In this part, we design a hybrid multi-layer control architecture. As shown in Fig. 5.16, the architecture can be divided into three layers: a super control layer, a user control layer, and a virtual network function placement layer. The user control layer is responsible for selecting the deployment action of SFC. And the virtual network function placement layer will execute the decisions of the multi-user controller. The super control layer is responsible for gathering and learning

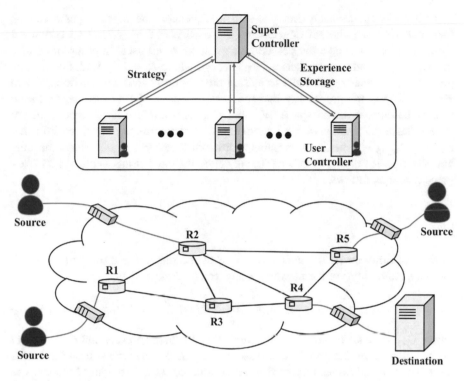

Fig. 5.16 The multi-layer control architecture

from network environment information and all controller behaviors of the user. And it is responsible for the user controller to optimize the execution strategy.

Consider a scenario that multiple network users want to establish service function chains from source to destination. Each user controller will make decisions according to its local observations. The user controller considers the problem from the aspect of the user and strives for the local maximum benefit. Multiple users compete with the network resource to satisfy their demands. The super controller will get the state information of each user controller. On account of the current policy of all user controllers, the super controller guides each user controller to optimize strategy until the game converging.

5.3.2.2   Markov Game Model

The SFC can be formulated as Markov decision process (MDP). When action a is executed in state s, the expected $R_{ss'}^a$ can be formulated as

$$R_{ss'}^a = E\{r_{t+1}|s_t = s, a_t = a, s_{t+1} = s'\}. \tag{5.75}$$

In Markov game, each user plays in a sequence. The Markov game can be formulated as: a finite set of users N that each user $i \in N$ has an action set W_i; a state space S as a the possible environment of all users; a group of actions $a_1, a_2, \ldots, a_N$ and observations $o_1, o_2, \ldots, o_N$, where $N = |N|$. With the Markov property, each user i will use a transition probability P to choose actions, where $P(s_i'|s_i, W_i)$ is the probability that the next state is in s_i' given the current state s_i and the current action space W_i, and obtains rewards r as a function of the state s and agent's action a. Each agent expects the maximum overall reward value $R_i = \sum_{t=0}^{T} \gamma^t r_i^t$ where γ represents a discount factor and T represents the time horizon. Specifically, in our scenario, we define the three components of a Markov game model as follows.

State

In the t-th time, we define the state $s_t \in S$ as the remaining resource (i.e., CPU, memory, bandwidth) of the network, which can be described as

$$s_t = (C_V, C_{\mathcal{E}}), \tag{5.76}$$

where C_V is the set of server node's remaining resource capacity and $C_{\mathcal{E}}$ is the set of the link's remaining transmission resources capacity. The state at time t indicates the environmental issues that need to be considered when providing SFC tasks to users.

Action

The agent has to decide how to deploy service function chain tasks. It includes two steps: assign server nodes resource to VNFs and choose suitable links to connect. As we mentioned above, there are n server nodes and m links in the network topology. The available resource of the network state can be obtained easily. Agent executes an action $a \in (A, L)$ at each step, where $A = 0, 1, 2, \ldots, n$ is the set of server node indexes. With the mapping of VNF, the action also considers forming a corresponding link L for the SFC task, where $L \in \mathcal{E}$.

Reward Function

In this paper, the optimization goal needs to consider both the load-balanced and service delay minimization. Thus, the reward function can be formulated as

$$r(a, s) = r_\alpha \cdot revenue(s, a) - r_\beta \cdot d(s, a), \tag{5.77}$$

where r_α and $r_\beta (r_\alpha, r_\beta > 0)$ are the custom constant, namely the reward coefficient. We can set the parameters according to the actual network environment. The larger the value of the function in Eq. (5.77), the smaller the total delay of SFC and the more balanced the network resource allocation.

5.3.2.3 Multi-agent Reinforcement Learning Approach

Multi-agent reinforcement learning can achieve excellent performance in the multi-agent system environment [68]. In this paper, we introduce a multi-agent reinforcement learning algorithm named MADDPG to solve our problem [69]. We design a centralized training and distributed implementation framework. In our framework, the centralized "critic" collects the data of all actors, while the actors can choose the actions guided by the critic. With such a learning process, SFC policy can be updated steadily and smoothly. The framework does not impose restrictions on the environment, in which each agent can have its reward function mechanism and determine to collaborate or compete. For each agent, only the current state data is required to make predictions during running, and the status of the environment is not persistently consistent.

As shown in Fig. 5.17, the actor executes the action after obtaining the observation from the environment. Each actor collects data (s, a, r, s') and stores it in replay buffer memory. When the number of buffer pools exceeds the warm-up threshold,

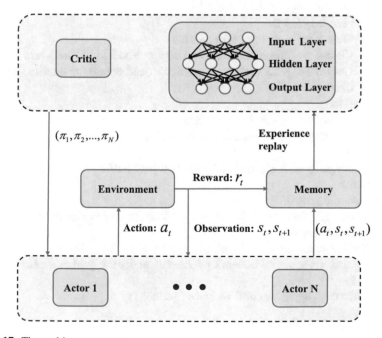

Fig. 5.17 The multi-agent system

the learning starts. The critic samples data from memory for training. The actor in each agent can interact with the environment independently, while the critic in the centralized controller can guide the actor to revise the strategy learning process. Note that each actor and each critic update their policy parameter separately. This training process will repeat until the policy converges. The specific algorithm is shown in Algorithm 5.4.

We denote the strategies as $\pi = \{\pi_1, \ldots, \pi_N\}$ of $N = |\mathcal{N}|$ agents, and let $\theta = \{\theta_1, \ldots, \theta_N\}$ represent the parameters of all agents. For users in the SFC deployment process, we define the policy gradient of the cumulative expected reward of the i-th agent as:

$$\nabla_{\theta_i} J(\delta_i) = E_{\ell,a \sim D}[\nabla_{\theta_i} \delta_i(a_i|o_i) \times$$
$$\nabla_{a_i} \partial_i^\delta(\ell, a_1, \ldots, a_n)|_{a_i = \delta_i(o_i)}], \tag{5.78}$$

Algorithm 5.4 The multi-agent reinforcement learning for SFC deployment

---

Input: observation o, reward r
Output: action a

1: Initialize the actor and critic neural network parameters
2: Setting replay buffer memory size and batch size
3: **for** $eachepisode = 1, 2, 3 \ldots$ **do**
4:     Initialize IoT network system network environment
5:     **for** $eachstep = 1, 2, 3 \ldots$ **do**
6:         **for** $each\ agent$ **do**
7:             Obtain IoT network observation o
8:             Choose an optimal action a and deploy VNFs in IoT network
9:             Get reward r for performing action a, and obtain the next IoT network observation o'
10:        **end for**
11:        **if** $replaysize \geq settingsize$ **then**
12:            Remove the oldest data in replay buffer
13:        **end if**
14:        Package (o, a, r, o') and add it to the replay buffer
15:    **end for**
16:    **for** $each\ agent$ **do**
17:        Randomly take out a batch of data from replay buffer according to the preset batch size
18:        Update critic neural network parameters with batch size data
19:        Update actor neural network parameters under the guidance of critic
20:    **end for**
21:    Update each agent's neural network parameters
22: **end for**

---

where the $\ell = \{o_1, o_2, \ldots, o_N\}$ represents the observation sets, $a_i, \forall i = 1 \ldots N$ represents the action of i-th agent, and θ_i represents the parameters of the actors. We set D as the replay buffer memory, where the data sample structure is $(\ell, a_1, \ldots, a_n, r_1, \ldots, r_n, \ell')$. Since ∂_i of each agent is learned separately, they have optimization objectives. The update function of ∂_i can be described as

$$L(\theta_i) = E_{\ell, a, r, \ell'}[(\partial_i^\delta(\ell, a_1, \ldots, a_n) - y)^2], \tag{5.79}$$

where

$$y = r_i + \gamma \overline{\partial_i}^{\delta'}(\ell', a_1', \ldots, a_n')|_{a_j' = \delta_j'(o_j)}, \tag{5.80}$$

where $\delta' = [\delta_1', \delta_2', \ldots, \delta_n']$ is the set of target policies with delayed parameter θ_j', and $\overline{\partial_i}^{\delta'}$ represents the target action-value function. The γ represents the discount factor.

The core method of this algorithm is that the critic of each agent will get the information of all agents. The critic is trained in a centralized model, and the actor is executed in distributing. The critic with global information can guide the training process of actors. In the running phase, only actors with local observations are used to perform actions. Centralized training is adopted off-line and decentralized execution is adopted online. The common point of the combination of online and off-line is that participants only need to use the observed local information.

In reality, the agent cannot always obtain the strategies of other agents timely. Therefore, each agent needs to maintain the strategy approximation function of other agents. We define $\hat{\delta}_{\phi_i^j}$ to represent the functional approximation of the i-th agent strategy to the j-th agent. Then, the cost function can be described as

$$L(\phi_i^j) = -E_{o_j, a_j}[log\hat{\delta}_{\phi_i^j}(a_j|o_j) + \lambda H(\hat{\delta}_{\phi_i^j})], \tag{5.81}$$

where $H(\hat{\delta}_{\phi_i^j})$ is the entropy regularizer function. The approximate cost function is a logarithmic cost with an entropy regularizer. As long as minimizing the log probability of agent j 's cost function, other agent strategies will be obtained.

Therefore, with the approximate policies, the value y in Eq. (5.80) can be replaced as follows

$$y = r_i + \gamma \bar{\partial}_i^{\delta'}(\ell', \hat{\delta}'^1_{\phi_i^j}(o_1), \ldots, \hat{\delta}'^n_{\phi_i^j}(o_n)), \tag{5.82}$$

where γ is cumulative discount factor and $\hat{\delta}'^n_{\phi_i^j}$ uses neural networks to approximate $\hat{\delta}_{\phi_i^j}$. Before updating ∂_i^δ, the algorithm uses the sampled data of replay buffer to update the parameters of ϕ_i^j.

The unstable environment caused by the policy changes of agents is another problem in multi-agent reinforcement learning. The problems that arise in competitive tasks are particularly serious, which can lead to over-fitting of the strategy by the agent. When a competitor's strategy is updated and changed, it will reduce the generalization ability of the agent. To improve the generalization ability of the agent's strategy, a strategy set mechanism is proposed in the algorithm. The strategy of the i-th agent is composed of K sub-strategies, and only one of the sub-strategies $u_i^{(k)}$ is used in each training epoch. The maximum reward of the agent's strategy set is defined as

$$J_e(\delta_i) = E_{k\sim unif(1,K),s\sim\rho^\delta,a\sim\delta_i^{(k)}}\left[\sum_{t=0}^{\infty}\gamma^t r_{i,t}\right]. \tag{5.83}$$

We construct a replay buffer memory $D_i(k)$ for sub-strategy k of agent i. In order to optimize the overall performance, the gradient of each sub-strategy $\theta_i^{(k)}$ of agent i is updated to

$$\nabla_{\theta_i^{(k)}} J(\delta_i) = \frac{1}{K} E_{\ell,a\sim D_i^{(k)}}[\Phi], \tag{5.84}$$

where

$$\Phi = \nabla_{\theta_i^{(k)}}\delta_i^{(k)}(a_i|o_i)\nabla_{a_i}\partial^{\delta_i}(\ell, a_1, \ldots, a_n)|_{a_i=\delta_i^{(k)}(o_i)}. \tag{5.85}$$

As shown in Table 5.6, we list the important notations of this section.

Table 5.6 List of main notations

Parameter	Description
a_i	The action of the i-th agent
o_i	The observation of the i-th agent
ℓ	The observation sets of all agents
r_i	The reward of the i-th agent
θ_i	The parameters of the i-th agent
D	The replay buffer memory
∂_i	The target action-value function of the i-th agent
γ	The discount factor
δ	The set of target policies
$H(*)$	The entropy regularizer function
$\hat{\delta}_{\phi_i^j}$	The functional approximation of the i-th agent strategy to the j-th agent

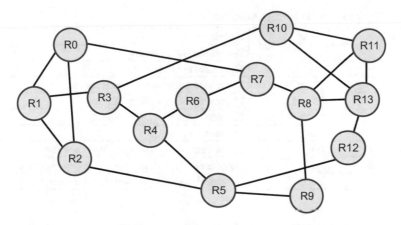

Fig. 5.18 The 14-nodes NSFNET topology

5.3.3 Experiments and Simulation Results

In this section, we will simulate the proposed algorithm and analyze its performance. The CPU uses an Intel (R) Core (TM) i5 8500 CPU @ 3.00 GHz. The software environment is python 3.7.6 and PyTorch 1.4.0.

5.3.3.1 Simulation Setup

We evaluate the proposed algorithm using the 14-nodes NSFNET topology [70], as shown in Fig. 5.18, which contains 44 fiber links. We assume that the service resource capacity of each server node ranges in [30, 50]. The latency between two server nodes is related to their distance. We assume that each server node can deploy all types of VNFs. If there is no special statement, we construct 5 different types of virtual network function, where each SFC with the number of VNFs randomly ranges from 3 to 5. The number of network users is $N = 4$, and all network users receive the same network service at the same moment. As shown in Table 5.7, we list the major parameter settings in the simulation.

5.3.3.2 Convergence Evaluation

First, we evaluate the training performance of our algorithm. In the simulations, we change the users' number and the number of VNFs contained in each SFC. Figure 5.19a shows rewards of the proposed SFC deployment algorithm with the different number of agents, where the numbers are 2 and 4, respectively. And the number of VNFs contained in each SFC is randomly ranged from 2 to 3 and from 3 to 5. Each episode in our algorithm includes 100 SFC deployment processes.

Table 5.7 The parameter settings in the simulation

Parameter	Value	Description
$|\mathcal{V}|$	14	The number of server nodes
$|\mathcal{E}|$	44	The number of fiber links
$C(v)$	[30, 50]	The resource capacity of server nodes
$C(e)$	[80, 100]	The transmission capacity of links
$|\mathcal{F}|$	[2, 5]	The number of VNFs in one SFC
N	[2, 4]	The number of users
$batchsize$	100	The number of training examples in one forward
$critic - lr$	$1 * 10^{-3}$	How much adjusting the weights of critic network
$actor - lr$	$1 * 10^{-4}$	How much adjusting the weights of actor network

To demonstrate the result more clearly, we use a simple 5-step moving average of the data. As shown in Fig. 5.19a, when the number of users is large, the network convergence speed will obviously slow down. For the same number of users, the less the number of VNFs, the more stable the network output result will be. However, no matter how the number of users and VNFs changes, the algorithm can always achieve a better convergence performance.

To evaluate the learning process, we present the loss value of one actor. Figure 5.19b shows the learning loss of the proposed multi-agent reinforcement learning SFC deployment algorithm with different batch sizes, in which the batch size values are 20, 100, and $1 * 10^3$, respectively. As mentioned earlier, to demonstrate the results clearly, we use a simple 100-step moving average of the data. As shown in the result, the batch size affects the value of the loss and learning efficiency with the increase of learning iterations. When the batch size is small, the convergence speed of the algorithm cannot reach an ideal result. With the increase of batch size, the convergence effect of the neural network is better and faster. Although the resource consumption of $batchsize = 1 * 10^3$ is much larger than that of $batchsize = 100$, the training speed is not significantly improved.

Similarly, we compare the effect of learning rate on convergence as shown in Fig. 5.19c. When the learning rate of actor and critic is set to $1 * 10^{-4}$ and $1 * 10^{-5}$, the training situation will not achieve the expected results. However, when they are set to $1 * 10^{-2}$ and $1 * 10^{-3}$, respectively, neural networks may cause non-convergence. In the following experiments, we will set the batch size and learning rate of the multi-agent reinforcement learning algorithm to the optimal value, which are the parameters listed in Table 5.7.

5.3.3.3 Performance Evaluation

We take the deep Q-network (DQN) algorithm and deep deterministic policy gradient (DDPG) algorithm as the benchmark algorithm. DQN uses neural networks to approximate the target strategy value function. After calculating the target

Fig. 5.19 Convergence
evaluation. (**a**) Convergence
of proposed algorithm. (**b**)
The impact of loss in different
batch size. (**c**) The impact of
loss in different learning rate

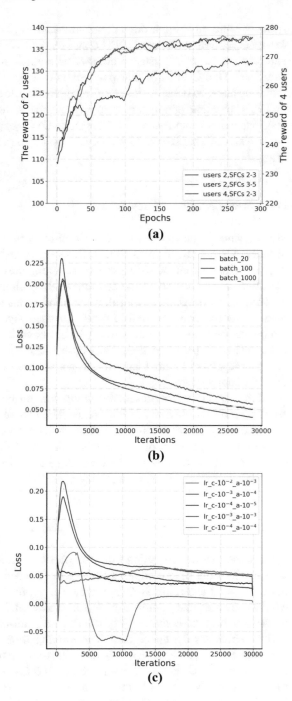

(a)

(b)

(c)

strategy value function through the neural network, DQN adopts a ε−greedy policy to execute action. Although DDPG draws on the main idea of DQN (memory replay and target network). It uses an actor-critic algorithm based on determining action strategies. DDPG ensures that parameters can be updated slowly through "soft" target updates, to achieve the effect of improving learning stability similar to that of DQN updating parameters regularly. At the same time, we also implement a distributed algorithm based on DQN (DQN-D) for experimental comparison. There are two sets of data selected for analysis: users' average reward and the calculation time.

Performance Analysis of Reward

We compare the performance of our algorithm with the benchmark methods as the number of SFCs ranges from 50 to 150. As shown in Fig. 5.20a, the performance of distributed algorithm DQN-D is the worst among the experimental results. Although the effect is not as well as the centralized algorithm (DQN algorithm and DDPG algorithm), the distributed algorithm can improve the response speed and the robustness of the placement system. In our proposed algorithm, a centralized platform simplifies the learning process with global network information whilst agents can make decisions based on their local observations in a distributed environment. It can get the global optimal solution, to get a more efficient deployment scheme in the SFC deployment.

At the same time, we present the simulation results of average revenue and delay. According to our previous definition, average revenue can be used to represent the utilization of network resources. The more the average revenue, the more balanced the resource utilization of the network. As shown in Fig. 5.20b, compared with other existing methods, MADDPG can make uniform use of network resources and achieve better results in reducing delay.

Complexity and Performance Analysis of Runtime

We assume that the execution time complexity of a single actor neural network is $O(T_e)$. Then the time complexity of the centralized algorithm is $O(N * T_e)$ and the distributed algorithm is $O(T_e)$, where N represents the number of agents. For MADDPG, the time complexity of training is $O(E * BT_c * N)$, where $O(BT_c)$ represents the time complexity of training an actor-critic network in batch size B, and E represents the epoch size. The training time complexity of DDPG is also approximately equal to $O(E * BT_c * N)$. The training time complexity of centralized DQN method is $O(E * BT_a * N)$, where $O(BT_a)$ represents the time complexity of training a single neural network in batch size B. Similarly, the training time complexity of distributed DQN method is $O(E * BT_a)$.

In the SFC deployment process, we only need to consider the execution process, rather than the training process. In our algorithm, the neural networks of actor

Fig. 5.20 Performance evaluation. (**a**) Average service reward. (**b**) Average revenue and delay. (**c**) Average service runtime

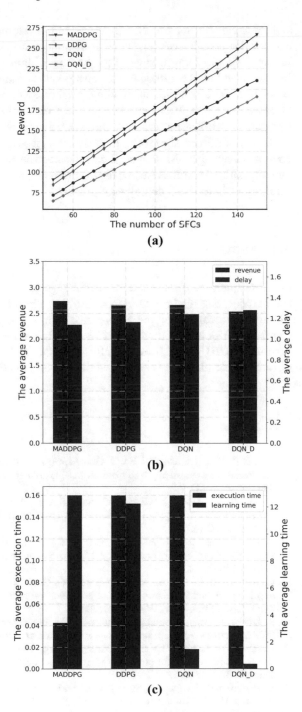

and critic can run separately. Therefore, in the runtime comparison part, we only consider the problem of choosing the substrate network for the execution algorithm. Figure 5.20c plots the runtime of the four algorithms to analyze their performance. As mentioned above, although the reward value obtained by distributed algorithm DQN-D cannot achieve outstanding performance, it has the best performance of response speed. From another perspective, centralized algorithms can better integrate network resources and improve network resource utilization. Compared with the centralized algorithm, the response speed of the proposed algorithm has obvious advantages. Moreover, it can observe the overall situation and achieve a globally optimal result. Based on the experimental results, our algorithm can achieve ideal results in the SFC deployment process.

5.4 Summary

In this chapter, we discuss the main challenge of IoT network resource scheduling and introduced several reinforcement learning algorithms. We first design a network slicing architecture over the SDN-based long-range wide area network. Then, a Continuous-Decision virtual network embedding scheme relying on Reinforcement Learning (CDRL) is proposed. Finally, we propose a hybrid intelligent control architecture, which adopts the centralized training and distributed execution paradigm.

References

1. T. Mai, H. Yao, N. Zhang, W. He, D. Guo, M. Guizani, Transfer reinforcement learning aided distributed network slicing optimization in industrial IoT. IEEE Trans. Ind. Inform. **18**(6), 4308–4316 (2021)
2. H. Yao, S. Ma, J. Wang, P. Zhang, C. Jiang, S. Guo, A continuous-decision virtual network embedding scheme relying on reinforcement learning. IEEE Trans. Netw. Service Manag. **17**(2), 864–875 (2020)
3. Y. Zhu, H. Yao, T. Mai, W. He, N. Zhang, M. Guizani, Multi-agent reinforcement learning aided service function chain deployment for Internet of Things. IEEE Internet Things J. **9**(17), 15674–15684 (2022)
4. G. Han, J. Tu, L. Liu, M. Martinez-Garcia, C. Choi, An intelligent signal processing data denoising method for control systems protection in the industrial Internet of Things. IEEE Trans. Ind. Inform. **18**(4), 2684–2692 (2021)
5. A. Lavric, V. Popa, Internet of things and LoRa low-power wide-area networks: a survey, in *2017 International Symposium on Signals, Circuits and Systems (ISSCS)* (IEEE, 2017), pp. 1–5
6. G. Han, J. Tu, L. Liu, M. Martínez-García, Y. Peng, Anomaly detection based on multidimensional data processing for protecting vital devices in 6g-enabled massive IIoT. IEEE Internet Things J. **8**(7), 5219–5229 (2021)
7. S. Wijethilaka, M. Liyanage, Survey on network slicing for internet of things realization in 5g networks. IEEE Commun. Surveys Tutorials **23**(2), 957–994 (2021)
8. C. Qiu, H. Yao, F.R. Yu, F. Xu, C. Zhao, Deep q-learning aided networking, caching, and computing resources allocation in software-defined satellite-terrestrial networks. IEEE Trans. Veh. Technol. **68**(6), 5871–5883 (2019)

9. M. Bor, J.E. Vidler, U. Roedig, Lora for the Internet of Things (2016)
10. J. Wang, C. Jiang, H. Zhang, Y. Ren, K.-C. Chen, L. Hanzo, Thirty years of machine learning: the road to pareto-optimal wireless networks. IEEE Commun. Surv. Tutorials **22**(3), 1472–1514 (2020)
11. S.J. Pan, Q. Yang, A survey on transfer learning. IEEE Trans. Knowl. Data Eng. **22**(10), 1345–1359 (2009)
12. B. K. Al-Shammari, N. Al-Aboody, H. S. Al-Raweshidy, IoT traffic management and integration in the QoS supported network. IEEE Internet Things J. **5**(1), 352–370 (2017)
13. K. Benzekki, A. El Fergougui, A. Elbelrhiti Elalaoui, Software-defined networking (SDN): a survey. Security Communication Networks **9**(18), 5803–5833 (2016)
14. C. Qiu, H. Yao, C. Jiang, S. Guo, F. Xu, Cloud computing assisted blockchain-enabled Internet of Things. IEEE Trans. Cloud Comput. **10**(1), 247–257. (2019)
15. O. Georgiou, U. Raza, Low power wide area network analysis: Can LoRa scale? IEEE Wirel. Commun. Lett. **6**(2), 162–165 (2017)
16. S. Dawaliby, A. Bradai, Y. Pousset, Distributed network slicing in large scale IoT based on coalitional multi-game theory. IEEE Trans. Netw. Service Manag. **16**(4), 1567–1580 (2019)
17. K. Xue, B. Zhu, Q. Yang, N. Gai, D. S. Wei, N. Yu, InPPTD: a lightweight incentive-based privacy-preserving truth discovery for crowdsensing systems. IEEE Internet Things J. **8**(6), 4305–4316 (2020)
18. T. Mai, H. Yao, N. Zhang, L. Xu, M. Guizani, S. Guo, Cloud mining pool aided blockchain-enabled internet of things: An evolutionary game approach. IEEE Trans. Cloud Comput. (2021)
19. Y. He, G. Han, J. Jiang, H. Wang, M. Martinez-Garcia, A trust update mechanism based on reinforcement learning in underwater acoustic sensor networks. IEEE Trans. Mobile Comput. **21**(3), 811–821 (2020)
20. B. Eysenbach, R. Salakhutdinov, S. Levine, Search on the replay buffer: bridging planning and reinforcement learning (2019). arXiv preprint arXiv:1906.05253
21. M.E. Taylor, P. Stone, Transfer learning for reinforcement learning domains: a survey. J. Mach. Learn. Res. **10**(7) (2009)
22. A. Fischer, J.F. Botero, M.T. Beck, H. De Meer, X. Hesselbach, Virtual network embedding: a survey. IEEE Commun. Surv. Tutorials **15**(4), 1888–1906 (2013)
23. N.M.M.K. Chowdhury, R. Boutaba, A survey of network virtualization. Comput. Netw. **54**(5), 862–876 (2010)
24. D. Drutskoy, E. Keller, J. Rexford, Scalable network virtualization in software-defined networks. IEEE Internet Comput. **17**(2), 20–27 (2013)
25. P. Zhang, X. Pang, Y. Bi, H. Yao, H. Pan, N. Kumar, DSCD: delay sensitive cross-domain virtual network embedding algorithm. IEEE Trans. Netw. Sci. Eng. **7**(4), 2913–2925 (2020)
26. Y. Zeng, R. Zhang, *Efficient Mapping of Virtual Networks onto a Shared Substrate* (Washington University in St Louis, 2006)
27. Y. Zhu, M.H. Ammar, Algorithms for assigning substrate network resources to virtual network components, in *25th IEEE International Conference on Computer Communications*, Barcelona, Spain, 2006, pp. 23–29
28. Z. Liu, M. Wu, Exact solutions of VNE: a survey. China Commun. **13**(6), 48–62 (2016)
29. M. Yu, Y. Yi, J. Rexford, M. Chiang, Rethinking virtual network embedding: substrate support for path splitting and migration. Comput. Commun. Rev. **38**(2), 17–29 (2008)
30. A. Razzaq, M.S. Rathore, An approach towards resource efficient virtual network embedding, in *International Conference on Evolving Internet*, Valencia, Spain, 2010, pp. 68–73
31. X. Cheng, S. Su, Z. Zhang, H. Wang, F. Yang, Y. Luo, J. Wang, Virtual network embedding through topology-aware node ranking. Comput. Commun. Rev. **41**(2), 38–47 (2011)
32. X. Hesselbach, J.R. Amazonas, S. Villanueva, J.F. Botero, Coordinated node and link mapping VNE using a new paths algebra strategy. J. Netw. Comput. Appl. **69**, 14–26 (2016)
33. J. Wang, C. Jiang, H. Zhang, Y. Ren, K.-C. Chen, L. Hanzo, Thirty years of machine learning: the road to pareto-optimal wireless networks. IEEE Commun. Surv. Tutorials (2020) https://doi.org/10.1109/COMST.2020.2965856

34. D.E. Goldberg, Genetic algorithms in search, optimization, and machine learning, in *Ethnographic Praxis in Industry Conference*, Portland, US, 1988, pp. 3104–3112
35. L.P. Kaelbling, M.L. Littman, A.W. Moore, Reinforcement learning: a survey. J. Artif. Intell. Res. **4**(1), 237–285 (1996)
36. R.S. Sutton, A.G. Barto, Reinforcement learning: an introduction. IEEE Trans. Neural Netw. **9**(5), 1054–1054 (1998)
37. Y. Lecun, Y. Bengio, G.E. Hinton, Deep learning. Nature **521**(7553), 436–444 (2015)
38. K. Cho, B. Van Merrienboer, C. Gulcehre, D. Bahdanau, F. Bougares, H. Schwenk, Y. Bengio, Learning phrase representations using RNN encoder–decoder for statistical machine translation, in *Conference on Empirical Methods in Natural Language Processing*, Doha, Qatar, 2014, pp. 1724–1734
39. I. Sutskever, O. Vinyals, Q. V. Le, Sequence to sequence learning with neural networks, in *Annual Conference on Neural Information Processing Systems*, Montreal, Canada, 2014, pp. 3104–3112
40. H. Yao, C. Xu, M. Li, P. Zhang, L. Wang, A novel reinforcement learning algorithm for virtual network embedding. Neurocomputing **284**, 1–9 (2018)
41. H. Yao, B. Zhang, L. Maozhen, P. Zhang, L. Wang, RDAM: a reinforcement learning based dynamic attribute matrix representation for virtual network embedding. IEEE Trans. Emer. Topics Comput. **PP**(99), 1–1 (2019)
42. R. Sutton, A. Barto, *Reinforcement Learning: An Introduction*, 2nd edn. a Bradford book (2018)
43. C. Watkins, P. Dayan, Q-learning[J]. Mach. Learn. **8**(3), 279–292 (1992)
44. R.J. Williams, Simple statistical gradient-following algorithms for connectionist reinforcement learning. Mach. Learn. **8**(3–4), 229–256 (1992)
45. D. Silver, G. Lever, N. Heess, T. Degris, D. Wierstra, M. Riedmiller, Deterministic policy gradient algorithms, in *31st International Conference on Machine Learning, ICML 2014*, vol. 1 (2014)
46. S. Hougardy, The Floyd-Warshall algorithm on graphs with negative cycles. Inform. Process. Lett. **110**(8), 279–281 (2010)
47. P. Koehn, Pharaoh: a beam search decoder for phrase-based statistical machine translation models. (2004), pp. 115–124
48. E.Z.M. Thomas, Generation and analysis of random graphs to model internetworks. College Comput. Georgia Institute Technol. **63**(4), 413–442 (1994)
49. M. Abadi, P. Barham, J. Chen, Z. Chen, A. Davis, J. Dean, M. Devin, S. Ghemawat, G. Irving, M. Isard, TensorFlow: a system for large-scale machine learning, in *25th IEEE International Conference on Computer Communications*, Georgia, USA, 2016, pp. 265–283
50. Wilcoxon, F., Individual comparisons of grouped data by ranking methods. J. Econ. Entomol. **39**(2), 269–270
51. R. Duan, J. Wang, C. Jiang, H. Yao, Y. Ren, Y. Qian, Resource allocation for multi-UAV aided IoT NOMA uplink transmission systems. IEEE Internet Things J. **6**(4), 7025–7037 (2019)
52. L. Cui, F.P. Tso, S. Guo, W. Jia, K. Wei, W. Zhao, Enabling heterogeneous network function chaining. IEEE Trans. Parallel Distrib. Syst. **30**(4), 842–854 (2019)
53. L. Qu, C. Assi, M.J. Khabbaz, Y. Ye, Reliability-aware service function chaining with function decomposition and multipath routing. IEEE Trans. Netw. Serv. Manag. **17**(2), 835–848 (2020)
54. S. DarOro, L. Galluccio, S. Palazzo, G. Schembra, Exploiting congestion games to achieve distributed service chaining in NFV networks. IEEE J. Sel. Areas Commun. **35**(2), 407–420 (2017)
55. J. Liu, W. Lu, F. Zhou, P. Lu, Z. Zhu, On dynamic service function chain deployment and readjustment. IEEE Trans. Netw. Serv. Manag. **14**(3), 543–553 (2017)
56. H. Hawilo, M. Jammal, A. Shami, Network function virtualization-aware orchestrator for service function chaining placement in the cloud. IEEE J. Sel. Areas Commun. **37**(3), 643–655 (2019)
57. A.M. Medhat, T. Taleb, A. Elmangoush, G.A. Carella, S. Covaci, T. Magedanz, Service function chaining in next generation networks: state of the art and research challenges. IEEE Commun. Mag. **55**(2), 216–223 (2017)

58. J. Pei, P. Hong, K. Xue, D. Li, Efficiently embedding service function chains with dynamic virtual network function placement in geo-distributed cloud system. IEEE Trans. Parallel Distrib. Syst. **30**(10), 2179–2192 (2019)
59. S. Bian, X. Huang, Z. Shao, X. Gao, Y. Yang, Service chain composition with resource failures in NFV systems: a game-theoretic perspective. IEEE Trans. Netw. Serv. Manag. **18**(1), 224–239 (2021)
60. J. Wang, H. Qi, K. Li, X. Zhou, PRSFC-IoT: a performance and resource aware orchestration system of service function chaining for Internet of Things. IEEE Internet Things J. **5**(3), 1400–1410 (2018)
61. W. Ren, Y. Sun, H. Luo, M.S. Obaidat, A new scheme for IoT service function chains orchestration in SDN-IoT network systems. IEEE Syst. J. **13**(4), 4081–4092 (2019)
62. T. Mai, H. Yao, N. Zhang, W. He, D. Guo, M. Guizani, Transfer reinforcement learning aided distributed network slicing resource optimization in industrial IoT. IEEE Trans. Ind. Inform. **18**(6), 4308–4316 (2021)
63. Y. He, F. R. Yu, N. Zhao, V.C.M. Leung, H. Yin, Software-defined networks with mobile edge computing and caching for smart cities: a big data deep reinforcement learning approach. IEEE Commun. Mag. **55**(12), 31–37 (2017)
64. L. Zhao, J. Wang, J. Liu, N. Kato, Routing for crowd management in smart cities: a deep reinforcement learning perspective. IEEE Commun. Mag. **57**(4), 88–93 (2019)
65. C. Qiu, H. Yao, C. Jiang, S. Guo, F. Xu, Cloud computing assisted blockchain-enabled internet of things. IEEE Trans. Cloud Comput. **10**(1), 247–257 (2019)
66. L. Gu, D. Zeng, W. Li, S. Guo, A.Y. Zomaya, H. Jin, Intelligent VNF orchestration and flow scheduling via model-assisted deep reinforcement learning. IEEE J. Sel. Areas Commun. **38**(2), 279–291 (2020)
67. T.A.Q. Pham, J.-M. Sanner, C. Morin, Y. Hadjadj-Aoul, Virtual network function–forwarding graph embedding: a genetic algorithm approach. Int. J. Commun. Syst. **33**(10), e4098 (2020)
68. A.S. Kumar, L. Zhao, X. Fernando, Mobility aware channel allocation for 5g vehicular networks using multi-agent reinforcement learning, in *ICC 2021—IEEE International Conference on Communications* (2021), pp. 1–6
69. Y. Xiao, Q. Zhang, F. Liu, J. Wang, M. Zhao, Z. Zhang, J. Zhang, NFVdeep: adaptive online service function chain deployment with deep reinforcement learning, in *Proceedings of the International Symposium on Quality of Service* (2019), pp. 1–10
70. W. Lu, Z. Zhu, Dynamic service provisioning of advance reservation requests in elastic optical networks. J. Lightwave Technol. **31**(10), 1621–1627 (2013)

Chapter 6
Mobile Edge Computing Enabled Intelligent IoT

Abstract The proliferation of the number of IoT devices, the ever-increasing computation intensive applications pose great challenges on resource allocation and offloading. In this chapter, to address spectrum sharing and edge computation offloading problems in SDN-based ultra dense networks, we propose a second-price auction scheme for ensuring the fair bidding for spectrum rent, which enables the MBS edge cloud and SBS edge cloud to occupy the channel in cooperative and competitive modes. Then, a novel deep reinforcement learning (DRL)-based network structure is proposed to jointly optimize task offloading and resource allocation. Finally, we propose two pervasive scenarios including single edge scene and multiple edge scenes. In the single edge scenario, a novel deep reinforcement learning (DRL)-based framework is invoked for collaboratively optimizing the task scheduling, transmission power, and CPU cycle frequency under metabolic channel conditions. Meanwhile, we propose a multi-agent aided deep deterministic policy gradient (MADDPG) algorithm to alleviate interference in multiple edge scenarios.

Keywords Mobile edge computing · Computing offloading · Ultra-dense networks · Multi-agent aided deep deterministic policy gradient

The proliferation of the number of IoT devices, the ever-increasing computation intensive applications pose great challenges on resource allocation and offloading. In this chapter, to address spectrum sharing and edge computation offloading problems in SDN-based ultra dense networks, we propose a second-price auction scheme for ensuring the fair bidding for spectrum rent, which enables the MBS edge cloud and SBS edge cloud to occupy the channel in cooperative and competitive modes [1]. Then, a novel deep reinforcement learning (DRL)-based network structure is proposed to jointly optimize task offloading and resource allocation [2]. Finally, we propose two pervasive scenarios including single edge scene and multiple edge scenes. In the single edge scenario, a novel deep reinforcement learning (DRL)-based framework is invoked for collaboratively optimizing the task scheduling, transmission power and CPU cycle frequency under metabolic channel conditions [3]. Meanwhile, we propose a multi-agent aided deep deterministic

© The Author(s), under exclusive license to Springer Nature Switzerland AG 2023
H. Yao, M. Guizani, *Intelligent Internet of Things Networks*, Wireless Networks,
https://doi.org/10.1007/978-3-031-26987-5_6

policy gradient (MADDPG) algorithm to alleviate interference in multiple edge scenarios.

6.1   Auction Design for Edge Computation Offloading

With the rapid growth of wireless communication demand [4, 5], the transmission rate and network capacity of traditional networks are facing unprecedented challenges. In addition, novel increased business scenarios in the next-generation networks (5G) [6, 7], e.g., vehicular networking, augmented virtual reality, and industrial Internet of things [8, 9], propose a higher requirement for the delay, energy efficiency, and other performance. In order to cope with the increasingly severe challenges above, ultra dense networks (UDNs) [10] empower 5G tremendous access capability, composed of extensive macro base stations (MBSs) and small-cell base stations (SBSs). Additionally, edge computing [11] technology promises the potential to provide available computation service ability for countless devices. It can effectively shorten the data transmission distance between the user equipment (UEs) and the data center as well as avoid the network congestion. With the assistance of edge computing, UDNs are capable of providing computation service for UEs, which is implemented by the MBS edge cloud and SBS edge cloud. Considering the severe channel interference of computation offloading in UDNs [12], therefore, cooperative and incentive spectrum management plays a significant influence in supporting of computation offloading between MBS edge cloud and SBS edge cloud.

To achieve rapid configuration as well as effective management in ultra dense networks, software-defined networking (SDN) has been considered as an efficient network architecture to promise the potential to realize flexible network control and management. Recently, the concept of SDN has been applied into UDNs, which is termed as SDN-based ultra dense networks [13]. In this case, the primary computation and control functions are decoupled from the distributed SBS edge cloud and MBS edge cloud. Specifically, the control function is integrated at the centralized SDN controller [14]. The SDN controller [15] is capable of collecting information from UEs and edge clouds, as well as perceive network state from a global perspective. There is a technical challenge for MBS edge cloud working in the unlicensed spectrum that can degrade service quality without appropriate cooperate channel interference management. In LTE networks, two main mechanisms are focusing on this issue: carrier-sensing adaptive transmission (CSAT) scheme and listen-before-talk (LBT) scheme. However, CSAT cannot deal with the response to on–off cycling, and LBT is difficult to assign proper backoff time and transmission length. Therefore, it is an emergency to explore an effective spectrum management mechanism for the cooperation between MBS edge cloud and SBS edge cloud [16]. As a result, according to the decision instruction of the SDN controller, the channel is allocated to the MBS edge cloud or SBS edge cloud for providing computation offloading service for multiple users.

With the assistance of SDN controller, spectrum management and computation offloading for the MBS edge cloud and SBS edge cloud can be effectively dealt with. Besides, we focus on the issue on how to achieve an efficient negotiation between the MBS edge cloud and the SBS edge cloud with competition mode and cooperation mode in this network architecture. Mitsis et al. [17] employed SDN and mobile edge computing technology to manage end-users computing demands in 5G networks. A non-cooperative game model among the end-users is formulated and the Nash Equilibrium is verified. However, they did not consider the scenario of MBS edge cloud and multiple edge clouds. Game theory has been applied into spectrum sharing recently. Duan et al. [18] determined the prices of femtocell and macrocell services and modeled it as a Stackelberg game. Duan et al. [19] provided the analysis of cooperative spectrum sharing between primary user and secondary user by contract theory. However, both references do not focus on the models in computing offloading scenario. To avoid the malicious bidding in the market and guarantee fair and efficient spectrum resource sharing, a second auction theory [20, 21] is employed to provide an appropriate allocation scheme for spectrum management in this paper. Specifically, the MBS edge cloud is denoted as the auctioneer (the buyer), and the SBS edge clouds are set as the channel owners (the sellers). In this paper, we only consider communication resource in computation offloading. Moreover, we analyze the SBS edge clouds' equilibrium strategies under the MBS edge cloud's offloading rate.

6.1.1   Architecture of SDN-Based Ultra Dense Networks

In traditional network architecture, control function and forwarding function are integrated at nearby network nodes. To overcome the high complexity of network management, researchers at Stanford University proposed the concept of SDN [22–24]. The idea of SDN separates control function from data forwarding layer, and the controller is capable of perceiving network topology, computing forward path, etc. Consequently, SDN greatly simplifies the infrastructure and enables network operators to manage and control the overall nodes more effectively. Recently, SDN technology is applied into wireless networks, which is termed as software-defined wireless networks (SDWN) [25]. SDWN consists of software-defined cellular network [26], software-defined mobile network [27], SDN-Wi-Fi [28], and SDN-based ultra dense networks.

In general, SDWN architecture is divided into three planes: application plane, control plane, and infrastructure plane. Specifically, as shown in Fig. 6.1, in this paper we focus on introducing the application plane, control plane, infrastructure plane, and interfaces. Afterwards, we will introduce the architecture of SDN-based ultra dense networks as follows.

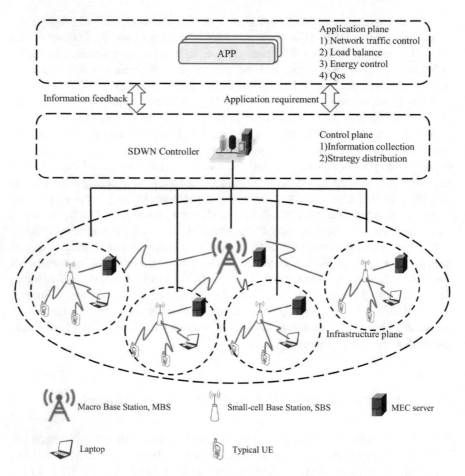

Fig. 6.1 Architecture of SDN-based Ultra dense networks

Application Plane   Service providers are capable of developing various applications on the application plane, as well as realizing the different requirements of users, e.g., network traffic control, load balance, energy control, etc.

Control Plane   The function of the control plane includes flow table control, strategy distribution, and the acquisition of network-wide information. After receiving the requirement from the application plane, the control plane transforms them into instructions that can be executed by the infrastructure plane, as well as sends them to the infrastructure plane through the flow table. Control plane connects to infrastructure plane by southbound interface [29, 30] and connects to application plane [31].

Infrastructure Plane   The infrastructure plane is composed of the MBS, SBS, and UE. Moreover, MBS and SBS connect with mobile edge computing server, termed

Table 6.1 Description of notations

Notation	Definition
$\mathcal{N} = \{1, 2, \ldots, N\}$	Set of the SBS edge clouds
N	Number of the SBS edge clouds
N	Number of channels for the SBS edge clouds
$r_n(t)$	Transmission rate for SBS edge cloud n at time t
$f(r)$	Probability distribution function of transmission rate r
$F(r)$	Cumulative distribution function of transmission rate r
σ^{MBS}	Discounting factor of the MBS edge cloud in competition mode
σ^{SBS}	Discounting factor of SBS edge cloud in competition mode
$r_{MBS}(t)$	Transmission rate of the MBS edge cloud
R	Maximum data offloading rate of the MBS edge cloud
D_j	Available offloading rate in channel j
$r_{compensation}(t)$	Guaranteed offloading rate
$b_n(t)$	Bidding strategy of SBS edge cloud n
J	Minimum bidding strategy
$U^{MBS}(b, R, t)$	The expected utility of the MBS edge cloud
$U_n^{MBS}(b, R, t)$	The expected utility of SBS edge cloud n

as the MBS edge cloud and the SBS edge cloud, respectively. By contrast, the SBS edge cloud is closer to UEs and can provide faster computation service via wireless link. UEs include smart devices (e.g., laptops and cell phones) connected to different application scenarios in the wireless networks. In addition, communication model in edge computing is from the resource allocation model in wireless access networks [32].

6.1.2   System Model

In this SDN-based ultra dense networks scenario, we consider one MBS edge cloud and a set $\mathcal{N} = \{1, 2, \ldots, N\}$ of SBS edge cloud, which provides computation offloading service for users. The n-th SBS edge cloud exclusively occupied channel n, $(n \in \mathcal{N})$. The MBS edge cloud is capable of providing a larger service area including different SBS edge clouds. In this case, it can work in channel n, which causes interference to the corresponding channel of the SBS edge cloud. Furthermore, we consider the auction between the MBS edge cloud and SBS edge cloud in timeslot $[t_1, t_2]$. Each auction is conducted after the last timeslot relying on the offloading rate in the last timeslot. As shown in Table 6.1, we list the notations of this section.

6.1.2.1 SBS Edge Clouds' Transmission Rate

In this paper, we consider a full-offloading mechanism in SBS edge clouds. Moreover, SBS edge cloud n occupies channel n, and the number of channels is equal to the number of the SBS edge clouds. Specifically, $r_n(t)$ represents the value of the transmission rate at time t, which is the private information of SBS edge cloud n. In addition, for the following timeslot, the transmission rate dynamically changes with time. The other $N - 1$ SBS edge clouds and MBS edge cloud only obtain the probability distribution of r_n. To be specific, r_n is assumed as a continuous random variable which generate in the range $[r_{min}, r_{max}]$, and r_{min} is the minimal value of r_n and r_{max} is the maximal value of r_n. Additionally, it obeys a probability distribution function $f(r)$ as well as a cumulative distribution function $F(r)$. In this case, all r_n is assumed to follow the same distribution.

6.1.2.2 MBS Edge Cloud's Cooperative and Competitive Modes

In this system, the MBS edge cloud should provide its computation service by occupying one of the channel N. Each SBS edge cloud has only one channel for its computation offloading service, but it cannot always be working which causes the consumption of channel. Besides, this scheme helps MBS edge cloud and SBS edge cloud cooperate with the channel and makes the transmission channel be utilized in an appropriate way. Furthermore, the competitive mode motivates the SBS to cooperate because each edge cloud will earn more profit in this mode. Specifically, the computation offloading service is operated in the following modes:

Competition Mode In this mode, the MBS edge cloud will choose a random channel with an equal probability. As a result, the MBS edge cloud will provide service in the channel at the case of SBS edge cloud n. Meanwhile, this will cause interference between the MBS edge cloud and SBS edge cloud n. We assume the original edge cloud in this channel will suffer more serious interference, which decreases the service quality of this edge cloud. In this case the transmission rate decreases by a certain discount. Because the discounting factors are not easy to be acquired in the real world, we denote $\sigma^{MBS} \in (0, 1)$ as the discounting factor of the MBS edge cloud and $\sigma^{SBS} \in (0, 1)$ as the discounting factor of the SBS edge cloud, respectively. In this mode, the computational complexity is $O(N)$, and N is the maximal number of SBS edge clouds.

Cooperation Mode In this mode, the MBS edge cloud will achieve the agreement with SBS edge cloud n, the transmission channel n will be occupied by the MBS edge cloud and SBS edge cloud n. Specifically, there is no interference in channel n and the transmission rate of the MBS edge cloud is set as r_{MBS}. As a compensation, the MBS edge cloud will serve the UEs of the SBS edge cloud in a timeslot with the guaranteed offloading rate $r_{compensation}(t) \in [0, r_{MBS}]$. In addition, the other $N - 1$ SBS edge clouds are not interfered by this channel occupied by the MBS edge cloud. In this mode, the computational complexity is $O(1)$, the MBS edge cloud will choose the SBS edge cloud with agreement. In the case of two modes, edge clouds prefer to choose cooperation mode when the channel is available. Nevertheless, when the channel is occupied, the competition mode is a reasonable way to assist edge computing.

6.1.2.3   Second-Price Auction Design

In the real world, different SBS edge clouds belong to different operators, and it is difficult to coordinate with each other. From the system model, the SDN controller has the ability to control the spectrum allocated to different SBS edge clouds. This can make it possible to complete the spectrum sharing in this architecture. The rules of the second-price auction are basically the same as the traditional bidding. The only difference is that the price paid by the winner is no longer his bid, but the second highest bid, so it is also termed as the "sub-highest price bidding method." Jehiel and Moldovanu [33] gave a comprehensive research of second-price forward auction, which characterized bidding strategies for general payoff functions. Bagwell et al. [34] researched auction bidding strategies in the WTO system. Nevertheless, both references only consider two bidders and not apply to the scenario with multiple bidders.

As shown in Fig. 6.2, a second-price auction mechanism is designed, which the MBS edge cloud is the buyer and the SBS edge clouds are denoted as the sellers. Each SBS edge cloud owns only one channel and tries to sell it. In addition, the bidding price is changing with different timeslots, and we assume that in a timeslot the bidding price does not change. When the SBS edge clouds are interested in the auction, they send their intentional bids to the controller. Afterwards, the controller determines the lowest price of the SBS edge cloud with the auction rule and delivers this information to the MBS edge cloud and the SBS edge clouds. With the assistance of the controller, two kinds of edge clouds need not communicate with each other directly. Moreover, we consider this auction in a timeslot and it can be termed as a differential game problem. In this case, we assume that the MBS edge cloud cannot occupy more than one channel simultaneously, therefore the MBS edge cloud is only interested in the winning seller of the SBS edge cloud. The MBS edge cloud will provide the offloading rate $r_{compensation}(t) \in [0, r_{MBS}]$ as the compensation.

The auction operation includes two stages: In the first stage of the auction, the MBS edge cloud announces its maximum data offloading rate R, which serves the winning the SBS edge cloud's users. In addition, in the second stage of the auction, after obtaining the data offloading rate R, SBS edge cloud n submits a bid $b_n(t) \in [0, R] \cup \emptyset$. Meanwhile, $b_n(t) \in [0, R]$ represents the data offloading rate that SBS edge cloud n requests the MBS edge cloud to serve SBS edge cloud n's users at time t. In addition, $b_n(t) = \emptyset$ indicates that at time t SBS edge cloud n does not provide sell service.

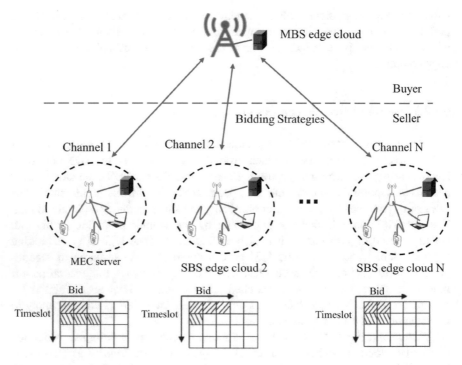

Fig. 6.2 Auction model of computation offloading

6.1.2.4 Auction Outcomes

In the following we will discuss the outcomes of auction for different values of b and R. Then the minimum bid $\mathcal{J}$ is defined as the following equation $\mathcal{J} = \{j \in \mathcal{N} : j = argmin_{n \in \mathcal{N}} \int_{t_1}^{t_2} b_n(t)\, dt\}$ and it has the following three possible outcomes:

(1) $|\mathcal{J}| = 1$. In this case, SBS edge cloud j is the winner and channel j is sold to the MBS edge cloud. The MBS edge cloud works in the cooperation mode. Then according to the principle of second-price auction theory, the transmission rate of the MBS edge cloud served SBS edge cloud j is $r_{compensation} = \min\{R, \int_{t_1}^{t_2} b_1(t)\, dt, \ldots, \int_{t_1}^{t_2} b_{j-1}(t)\, dt, \int_{t_1}^{t_2} b_{j+1}(t)\, dt, \ldots, \int_{t_1}^{t_2} b_N(t)\, dt\}$. Specifically, allocated transmission rate $r_{compensation}$ is larger than the minimum SBS edge cloud's bid.

(2) $|\mathcal{J}| \geq 2$. In this case, the MBS edge cloud works in the cooperation mode and it will choose a channel for minimum bid $\mathcal{J}$ with possibility $\frac{D_j}{\Sigma_{i=1}^{|\mathcal{J}|} D_i}$, where D_j is the available offloading rate in channel j. In the real scenario, the offloading rate plays an important role in the chosen possibility. The MBS edge cloud prefers to choose a channel with a higher service quality. Therefore, we define the chosen possibility with different values according to their offloading rates.

The more offloading rate can be provided, the higher the service quality will be, and the larger possibility the channel will be chosen. To be specific, the MBS edge cloud serves SBS edge cloud j's users with the data offloading rate $r_{compensation} = \min_{n \in N} \int_{t_1}^{t_2} b_n(t)\, dt$. Allocated transmission rate $r_{compensation}$ equals the minimum SBS edge cloud's bid.

(3) $|\mathcal{J}| = 0$. In this case, there is no SBS edge cloud that is willing to sell the channel to the MBS edge cloud, the MBS edge cloud chooses the competition mode and it will occupy a random channel with probability $\frac{1}{K}$. Specifically, the chosen channel will be shared by both providers.

Based on the above outcomes of three different cases, the $r_{compensation}$ can be given as

$$
r_{compensation}(b, R, t)
$$

$$
= \begin{cases}
A, & \text{if } |\mathcal{J}| = 1,\ j = \min_{n \in N} \int_{t_1}^{t_2} b_n(t)dt, \\[2mm]
\min\left\{ R, \min_{n \in N} \int_{t_1}^{t_2} b_n(t)dt \right\}, & \text{if } |\mathcal{J}| \geq 2, \\[2mm]
0, & \text{if } |\mathcal{J}| = 0,
\end{cases}
\tag{6.1}
$$

where A denotes $\min\{R, \min_{n \in N \setminus \{j\}} \int_{t_1}^{t_2} b_n(t)dt\}$. The utility of the MBS edge cloud obtained from offloading is defined as

$$
U^{MBS}(b, R, t)
$$

$$
= \begin{cases}
R - r_{compensation}(b, R, t), & \text{if } |\mathcal{J}| \geq 1, \\[2mm]
\sigma^{MBS} R, & \text{if } |\mathcal{J}| = 0,
\end{cases}
\tag{6.2}
$$

where if $|\mathcal{J}| \geq 1$, the MBS edge cloud works in the cooperation mode and its utility is denoted as $R - r_{compensation}(b, R, t)$. In addition, if $|\mathcal{J}| = 0$, the MBS edge cloud works in the competition mode, and then the utility is influenced by the interference of channel and it can be denoted as $\sigma^{MBS} R$.

Then relying on the analysis above, the expected utility of SBS edge cloud n can be formulated as

$$
U_n^{SBS}(b, R, t)
$$

$$
= \begin{cases}
\int_{t_1}^{t_2} r_n(t)dt, & \text{if } \int_{t_1}^{t_2} b_n(t)dt > \min_{j \in N} \int_{t_1}^{t_2} b_j(t)dt, \\[2mm]
B, & \text{if } \int_{t_1}^{t_2} b_n(t)dt = \min_{j \in N} \int_{t_1}^{t_2} b_j(t)dt, \\[2mm]
C, & \text{if } \min_{j \in N} \int_{t_1}^{t_2} b_j(t)dt = \emptyset,
\end{cases}
\tag{6.3}
$$

where B represents $\dfrac{D_j}{\Sigma_{i=1}^{|\mathcal{J}|} D_i} r_{compensation}(b, R, t) + \left(1 - \dfrac{D_j}{\Sigma_{i=1}^{|\mathcal{J}|} D_i}\right) \int_{t_1}^{t_2} r_n(t)dt$ and C denotes $\frac{1}{N}\sigma^{SBS} \int_{t_1}^{t_2} r_n(t)dt + \left(1 - \frac{1}{N}\right) \int_{t_1}^{t_2} r_n(t)dt$. In the case of $b_n >$

$\min_{j \in N} \int_{t_1}^{t_2} b_j(t) \, dt$, the MBS edge cloud occupies the other channel except channel n and SBS edge cloud n can provide its users with original transmission rate $\int_{t_1}^{t_2} r_n(t) \, dt$. In the case of $b_n = \min_{j \in N} \int_{t_1}^{t_2} b_j(t) \, dt$, the MBS edge cloud occupies channel n in the cooperation mode. Therefore, the SBS edge cloud's users can obtain the transmission rate $\frac{D_j}{\Sigma_{i=1}^{|\mathcal{J}|} D_i} r_{compensation}(b, R, t) + (1 - \frac{D_j}{\Sigma_{i=1}^{|\mathcal{J}|} D_i}) \int_{t_1}^{t_2} r_n(t) \, dt$. In the case of $b_n = \min_{j \in N} \int_{t_1}^{t_2} b_j(t) \, dt = \emptyset$, there is no SBS edge cloud that is willing to sell its channel and the MBS edge cloud will chose a random channel in competition mode. As a result, the transmission rate of computation offloading service is $\frac{1}{N} \sigma^{SBS} \int_{t_1}^{t_2} r_n(t) \, dt + (1 - \frac{1}{N}) \int_{t_1}^{t_2} r_n(t) \, dt$.

6.1.3 SBSs' Equilibrium Bidding Strategies

6.1.3.1 Definition of the Symmetric Bayesian Nash Equilibrium

Assume the maximum data offloading rate R of the MBS edge cloud in Stage I is given, and the SBS edge clouds' equilibrium bidding strategies will be analyzed and discussed. In the following subsections, we analyze the SBS edge clouds' equilibrium bidding strategies by taking into account different intervals of R.

The definition of the symmetric Bayesian NASH equilibrium (SBNE) is first given in the following Definition 6.3.

Definition 6.3 Given data offloading rate of the MBS edge cloud, a bidding strategy $b^*(r_n(t), R, t), r_n(t) \in [r_{\min}, r_{\max}], t \in [t_1, t_2]$ constitutes the SBNE if $s_n \in [0, R] \cup \emptyset, \forall r_n(t) \in [r_{\min}, r_{\max}]$ at time t, it holds that:

$$E_{r_{-n}} \{ U_n^{SBS}(b * (r_1(t), R, t), \ldots, b * (r_{n-1}(t), R, t),$$

$$b * (r_n(t), R, t), b * (r_{n+1}(t), R, t), \ldots, b * (r_N(t), R, t)$$

$$|r_n(t))\} \tag{6.4}$$

$$\geq E_{r_{-n}} \{ U_n^{SBS}(b * (r_1(t), R, t), \ldots, b * (r_{n-1}(t), R, t),$$

$$s_n, b * (r_{n+1}(t), R, t), \ldots, b * (r_N(t), R, t) | r_n(t)) \} .$$

Inequality (6.4) shows the SBNE of the SBS edge clouds, and all the SBS edge clouds adopt the identical bidding strategy $b^*(r_n(t), R, t)$ owing to the symmetric equilibrium. The left side of (6.4) represents the expected utility of SBS edge cloud n. Moreover, all the other SBS edge clouds' types are unknown to SBS edge cloud n. This inequality implies that SBS edge cloud n is not capable of obtaining a better utility when changing its strategy from $b^*(r_n(t), R, t)$ to $\forall s_n \in [0, R] \cup \emptyset$.

In the following, we will analyze the symmetric Bayesian NASH equilibrium for bidding strategies when offloading rate R in different intervals. The intervals are constituted of $[0, \frac{N-1+\sigma^{SBS}}{N} r_{\min}], (\frac{N-1+\sigma^{SBS}}{N} r_{\min}, r_{\min}), [r_{\min}, r_{\max}), [r_{\max}, +\infty)$.

First, we will introduce the case in $R \in [r_{\min}, r_{\max})$, and a detailed proof will be described. Similar to this case, the proof in other cases will be presented in general.

6.1.3.2   Equilibrium for $R \in [r_{\min}, r_{\max})$

First of all, we consider the most complex equilibrium analysis for the SBS edge clouds' equilibrium bidding strategies in the case of $R \in (r_{\min}, r_{\max}]$. To be specific, other cases are capable of being discussed in the same method. The following Lemma 6.1 is introduced to help analyze the SBS edge clouds' equilibrium bidding strategies.

Lemma 6.1 *There exists at least one solution $r(t)$ in the range $(R, r_{\max}]$ meeting the following equation:*

$$\Sigma_{n=1}^{N} C_{N-1}^{n} \left(\int_{t_1}^{t_2} \int_{R}^{r(t)} f\left(r(t)\right) d\left(r(t)\right) dt \right)^{n}$$

$$\times \left(\int_{t_1}^{t_2} \int_{r(t)}^{\infty} f(r(t)) d(r(t)) dt \right)^{N-1-n} \frac{\int_{t_1}^{t_2} [R - r(t)] dt}{n+1} \quad (6.5)$$

$$+ \left(\int_{t_1}^{t_2} \int_{r(t)}^{\infty} f(r(t)) d(r(t)) dt \right)^{N-1} \left(R - \frac{N-1+\sigma^{SBS}}{N} r(t) \right) = 0.$$

Specifically, we denote $F(r(t))$ as the CDF of random variable $r_n(t)$ at time t. Furthermore, the solutions $r_n(t)$ in $(R, r_{\max}]$ are denoted as $\tilde{r}_1(R, t), \tilde{r}_2(R, t), \ldots, \tilde{r}_K(R, t)$, where $K = \{1, 2, \ldots, K_{\max}\}$ represents the number of solutions and $K_{\max}$ is the maximal number of solutions. To be specific, the proof of Lemma 6.1 is provided in the following.

Proof In the following we will give the proof that there is at least one solution $r(t)$ that satisfies (6.5). First, the function of left hand side of equation is defined as $\mathcal{L}(r(t))$, and then we can obtain that

$$\mathcal{L}(r(t))$$

$$= \Sigma_{n=1}^{N} C_{N-1}^{n} \left(\int_{t_1}^{t_2} [F(r(t)) - F(R)] dt \right)^{n}$$

$$\times \left(\int_{t_1}^{t_2} [1 - F(r(t))] dt \right)^{N-1-n} \frac{\int_{t_1}^{t_2} [R - r(t)] dt}{n+1} \quad (6.6)$$

$$+ \left(\int_{t_1}^{t_2} [1 - F(r(t))] dt \right)^{N-1} \left(\int_{t_1}^{t_2} \left[R - \frac{N-1+\sigma^{SBS}}{N} r(t) \right] dt \right),$$

where function $\mathscr{L}(r(t))$ is continuous for $r_n(t)$ in $(R, r_{\max}]$. Then we can obtain that

$$
\begin{aligned}
\mathscr{L}(R) \\
= \left(\int_{t_1}^{t_2} [1 - F(R)] dt \right)^{N-1} \left(\int_{t_1}^{t_2} \left[R - \frac{N - 1 + \sigma^{SBS}}{N} R \right] dt \right) \\
= \left(\int_{t_1}^{t_2} [1 - F(R)] dt \right)^{N-1} \left(\int_{t_1}^{t_2} \left[R - \frac{1 - \sigma^{SBS}}{N} R \right] dt \right).
\end{aligned}
\tag{6.7}
$$

And since $F(r_{\max}) = 1$, we can obtain that

$$
\begin{aligned}
\mathscr{L}(r_{\max}) \\
= \Sigma_{n=1}^{N} C_{N-1}^{n} \left(\int_{t_1}^{t_2} [F(r_{\max}) - F(R)] dt \right)^{n} \\
\times \left(\int_{t_1}^{t_2} [1 - F(r_{\max})] dt \right)^{N-1-n} \frac{\int_{t_1}^{t_2} [R - r_{\max}] dt}{n + 1} \\
+ \left(\int_{t_1}^{t_2} [1 - F(r_{\max})] dt \right)^{N-1} \left(\int_{t_1}^{t_2} \left[R - \frac{N - 1 + \sigma^{SBS}}{N} r_{\max} \right] dt \right) \\
= \left(\int_{t_1}^{t_2} [1 - F(r_{\max})] dt \right)^{N-1} \int_{t_1}^{t_2} \frac{R - r_{\max}}{N} dt.
\end{aligned}
\tag{6.8}
$$

According to the property of cumulative distribution function, we have $F(R) \leq 1$. Then we will give the proof of $F(R)$ is not equal to 1.

Assumed that $F(R) = 1$, and since $R \in [r_{\min}, r_{\max})$, there can be found a ζ definitely, which holds $R + \zeta \in [r_{\min}, r_{\max}]$. Moreover, we have $F(R + \zeta) \leq 1$. Because $F(R) = 1$, $F(R + \zeta)$ only is equal to 1.

In conclusion, $F(R + \zeta) = F(R) = 1$. Since $F(r)$ is a cumulative distribution function, then $F(R + \zeta) - F(R) = 0$ contracts with the property of this function. Ultimately, we can obtain that $F(R) < 1$.

Since $F(R) \leq 1$ and $R \in [r_{\min}, r_{\max})$, we can conclude that $\mathscr{L}(r) > 0$ and $\mathscr{L}(r_{\max}) < 0$. Relying on the intermediate value theorem, there is at least one solution $r(t)$ in $(R, r_{\max}]$ satisfying (6.5). This completes the proof.

Relying on Lemma 6.1, the SBS edge clouds' equilibrium bidding strategies can be provided in Theorem 6.1.

Theorem 6.1 *Consider that there is a $\tilde{r}_x(R, t)$ submitted to $\{\tilde{r}_1(R, t), \tilde{r}_2(R, t)\}, \ldots,$ $\tilde{r}_K(R, t)$, then we can obtain the following bidding strategies $b^*(r_n(t), R, t)$ that constitute the SBNE for SBS edge cloud n.*

$$b^*(r_n(t), R, t)$$

$$= \begin{cases} any\ value \in [0, r_{\min}], & r_n(t) = r_{\min}; \\ r_n(t), & r_n(t) \in (r_{\min}, R]; \\ R, & r_n(t) \in (R, \tilde{r}_x(R, t)); \\ R\ or\ \emptyset, & r_n(t) = \tilde{r}_x(R, t); \\ \emptyset, & r_n(t) \in (\tilde{r}_x(R, t), r_{\max}]. \end{cases} \tag{6.9}$$

From (6.9), in the case of $r_n(t) = r_{\min}$, the optimal bidding strategy for SBS edge cloud n is to choose any value in the range $[0, r_{\min}]$. When $r_n(t) \in (r_{\min}, R]$, $b^*(r_n(t), R, t)$ should be $r_n(t)$. In addition, when $r_n(t) \in (R, \tilde{r}_x(R, t))$, the best strategy for SBS edge cloud n is R. When $r_n(t) = \tilde{r}_x(R, t)$, it may be R or $\emptyset$. Ultimately, when $r_n(t) \in (\tilde{r}_x(R, t), r_{\max}]$, the optimal bidding strategy for SBS edge cloud n should be $\emptyset$.

In conclusion, when the other SBS edge clouds choose their strategies in (6.9), the optimal strategy for SBS edge cloud n is to adopt $b^*(r_n(t), R, t)$ in (6.9). To be specific, the proof of Theorem 6.1 will be given in the following.

Proof In the following we will give the proof that the bidding strategies $b^*(r_n(t), R, t)$ constitute the SBNE for SBS edge cloud n. For SBS edge cloud n, suppose that all the other SBS edge clouds choose strategy $b^*(r_n(t), R, t)$, and then the following proof will give the maximum utility of SBS edge cloud n, which consists of four cases.

Case I $r_n(t) \in [r_{\min}, R]$ Supposed that when time is t, the data offloading rate at SBS edge cloud n satisfies $r_n(t) \in [r_{\min}, R]$, then we can obtain the following two situations.

(1) $b_{\min}^{-n} \in [0, R]$. In this case, the MBS edge cloud can always find a SBS edge cloud to cooperate with offloading data. If $b_{\min}^{-n} < r_n(t)$, then the expectation utility of SBS edge cloud n is $\int_{t_1}^{t_2} r_n(t)dt$. Moreover, when $b_{\min}^{-n} = r_n(t)$, the expectation utility of SBS edge cloud n is $\omega b_{\min}^{-n} + (1 - \omega)\int_{t_1}^{t_2} r_n(t)dt = \int_{t_1}^{t_2} r_n(t)dt$. Hence, bidding $\int_{t_1}^{t_2} r_n(t)dt$ is the optimal strategy of SBS edge cloud n.

To be specific, when $r_n(t) = r_{\min}$, we can obtain that $b_{\min}^{-n} > r_{\min}$ with possibility one. Therefore, for SBS edge cloud n, bidding any value in $[0, r_{\min})$ has the equivalent utility with bidding $r_{\min}$. In other words, bidding any value in $[0, r_{\min}]$ is the optimal strategy of SBS edge cloud n.

(2) $b_{\min}^{-n} = \emptyset$. In this case, If SBS edge cloud n bids, its utility will be $\frac{N-1+\sigma^{SBS}}{N}\int_{t_1}^{t_2} r_n(t)dt$. Since $\frac{N-1+\sigma^{SBS}}{N}\int_{t_1}^{t_2} r_n(t)dt < \int_{t_1}^{t_2} r_n(t)dt$, bidding $r_n(t)$ is one of the optimal strategy of SBS edge cloud n. In addition, when $r_n(t) = r_{\min}$, bidding any value in $[0, r_{\min}]$ is the optimal strategy of SBS edge cloud n.

As a result, when the other SBS edge clouds choose their strategies in (6.9), the optimal strategy for SBS edge cloud n is to adopt $b^*(r_n(t), R, t)$ in (6.9). To be specific, when $r_n(t) = r_{\min}$, the optimal bidding price for SBS edge cloud n is any value in the range $[0, r_{\min}]$. In addition, in the case of $r_n(t) \in (r_{\min}, R]$, the optimal bidding price should be the value of $r_n(t)$.

Case II $r_n(t) \in [R, \tilde{r}_x(R, t)]$ We assume that the data offloading rate at SBS edge cloud n satisfies $r_n(t) \in [R, \tilde{r}_x(R, t)]$. To be specific, we will analyze this case with the following two situations.

(1) Comparison between R and $\emptyset$. When bid R, the utility of SBS edge cloud n can be obtained as follows.

$$
\begin{aligned}
U_n^{SBS}&(b_n = R, R, t) \\
&= \int_{t_1}^{t_2} (1 - (1 - F(R))^{N-1}) r_n(t) dt + \int_{t_1}^{t_2} (1 - F(\tilde{r}_x(R, t)))^{N-1} R \, dt \\
&\quad + \Sigma_{n=1}^{N-1} \int_{t_1}^{t_2} C_{N-1}^n (F(\tilde{r}_x(R, t)) - F(R))^n \\
&\quad \times (1 - F(\tilde{r}_x(R, t)))^{N-1-n} \frac{R + n r_n(t)}{n+1} dt.
\end{aligned}
\tag{6.10}
$$

When bid $\emptyset$, the utility of SBS edge cloud n can be obtained as follows:

$$
\begin{aligned}
U_n^{SBS}&(b_n = \emptyset, R, t) \\
&= \int_{t_1}^{t_2} (1 - (1 - F(R))^{N-1}) r_n(t) dt + \int_{t_1}^{t_2} (1 - F(\tilde{r}_x(R, t)))^{N-1} \\
&\quad \times \frac{N - 1 + \sigma^{SBS}}{N} r_n(t) dt + \Sigma_{n=1}^{N-1} \int_{t_1}^{t_2} C_{N-1}^n (F(\tilde{r}_x(R, t)) \\
&\quad - F(R))^n (1 - F(\tilde{r}_x(R, t)))^{N-1-n} r_n(t) dt.
\end{aligned}
\tag{6.11}
$$

Then we can conclude that

$$
\begin{aligned}
U_n^{SBS}&(b_n = R, R, t) - U_n^{SBS}(b_n = \emptyset, R, t) \\
&= \int_{t_1}^{t_2} (1 - F(\tilde{r}_x(R, t)))^{N-1} \left(R - \frac{N - 1 + \sigma^{SBS}}{N} r_n(t) \right) dt \\
&\quad + \Sigma_{n=1}^{N-1} \int_{t_1}^{t_2} C_{N-1}^n \left(F(\tilde{r}_x(R, t)) - F(R))^n (1 - F(\tilde{r}_x(R, t))) \right)^{N-1-n} \\
&\quad \times \frac{R - r_n(t)}{n+1} dt.
\end{aligned}
\tag{6.12}
$$

It is simple to realize that $U_n^{SBS}(b_n = R, R, t) - U_n^{SBS}(b_n = \emptyset, R, t)$ is a decreasing function and it is larger than 0. Hence, choose to bid R can obtain a higher utility than biding $\emptyset$.

(2) Comparison between R and any value belongs to $[0, R)$. Assume that there is a value $\xi \in [0, R)$, we will analyze different cases, which are $b_{min}^{-n} \in (\xi, R)$ and $b_{min}^{-n} = R$. The concreted description will be presented in the following.

- $b_{min}^{-n} \in (\xi, R)$. If SBS edge cloud n choose to bid R, the utility is $\int_{t_1}^{t_2} r_n(t)dt$. Moreover, if SBS edge cloud n choose to bid $temp\_a$, the utility is $\int_{t_1}^{t_2} b_{min}^{-n}dt$. Since $b_{min}^{-n} < R \le r_n(t)$, the strategy to bid R is the optimal choice.

- $b_{min}^{-n} = R$. If SBS edge cloud n choose to bid R, the utility is belonged to $(R, r_n(t))$. Furthermore, if SBS edge cloud n choose to bid $temp\_a$, the utility is belonged to R.

- $b_{min}^{-n} = \xi$. If SBS edge cloud n choose to bid R, the utility is $\int_{t_1}^{t_2} r_n(t)dt$. Moreover, if SBS edge cloud n choose to bid $temp\_a$, the utility is $\int_{t_1}^{t_2} b_{min}^{-n}dt$. Since $b_{min}^{-n} < R \le r_n(t)$, the strategy to bid R is the optimal choice.

Therefore, when the other SBS edge clouds choose their strategies in (6.9), the optimal strategy for SBS edge cloud n is to adopt $b^*(r_n(t), R, t)$ in (6.9). Specifically, when $r_n(t) \in (R, \tilde{r}_x(R, t))$, the optimal bidding price for SBS edge cloud n should be the value of R. In the following cases, we will introduce the case when $r_n(t) = \tilde{r}_x(R, t)$, and the analysis is as same as **Case II**, the detailed proof will be neglected.

Case III $r_n(t) = \tilde{r}_x(R, t)$ Similar like the analysis in **Case II**, we can obtain that bidding R has the same utility with bidding $\emptyset$. Consequently, when the other SBS edge clouds choose their strategies in (6.9), the optimal strategy for SBS edge cloud n is to adopt $b^*(r_n(t), R, t)$ in (6.9). To be specific, when $r_n(t) = \tilde{r}_x(R, t)$, the optimal bidding price for SBS edge cloud n may be the value of R or not participating in this bidding.

Case IV $r_n(t) \in [\tilde{r}_x(R, t), r_{max}]$ Similar like the analysis in **Case II**, we will consider the two situations when bid R and bid $\emptyset$. We assume that the data offloading rate at SBS edge cloud n satisfies $r_n(t) \in [\tilde{r}_x(R, t), r_{max}]$. To be specific, we will analyze this case with the following situation.

Comparison between R and $\emptyset$. When bid R, the utility of SBS edge cloud n can be obtained as follows.

$$U_n^{SBS}(b_n = R, R, t)$$

$$= \int_{t_1}^{t_2} \left(1 - \left(\int_{\tilde{r}_x(R,t)}^{\infty} f(r_n(t)) d(r_n(t)) \right)^{N-1} \right) r_n(t) dt$$

$$+ \int_{t_1}^{t_2} \left(\int_{r_{\max}}^{\infty} f(r_n(t)) d(r_n(t)) \right)^{N-1} R dt + \Sigma_{n=1}^{N-1} \int_{t_1}^{t_2} C_{N-1}^n$$

$$\times \left(\int_{\tilde{r}_x(R,t)}^{r_{\max}} f(r_n(t)) d(r_n(t)) \right)^{n} \left(\int_{r_{\max}}^{\infty} f(r_n(t)) d(r_n(t)) \right)^{N-1-n}$$

$$\frac{R + n r_n(t)}{n+1} dt$$

$$= \int_{t_1}^{t_2} (1 - (1 - F(\tilde{r}_x(R, t)))^{N-1}) r_n(t) dt$$

$$+ \int_{t_1}^{t_2} (1 - F(r_{\max}))^{N-1} R dt + \Sigma_{n=1}^{N-1} \int_{t_1}^{t_2} C_{N-1}^n (F(r_{\max}) - F(\tilde{r}_x(R, t)))^n$$

$$\times (1 - F(r_{\max}))^{N-1-n} \frac{R + n r_n(t)}{n+1} dt.$$

(6.13)

When bid Ø, the utility of SBS edge cloud n can be obtained as follows.

$$U_n^{SBS}(b_n = \emptyset, R, t)$$

$$= \int_{t_1}^{t_2} \left(1 - \left(\int_{\tilde{r}_x(R,t)}^{\infty} f(r_n(t)) d(r_n(t)) \right)^{N-1} \right) r_n(t) dt$$

$$+ \int_{t_1}^{t_2} \left(\int_{r_{\max}}^{\infty} f(r_n(t)) d(r_n(t)) \right)^{N-1} \frac{N - 1 + \sigma^{SBS}}{N} r_n(t) dt$$

$$+ \Sigma_{n=1}^{N-1} \int_{t_1}^{t_2} C_{N-1}^n \left(\int_{\tilde{r}_x(R,t)}^{r_{\max}} f(r_n(t)) d(r_n(t)) \right)^n$$

$$\times \left(\int_{r_{\max}}^{\infty} f(r_n(t)) d(r_n(t)) \right)^{N-1-n} r_n(t) dt$$

$$= \int_{t_1}^{t_2} (1 - (1 - F(\tilde{r}_x(R, t)))^{N-1}) r_n(t) dt$$

$$+ \int_{t_1}^{t_2} (1 - F(r_{\max}))^{N-1} \frac{N - 1 + \sigma^{SBS}}{N} r_n(t) dt + \Sigma_{n=1}^{N-1} \int_{t_1}^{t_2} C_{N-1}^n (F(r_{\max})$$

$$- F(\tilde{r}_x(R, t)))^n (1 - F(r_{\max}))^{N-1-n} r_n(t) dt.$$

(6.14)

Then we can conclude that

$$U_n^{SBS}(b_n = R, R, t) - U_n^{SBS}(b_n = \emptyset, R, t)$$

$$= \int_{t_1}^{t_2} (1 - F(r_{\max}))^{N-1} \left(R - \frac{N - 1 + \sigma^{SBS}}{N} r_n(t) \right) dt$$

$$+ \Sigma_{n=1}^{N-1} \int_{t_1}^{t_2} C_{N-1}^n (F(r_{\max}) - F(\tilde{r}_x(R, t)))^n$$

$$\times (1 - F(r_{\max}))^{N-1-n} \frac{R - r_n(t)}{n + 1} dt.$$

(6.15)

According to (6.15), we can obtain that $U_n^{SBS}(b_n = R, R, t) < U_n^{SBS}(b_n = \emptyset, R, t)$. Therefore, choose to bid $\emptyset$ will obtain a higher utility for SBS edge cloud n.

Relying on the four cases above, when the other SBS edge clouds choose their strategies in (6.9), the optimal strategy for SBS edge cloud n is to adopt $b^*(r_n(t), R, t)$ in (6.9). To conclude, we have provided the concrete proof of Theorem 6.1.

6.1.3.3 Equilibrium for $R \in [0, \frac{N-1+\sigma^{SBS}}{N} r_{\min}]$

Second, we analyze that when the offloading rate $R \in [0, \frac{N-1+\sigma^{SBS}}{N} r_{\min}]$, the optimal bidding strategy is provided in the following theorem.

Theorem 6.2 *In the case of $R \in [0, \frac{N-1+\sigma^{SBS}}{N} r_{\min}]$, the optimal strategy for SBS edge cloud n is to adopt $b^*(r_n(t), R, t) = \emptyset$. In addition, when $R = \frac{N-1+\sigma^{SBS}}{N} r_{\min}$, the optimal strategy form is presented as follows.*

$$b^*(r_n(t), R, t)$$

$$= \begin{cases} any \ value \in [0, R] \ or \ \emptyset, & r_n(t) = r_{\min}; \\ \emptyset, & r_n(t) \in (r_{\min}, r_{\max}]. \end{cases}$$

(6.16)

To conclude, when the other SBS edge clouds choose their strategies in (6.16), the optimal strategy for SBS edge cloud n is to adopt $b^*(r_n(t), R, t)$ in (6.16). To be specific, when $r_n(t) = r_{\min}$, the optimal strategy for SBS edge cloud n is to choose any value in the range $[0, R]$ or not participate in this bid. In the case of $r_n(t) \in (r_{\min}, r_{\max}]$, it is the optimal strategy not to participate in this bid.

6.1.3.4 Equilibrium for $R \in (\frac{N-1+\sigma^{SBS}}{N} r_{\min}, r_{\min})$

Third, we discuss the optimal strategy for SBS edge cloud n when $R \in (\frac{N-1+\sigma^{SBS}}{N} r_{\min}, r_{\min})$. Then we introduce Lemma 6.2 for the following analysis.

Lemma 6.2 *There exists at least one solution $r_n(t) \in (r_{\min}, r_{\max})$ satisfying the following equation.*

$$\Sigma_{n=1}^{N-1} \int_{t_1}^{t_2} C_{N-1}^n F^n(r(t))(1-F(r(t)))^{N-1-n} \frac{R-r(t)}{n+1}$$

$$dt + \int_{t_1}^{t_2} (1-F(r(t)))^{N-1} \left(R - \frac{N-1+\sigma^{SBS}}{N} r(t) \right) dt \qquad (6.17)$$

$$= 0,$$

where $F(r_n(t))$ is the cumulative distribution function of random variable $r_n(t)$. Moreover, the solutions are denoted as $\tilde{r}_1(R,t), \tilde{r}_2(R,t), \ldots, \tilde{r}_L(R,t)$, where $L = \{1, 2, \ldots, L_{\max}\}$ which represents the number of solutions and $L_{\max}$ is the maximum number of solutions.

Proof The function of left hand side of equation is defined as $\mathscr{R}(r(t))$

$$\mathscr{R}(r(t))$$

$$\triangleq \Sigma_{n=1}^{N-1} \int_{t_1}^{t_2} C_{N-1}^n F^n(r(t))(1 - F(r(t)))^{N-1-n} \frac{R-r(t)}{n+1} dt \qquad (6.18)$$

$$+ \int_{t_1}^{t_2} (1 - F(r(t)))^{N-1} \left(R - \frac{N-1+\sigma^{SBS}}{N} r(t) \right) dt,$$

where function $\mathscr{R}(r(t))$ is continuous for $r_n(t)$ in $(r_{\min}, r_{\max})$. In particular, $F(r_{\min}) = 0$ for $r_n(t) \in (-\infty, r_{\min}]$, and $F(r_{\max}) = 1$ for $r_n(t) \in [r_{\max}, +\infty)$. Then we can obtain that

$$\mathscr{R}(r_{\min})$$

$$= \Sigma_{n=1}^{N-1} \int_{t_1}^{t_2} C_{N-1}^n F^n(r_{\min})(1 - F(r_{\min}))^{N-1-n} \frac{R-r(t)}{n+1} dt$$

$$+ \int_{t_1}^{t_2} (1-F(r_{\min}))^{N-1} \left(R - \frac{N-1+\sigma^{SBS}}{N} r_{\min} \right) dt \qquad (6.19)$$

$$\times \int_{t_1}^{t_2} \left(R - \frac{N-1+\sigma^{SBS}}{N} r_{\min} \right) dt.$$

And since $F(r_{\max}) = 1$, we can obtain that

$$\mathcal{R}(r_{\max})$$

$$= \Sigma_{n=1}^{N-1} \int_{t_1}^{t_2} C_{N-1}^n F^n(r_{\max})(1 - F(r_{\max}))^{N-1-n} \frac{R - r(t)}{n+1} dt$$

$$+ \int_{t_1}^{t_2} (1 - F(r_{\max}))^{N-1} \left(R - \frac{N-1+\sigma^{SBS}}{N} r_{\max} \right) dt \qquad (6.20)$$

$$\times \int_{t_1}^{t_2} \frac{R - r_{\max}}{N} dt.$$

Since $\frac{N-1+\sigma^{SBS}}{N} r_{\min} < R < r_{\min} < r_{\max}$, we can conclude that $\mathcal{R}(r_{\max}) > 0$ and $\mathcal{R}(r_{\max}) < 0$. Based on the intermediate value theorem, there is at least one solution $r(t)$ in $(\frac{N-1+\sigma^{SBS}}{N} r_{\min}, r_{\min})$ satisfying (6.17). The proof is completed.

To be specific, the proof of Lemma 6.2 is provided in the above. Based on Lemma 6.2, we can obtain the following theorem.

In the following we will give an analysis when $R \in (\frac{N-1+\sigma^{SBS}}{N} r_{\min}, r_{\min})$, first an optimal bidding strategy will be presented in Theorem 6.3, then its detailed proof will be described.

Theorem 6.3 *In the case of* $R \in (\frac{N-1+\sigma^{SBS}}{N} r_{\min}, r_{\min})$, *assuming that there is a* $\tilde{r}_y(R, t) \in (R, r_{\max})$ *subject to* $\{\tilde{r}_1(R, t), \tilde{r}_2(R, t), \ldots, \tilde{r}_L(R, t)\}$, *we can obtain the following bidding strategies* $b^*(r_n(t), R, t)$ *constitute the SBNE for SBS edge cloud* n,

$$b^*(r_n(t), R, t)$$

$$= \begin{cases} R, & r_n(t) \in [r_{\min}, \tilde{r}_y(R, t)]; \\ R \text{ or } \emptyset, & r_n(t) = \tilde{r}_y(R, t); \\ \emptyset, & r_n(t) \in (\tilde{r}_y(R, t), r_{\max}]. \end{cases} \qquad (6.21)$$

Similar to the analysis of Theorem 6.1, Theorem 6.3 presents that when the other SBS edge clouds choose their strategies in (6.21), the optimal strategy for SBS edge cloud n is to adopt $b^*(r_n(t), R, t)$ in (6.21). To be specific, the proof of Theorem 6.3 is provided in the following.

Proof

Case I $r_n(t) \in [r_{\min}, \tilde{r}_y(R, t)]$ We assume that the data offloading rate at SBS edge cloud n satisfies $r_n(t) \in [R, \tilde{r}_x(R, t)]$. To be specific, we will analyze this case with the following situation.

Comparison between R and $\emptyset$. When bid R, the utility of SBS edge cloud n can be obtained as follows.

$$U_n^{SBS}(b_n = R, R, t)$$

$$= \int_{t_1}^{t_2} \left(1 - \left(\int_{r_{\min}}^{\infty} f(r_n(t)) d(r_n(t)) \right)^{N-1} \right) r_n(t) dt$$

$$+ \int_{t_1}^{t_2} \left(\int_{\tilde{r}_y(R,t)}^{\infty} f(r_n(t)) d(r_n(t)) \right)^{N-1} R dt$$

$$+ \Sigma_{n=1}^{N-1} \int_{t_1}^{t_2} C_{N-1}^n \left(\int_{r_{\min}}^{\tilde{r}_y(R,t)} f(r_n(t)) d(r_n(t)) \right)^{n}$$

$$\times \left(\int_{\tilde{r}_y(R,t)}^{\infty} f(r_n(t)) d(r_n(t)) \right)^{N-1-n} \tag{6.21}$$

$$\frac{R + n r_n(t)}{n+1} dt$$

$$= \int_{t_1}^{t_2} (1 - (1 - F(r_{\min}))^{N-1}) r_n(t) dt + \int_{t_1}^{t_2} (1 - F(\tilde{r}_y(R,t)))^{N-1} R dt$$

$$+ \Sigma_{n=1}^{N-1} \int_{t_1}^{t_2} C_{N-1}^n (F(\tilde{r}_y(R,t)) - F(r_{\min}))^n$$

$$\times (1 - F(\tilde{r}_y(R,t)))^{N-1-n} \frac{R + n r_n(t)}{n+1} dt.$$

When bid Ø, the utility of SBS edge cloud n can be obtained as follows:

$$U_n^{SBS}(b_n = \emptyset, R, t)$$

$$= \int_{t_1}^{t_2} \left(1 - \left(\int_{r_{\min}}^{\infty} f(r_n(t)) d(r_n(t)) \right)^{N-1} \right) r_n(t) dt$$

$$+ \int_{t_1}^{t_2} \left(\int_{\tilde{r}_y(R,t)}^{\infty} f(r_n(t)) d(r_n(t)) \right)^{N-1} \frac{N - 1 + \sigma^{SBS}}{N} r_n(t) dt$$

$$+ \Sigma_{n=1}^{N-1} \int_{t_1}^{t_2} C_{N-1}^n \left(\int_{r_{\min}}^{\tilde{r}_y(R,t)} f(r_n(t)) d(r_n(t)) \right)^{n}$$

$$\times \left(\int_{\tilde{r}_y(R,t)}^{\infty} f(r_n(t)) d(r_n(t)) \right)^{N-1-n} r_n(t) dt$$

$$= \int_{t_1}^{t_2} (1-(1-F(r_{\min}))^{N-1}) r_n(t) dt + \int_{t_1}^{t_2} (1-F(\tilde{r}_y(R,t)))^{N-1}$$

$$\times \frac{N-1+\sigma^{SBS}}{N} r_n(t) dt + \Sigma_{n=1}^{N-1} \int_{t_1}^{t_2} C_{N-1}^n (F(\tilde{r}_y(R,t))) \qquad (6.21)$$

$$- F(r_{\min}))^n (1 - F(\tilde{r}_y(R,t)))^{N-1-n} r_n(t) dt.$$

Then we can conclude that

$$U_n^{SBS}(b_n = R, R, t) - U_n^{SBS}(b_n = \emptyset, R, t)$$

$$= \int_{t_1}^{t_2} (1 - F(\tilde{r}_x(R,t)))^{N-1} \left(R - \frac{N-1+\sigma^{SBS}}{N} r_n(t) \right) dt$$

$$+ \Sigma_{n=1}^{N-1} \int_{t_1}^{t_2} C_{N-1}^n (F(\tilde{r}_y(R,t)) - F(r_{\min}))^n (1 - F(\tilde{r}_y(R,t)))^{N-1-n}$$

$$\times \frac{R - r_n(t)}{n+1} dt.$$

$$(6.21)$$

It is simple to realize that $U_n^{SBS}(b_n = R, R, t) - U_n^{SBS}(b_n = \emptyset, R, t)$ is a decreasing function and it is larger than 0. Hence, choose to bid R can obtain a higher utility than biding $\emptyset$.

Case II $r_n(t) = \tilde{r}_y(R, t)$ Similar like the analysis in **Case I**, we can obtain that bidding R has the same utility with bidding $\emptyset$. Consequently, when the other SBS edge clouds choose their strategies in (6.21), the optimal strategy for SBS edge cloud n is to adopt $b^*(r_n(t), R, t)$ in (6.21).

Case III $r_n(t) \in [\tilde{r}_y(R, t), r_{\max}]$ Similar like the analysis in **Case I**, we will consider the two situations when bid R and bid $\emptyset$. We assume that the data offloading rate at SBS edge cloud n satisfies $r_n(t) \in [\tilde{r}_y(R, t), r_{\max}]$. To be specific, we will analyze this case with the following situation.

Case IV $r_n(t) > r_{\max}$ Similar like the analysis in **Case I**, we will consider the two situations when bid R and bid $\emptyset$. We assume that the data offloading rate at SBS edge cloud n satisfies $r_n(t) \in [\tilde{r}_y(R, t), r_{\max}]$. To be specific, we will analyze this case with the following situation.

Comparison between R and $\emptyset$. When bid R, the utility of SBS edge cloud n can be obtained as follows:

$$U_n^{SBS}(b_n = R, R, t)$$

$$= \int_{t_1}^{t_2} \left(1 - \left(\int_{\tilde{r}_y(R,t)}^{\infty} f(r_n(t)) d(r_n(t)) \right)^{N-1} \right) r_n(t) dt$$

$$+ \int_{t_1}^{t_2} \left(\int_{r_{\max}}^{\infty} f(r_n(t)) d(r_n(t)) \right)^{N-1} R dt + \Sigma_{n=1}^{N-1} \int_{t_1}^{t_2} C_{N-1}^n$$

$$\times \left(\int_{\tilde{r}_y(R,t)}^{r_{\max}} f(r_n(t)) d(r_n(t)) \right)^n \left(\int_{r_{\max}}^{\infty} f(r_n(t)) d(r_n(t)) \right)^{N-1-n}$$

$$\frac{R + n r_n(t)}{n+1} dt$$

$$= \int_{t_1}^{t_2} (1 - (1 - F(\tilde{r}_y(R,t)))^{N-1}) r_n(t) dt + \int_{t_1}^{t_2} (1 - F(r_{\max}))^{N-1} R \, dt \qquad (6.22)$$

$$+ \Sigma_{n=1}^{N-1} \int_{t_1}^{t_2} C_{N-1}^n (F(r_{\max}) - F(\tilde{r}_y(R,t)))^n$$

$$\times (1 - F(r_{\max}))^{N-1-n} \frac{R + n r_n(t)}{n+1} dt.$$

When bid Ø, the utility of SBS edge cloud n can be obtained as follows:

$$U_n^{SBS}(b_n = \text{Ø}, R, t)$$

$$= \int_{t_1}^{t_2} \left(1 - \left(\int_{\tilde{r}_y(R,t)}^{\infty} f(r_n(t)) d(r_n(t)) \right)^{N-1} \right) r_n(t) dt$$

$$+ \int_{t_1}^{t_2} \left(\int_{r_{\max}}^{\infty} f(r_n(t)) d(r_n(t)) \right)^{N-1} \frac{N - 1 + \sigma^{SBS}}{N} r_n(t) dt$$

$$+ \Sigma_{n=1}^{N-1} \int_{t_1}^{t_2} C_{N-1}^n \left(\int_{\tilde{r}_y(R,t)}^{r_{\max}} f(r_n(t)) d(r_n(t)) \right)^n$$

$$\times \left(\int_{r_{\max}}^{\infty} f(r_n(t)) d(r_n(t)) \right)^{N-1-n} r_n(t) dt$$

$$= \int_{t_1}^{t_2} (1 - (1 - F(\tilde{r}_y(R,t)))^{N-1}) r_n(t) dt$$

$$+ \int_{t_1}^{t_2} (1 - F(r_{\max}))^{N-1} \frac{N - 1 + \sigma^{SBS}}{N} r_n(t) dt + \Sigma_{n=1}^{N-1} \int_{t_1}^{t_2} C_{N-1}^n (F(r_{\max})$$

$$- F(\tilde{r}_y(R,t)))^n (1 - F(r_{\max}))^{N-1-n} r_n(t) dt.$$

$$(6.23)$$

Then we can conclude that

$$U_n^{SBS}(b_n = R, R, t) - U_n^{SBS}(b_n = \text{Ø}, R, t)$$

$$= \int_{t_1}^{t_2} (1 - F(r_{\max}))^{N-1} \left(R - \frac{N - 1 + \sigma^{SBS}}{N} r_n(t) \right) dt$$

$$+ \Sigma_{n=1}^{N-1} \int_{t_1}^{t_2} C_{N-1}^n (F(r_{\max}) - F(\tilde{r}_y(R,t)))^n$$

$$\times (1 - F(r_{\max}))^{N-1-n} \frac{R - r_n(t)}{n+1} dt. \tag{6.24}$$

According to (6.24), we can obtain that $U_n^{SBS}(b_n = R, R, t) < U_n^{SBS}(b_n = \emptyset, R, t)$. Therefore, choose to bid $\emptyset$ will obtain a higher utility for SBS edge cloud n. This completes the proof.

6.1.3.5 Equilibrium for $R \in [r_{\max}, +\infty)$

Ultimately, we analyze the case of offloading rate $R \in (r_{\max}, +\infty)$, and the form of SBNE is presented in Theorem 6.4.

Theorem 6.4 *When the offloading rate $R \in (r_{\max}, +\infty)$, the optimal bidding strategy $b^*(r_n(t), R, t)$ for SBS edge cloud n is provided as*

$$b^*(r_n(t), R, t)$$

$$= \begin{cases} any \ value \ in \ [0, r_{\min}], & r_n(t) = r_{\min}; \\ r_n(t), & r_n(t) \in [r_{\min}, r_{\max}]; \\ any \ value \ in \ [r_{\min}, R] \ or \ \emptyset, & r_n(t) = r_{\max}. \end{cases} \tag{6.25}$$

As a result, when the other SBS edge clouds choose their strategies in (6.25), the optimal strategy for SBS edge cloud n is to adopt $b^*(r_n(t), R, t)$ in (6.25). When $r_n(t) = r_{\min}$, the optimal price strategy for SBS edge cloud is any value in $[0, r_{\min}]$. When $r_n(t) \in [r_{\min}, r_{\max}]$, the optimal price strategy is $r_n(t)$. Furthermore, when $r_n(t) = r_{\max}$, the optimal price strategy for SBS edge cloud is any value in $r_{\min}, R$ or giving up bidding.

6.1.4 MBS Edge Cloud's Expected Utility Analysis

In this section, we will investigate the optimal expected utility of the MBS edge cloud based on the above analysis in Sect. 6.1.2. In addition, we assume that there is a unique solution in (6.5) and (6.17), respectively.

In the following we will prove (6.5) has one solution when $N = 2$. The cumulative distribution function is $F(r(t)) = \frac{r(t) - r_{\min}}{r_{\max} - r_{\min}}$, where $r(t) \in [r_{\min}, r_{\max}]$. Relying on the expression of (6.5) and the number of the SBS edge cloud is 2, then the equation can be

$$\frac{r(t) - R}{r_{\max} - r_{\min}} \frac{R - r(t)}{2} + \frac{r_{\max} - r}{r_{\max} - r_{\min}} \left(R - \frac{1 - \sigma^{SBS}}{2} r(t) \right) = 0. \tag{6.26}$$

After transformation of equation, we obtain that

$$\frac{\sigma^{SBS}}{2}r(t)^2-\left(\frac{1+\sigma^{SBS}}{2}\right)r_{\max}r(t)+r_{\max}R-\frac{R^2}{2}=0. \tag{6.27}$$

Then the left side of equation above can be defined as

$$H(r(t)) \triangleq \frac{\sigma^{SBS}}{2}r(t)^2-\left(\frac{1+\sigma^{SBS}}{2}\right)r_{\max}r(t)+r_{\max}R$$
$$-\frac{R^2}{2}. \tag{6.28}$$

Because this function is a quadratic, the derivative function of the equation above can be shown as

$$d(H(r(t)))/d(r(t))=\sigma^{SBS}r(t)-\left(\frac{1+\sigma^{SBS}}{2}\right)r_{\max}. \tag{6.29}$$

Let the derivative function above be equal to zero, we can obtain that in this case the solution of $r_0(t)$ can be

$$r_0(t) = \left(\frac{1+\sigma^{SBS}}{2\sigma^{SBS}}\right)r_{\max}. \tag{6.30}$$

When the solution $r(t) < r_0(t)$, i.e., the derivative function < 0, the quadratic function $H(r(t)))$ is a decreasing function. We will analyze the range of $r(t)$ as follows.

It is obvious that when $r(t) \in [r_{\min}, \frac{1+\sigma^{SBS}}{2\sigma^{SBS}}r_{\max})$, the quadratic function $H(r(t)))$ decreases with $r(t)$. Since $\frac{1+\sigma^{SBS}}{2\sigma^{SBS}} > \frac{\sigma^{SBS}+\sigma^{SBS}}{2\sigma^{SBS}} > 1$, we can obtain that $r_{\min} < r(t) < r_{\max} < \frac{1+\sigma^{SBS}}{2\sigma^{SBS}}r_{\max}$. Then it can be concluded that $H(r(t)))$ decreases with $r(t)$ in its range.

Moreover, when $r(t) = R$, $H(R) = \frac{1-\sigma^{SBS}}{2}(r_{\max} - R)R > 0$, and $r(t) = r_{\max}$, $H(r_{\max})) = -\frac{1}{2}(r_{\max} - R)^2 < 0$. Based on the above analysis, (6.5) has only one solution when $N = 2$. In addition, when $N > 2$, we will give the function curve in Sect. 6.1.5.1 to verify its uniqueness.

Remarks The uniqueness of solution for (6.17) is proved like (6.5), therefore the proof is omitted in detail.

For now, we have obtained the uniqueness of solution for (6.5) and (6.17), then in the following subsections the expected compensation $r_{compensation}$ and the expected utility of the MBS edge cloud $\int_{t_1}^{t_2} U^{MBS}(b, R, t)dt$ in different cases of R can be formulated.

6.1.4.1 Definition of MBS Edge Cloud's Expected Utility

Definition 6.4 First, we define the MBS edge cloud's expected utility as

$$U^{MBS}(b, R, t) \triangleq E\{U^{SBS}(b^*(r_1, R, t), b^*(r_2, R, t), \ldots,$$
$$b^*(r_N, R, t)), R\}, \tag{6.31}$$

where $b^*(r_n, R, t), n \in N$ represents the optimal bidding strategy for each SBS edge cloud under the offloading rate R. Based on the different intervals of offloading rate R, the MBS edge cloud's expected utility has variant forms.

6.1.4.2 MBS Edge Cloud's Optimal Expected Utility

The MBS edge cloud's optimal offloading rate should satisfy

$$\max \int_{t_1}^{t_2} U^{MBS}(b, R, t) dt;$$
$$s.t.\ b_{\max}(R) \leq r_{MBS}; \tag{6.32}$$
$$t \in t_1, t_2.$$

Case I $R \in [r_{\min}, r_{\max})$ Then we compute the probability distribution of $b_{\min}^{-k}$. The cumulative distribution function of $b_{\min}^{-k}$ is denoted as $H(\cdot)$ and it can be computed as

$$H(\cdot) = 1 - (1 - F(r_n(t)))^{N-1}, r_n(t) \in [r_{\min}, r_{\max}]. \tag{6.33}$$

Therefore, the probability distribution function of $b_{\min}^{-k}$ can be computed as

$$h(r_n(t)) = \frac{dH(\cdot)}{dr(t)} = (N - 1)f(r_n(t))(1 - F(r_n(t)))^{N-2},$$
$$r_n(t) \in [r_{\min}, r_{\max}]. \tag{6.34}$$

Then the expected compensation received by SBS edge cloud n. Specifically, SBS edge cloud n is capable of winning the auction under the following three cases:

(1) $r_n(t) \in [r_{\min}, R)$ and $b_{\min}^{-k} \in [r_n(t), R)$. In this case, SBS edge cloud n is capable of receiving $b_{\min}^{-k}$ from the MBS edge cloud.

(2) $r_n(t) \in [r_{\min}, R)$ and $b_{\min}^{-k} = R$ or $\emptyset$. In this case, SBS edge cloud n is capable of receiving R from the MBS edge cloud.

(3) $r_n(t) \in [\tilde{r}_x(t), R)$ and $b_{\min}^{-k} = R$ *or* $\emptyset$. In this case, SBS edge cloud n can receive the expected compensation from the MBS edge cloud depends on the number of the SBS edge clouds bidding R.

Relying on the analysis above, the expected compensation that SBS edge cloud n received is

$$
\begin{aligned}
r_{compensation} = &\int_{r_{\min}}^{R} r(t)g(r(t))F(r(t))dr(t) + RF(R)(1-G(R)) \\
&+ (F(\tilde{r}_x(R)) - F(R))\Sigma_{n=0}^{N-1}C_{N-1}^{n}(F(\tilde{r}_x(t)) \\
&- F(R))^{n}(1 - F(\tilde{r}_x(t)))^{N-1-n}\frac{R}{n+1}.
\end{aligned}
\tag{6.35}
$$

Furthermore, we can hold that

$$
\begin{aligned}
\frac{1}{N}&((1-F(R))^{N} - (1-F(\tilde{r}_x(t)))^{N}) \\
&= (F(\tilde{r}_x(R)) - F(R))\Sigma_{n=0}^{N-1}C_{N-1}^{n}(F(\tilde{r}_x(t)) - F(R))^{n} \\
&\times (1 - F(\tilde{r}_x(t)))^{N-1-n}\frac{R}{n+1}.
\end{aligned}
\tag{6.36}
$$

Relying on (6.33), (6.34), and (6.36), we can transfer (6.35) to the following equation:

$$
\begin{aligned}
r_{compensation} = &(N-1)\int_{r_{\min}}^{R} r(t)f(r(t))F(r(t))(1-F(r(t)))^{N-2}dr(t) \\
&+ RF(R)(1-F(R))^{N-1} + \frac{1}{N}R\left((1-F(R))^{N} \right. \\
&\left. -(1-F(\tilde{r}_x(t)))^{N}\right).
\end{aligned}
\tag{6.37}
$$

The maximal expected utility of the MBS edge cloud can be concluded as the minimal $r_{compensation}$. Then consider there is N the SBS edge clouds, the total expected compensation can be summarized as

$$
\begin{aligned}
\widetilde{r}_{compensation} = &N(N-1)\int_{r_{\min}}^{R} r(t)f(r(t))F(r(t))(1-F(r(t)))^{N-2}dr(t) \\
&+ NRF(R)(1-F(R))^{N-1} \\
&+ R\left((1-F(R))^{N} - (1-F(\tilde{r}_x(t)))^{N}\right).
\end{aligned}
\tag{6.38}
$$

Considering that the distribution of the SBS edge clouds offloading rate can be obtained by the MBS edge cloud, moreover, relying on the proof above, no matter how many numbers of the SBS edge clouds, the solution of (6.5) is only one. In addition, the utility of the MBS edge cloud is defined as (6.2). Then we can obtain the expected utility of the MBS edge cloud as

$$
E\left(\int_{t_1}^{t_2} U^{MBS}(b, R, t)dt\right) = \int_{t_1}^{t_2} \Bigg((F(\tilde{r}_x(R, t))^N \sigma^{MBS} R
$$

$$
+ (1 - F(\tilde{r}_x(R, t))^N)R - \tilde{r}_{compensation}\Bigg)dt.
$$

(6.39)

Case II $R \in [0, \frac{N-1+\sigma^{SBS}}{N}r_{\min}]$ In this case, the SBS edge clouds work with the MBS edge cloud in the competition mode, and expected utility of the MBS edge cloud is

$$
E\left(\int_{t_1}^{t_2} U^{MBS}(b, R, t)dt\right) = \int_{t_1}^{t_2} \sigma^{MBS} r_{MBS} dt.
$$

(6.40)

Case III $R \in (\frac{N-1+\sigma^{SBS}}{N}r_{\min}, r_{\min})$ In this case, the SBS edge clouds choose to bid R or $\emptyset$ with possibility, and expected utility of the MBS edge cloud is formulated as

$$
E\left(\int_{t_1}^{t_2} U^{MBS}(b, R, t)dt\right) = \int_{t_1}^{t_2} \Bigg((1 - F(\tilde{r}_y(R)))^N \sigma^{MBS} r_{MBS}
$$

(6.41)

$$
+ (1 - (1 - F(\tilde{r}_y(R)))^N)(r_{MBS} - R)\Bigg)dt.
$$

Case IV $R \in (r_{\max}, \infty)$ In this case, the SBS edge clouds choose to bid $[0, R]$ in the cooperation mode, and expected utility of the MBS edge cloud is formulated as

$$
E\left(\int_{t_1}^{t_2} U^{MBS}(b, R, t)dt\right) = \int_{t_1}^{t_2} \Bigg(r_{MBS} - N(N-1)\int_{r_{\min}}^{r_{\max}} r(t)f(r(t))F(r(t))
$$

$$
\times (1 - F(r(t)))^{N-2}d(r(t))\Bigg)dt.
$$

(6.42)

6.1.4.3 MBS Edge Cloud's Optimal Offloading Rate

The optimal expected utility of the MBS edge cloud is clarified in the following theorem, which satisfying the assumption in Sect. 6.1.2.

Theorem 6.5 *The optimal offloading rate of the MBS edge cloud is denoted as R^*, which has the following properties.*

Case I *If $r_{MBS} \leq \frac{N-1+\sigma^{SBS}}{N(1-\sigma^{MBS})}r_{\min}$, offloading rate R^* can be the value from range $[0, \frac{N-1+\sigma^{SBS}}{N}r_{\min}]$.*

Case II *If $\frac{N-1+\sigma^{SBS}}{N(1-\sigma^{MBS})}r_{\min} < r_{MBS} \leq r_{\max}$, offloading rate R^* can be any value from $(\frac{N-1+\sigma^{SBS}}{N}r_{\min}, r_{MBS}]$.*

Case III *If $r_{MBS} > MAX\{r_{\max}, \frac{N-1+\sigma^{SBS}}{N(1-\sigma^{MBS})}r_{\min}\}$, offloading rate R^* can be any value from $(\frac{N-1+\sigma^{SBS}}{N}r_{\min}, r_{\max}]$.*

In the case of $r_{MBS} \leq \frac{N-1+\sigma^{SBS}}{N(1-\sigma^{MBS})}r_{\min}$, the MBS edge cloud is not capable of providing enough service to the SBS edge cloud. To be specific, $r_{MBS} \leq \frac{N-1+\sigma^{SBS}}{N(1-\sigma^{MBS})}r_{\min} \Rightarrow (1 - \sigma^{MBS})r_{MBS} \leq \frac{N-1+\sigma^{SBS}}{N}r_{\min}$. Relying on the above analysis, $\frac{N-1+\sigma^{SBS}}{N}r_{\min}$ should be the lower bound of the computation offloading rate, which the SBS edge cloud can request from the MBS edge cloud. As a result, when $r_{MBS} \leq \frac{N-1+\sigma^{SBS}}{N(1-\sigma^{MBS})}r_{\min}$, the MBS edge cloud cannot meet the request from the SBS edge cloud under cooperation mode. Therefore, it chooses $R^* \in [0, \frac{N-1+\sigma^{SBS}}{N}r_{\min}]$ in the competition mode.

In the case of $\frac{N-1+\sigma^{SBS}}{N(1-\sigma^{MBS})}r_{\min} < r_{MBS} \leq r_{\max}$, the offloading rate capacity can hold the request from the SBS edge clouds. Therefore, the MBS edge cloud chooses $\frac{N-1+\sigma^{SBS}}{N}r_{\min}$ as the lowest bound of R^*. Moreover, the offloading rate should not be larger than R^*, otherwise, it does not satisfy the SBS edge cloud with the largest bidding value.

In the case of $r_{MBS} > MAX\{r_{\max}, \frac{N-1+\sigma^{SBS}}{N(1-\sigma^{MBS})}r_{\min}\}$, because the maximum bidding value from the SBS edge cloud is $r_{\max}$, the MBS edge cloud always provides enough service ability to meet the SBS edge cloud's request. Meanwhile, the offloading rate chooses R^* from the interval of $(\frac{N-1+\sigma^{SBS}}{N}r_{\min}, r_{\max}]$.

6.1.5 Experiments and Simulation Results

We will discuss the influence of parameters on the MBS edge cloud optimal offloading rate, the expected utility of the MBS edge cloud and the SBS edge cloud in this section. Specifically, we verify the effectiveness of our proposed scheme in SDN-based ultra dense networks. To be specific, we simulate our proposed scheme in Matlab 2013. We consider one MBS edge cloud and the number of the SBS edge clouds N is determined in the concrete experiment. The timeslot $t \in [0, 2]$ and the transmission rate $r_n(t)$ submit to the normal distribution with $\mathcal{N}(r_n(t))$, where the mean value is 125 Mbps and the standard deviation is set as 50 Mbps. In addition, $r_{\min} = 50$ and $r_{\max} = 200$. The discounting factor of the MBS edge cloud and the SBS edge cloud will be given in the following subsections.

6.1.5.1 Uniqueness of $\tilde{r}_x(t)$

In this subsection, we present the proof that the uniqueness of $\tilde{r}_x(t)$ for (6.5). It includes two part, the first experimental result gives the numerical curve of (6.5) at the case of different values of R. The second one presents the numerical curve of (6.5) for different values of N. Relying on the proof, we can conclude that the numerical curve decreases gradually and it has only one solution when this function is equal to zero.

From Fig. 6.3a, we choose $N = 3, \sigma^{MBS} = 0.3, \sigma^{SBS} = 0.8, t \in [0, 2]$ and $r_n(t)$ submits to the normal distribution with $\mathcal{N}(r_n(t)) \sim [125\,\text{Mbps}, 2500\,\text{Mbps}^2]$. Moreover, the computation offloading service rate of the MBS edge cloud R is set as $\{60, 80, 100\}$ Mbps, respectively, and it is plot with Fig. 6.3a. We can see that, with the changing values of R, there is an unique solution for (6.5). In other words, there is only one $\tilde{r}_x(t)$ for this equation at the different cases of R. To be specific, the increase of R results in a higher function value of (6.5), as well as the zero-point value of $r_x(t)$ increases with a larger R.

As shown in Fig. 6.3b, we choose $R = 60, \sigma^{MBS} = 0.3, \sigma^{SBS} = 0.8, t \in [0, 2]$ and $r_n(t)$ submits to the normal distribution with $\mathcal{N}(r_n(t)) \sim [125\,\text{Mbps}, 2500\,\text{Mbps}^2]$. In addition, the number of the SBS edge clouds N is denoted as $\{3, 4, 5\}$ respectively, and it is shown with Fig. 6.3b for (6.5) with different N. In other words, the number of the SBS edge clouds does not have any effect on the number of $\tilde{r}_x(t)$. Specifically, the zero-point values of three curves are in touching distance. The larger the numerical value of N, the faster the curve descends.

6.1.5.2 Impact on Offloading Rate R^*

In this subsection, we implement the experiment to verify the impact of different discounting factors for the SBS edge cloud and the MBS edge cloud. To be specific, we choose the number of the SBS edge clouds as 3, and the distribution of $r_n(t)$ is the same as Sect. 6.1.5.1. As shown in Fig. 6.4a, the discounting factor of the MBS edge cloud is set as 0.3. Moreover, the discounting factor of the SBS edge cloud is denoted as $\{0.1, 0.3, 0.5, 0.7\}$. For different pairs of discounting factors, the offloading rate for the MBS edge cloud R^* is increasing with the change of r_{MBS}.

From Fig. 6.4a, we can see that R^* is constant when r_{MBS} does not exceed $\frac{N-1+\sigma^{SBS}}{N}$. Afterwards, when r_{MBS} is above $\frac{N-1+\sigma^{SBS}}{N}$, R^* increases with r_{MBS}. Specifically, the higher the discount factor of the SBS edge cloud r_{MBS} is, the more R^* the MBS edge cloud is able to provide. It can prove that the difference of r_{MBS} has a great impact on R^*.

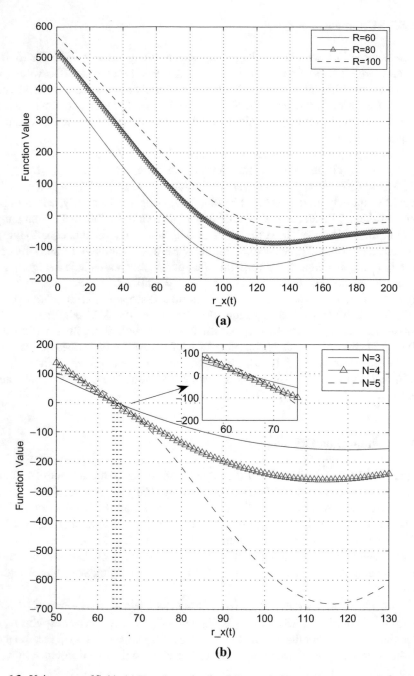

Fig. 6.3 Uniqueness of $\tilde{r}_x(t)$. (**a**) Function value for different R. (**b**) Function value for different N

Fig. 6.4 Performance
analysis. (**a**) Impact on
offloading rate. (**b**) Utility of
the MBS edge cloud. (**c**)
Giving up bidding rate of the
SBS edge cloud

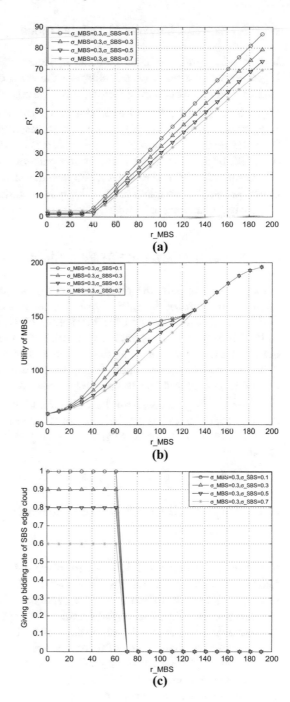

6.1.5.3   Expected Utility of MBS Edge Cloud

In this subsection, we investigate the expected utility in different parameters. In Fig. 6.4b, we plot the expected utility of the MBS edge cloud against r_{MBS} in the case of different discounting factors. First, we can obviously obtain that, at the first stage, the expected utility of MBS does not change with r_{MBS} is below $\frac{N-1+\sigma^{SBS}}{N}$. Specifically, in the latter stage, with an increase of r_{MBS} the expected utility of the MBS edge cloud increases rapidly. This is because that a higher r_{MBS} enables the MBS edge cloud to set a higher computation offloading rate, which results in a larger possibility of cooperation between the SBS edge cloud and the MBS edge cloud. Moreover, the increase of σ^{SBS} helps deteriorate the expected utility of the MBS edge cloud increases slightly.

6.1.5.4   Utility Analysis of SBS Edge Cloud

As shown in Fig. 6.4c, the giving up bidding rate of the SBS edge clouds against r_{MBS} is presented in the case of different parameters. We set the number of SBS as 10, the discount factor of the MBS edge cloud is 0.3, and the discount factor of the MBS edge cloud is chosen from $\{0.1, 0.3, 0.5, 0.7\}$. It is obviously obtained that, with the increase of r_{MBS} the giving up bidding rate of total SBS edge clouds decreases rapidly in the latter stage. In the initial stage, the SBS edge clouds choose to give up bidding because of the low offloading rate when r_{MBS} is below $\frac{N-1+\sigma^{SBS}}{N}$. A larger r_{MBS} helps the SBS edge clouds cooperate the bidding between the SBS edge clouds and the MBS edge cloud in a larger possibility. Furthermore, the increase of σ^{SBS} helps reduce the giving up bidding rate of total SBS edge clouds.

6.2   Edge Intelligence-Driven Offloading and Resource Allocation

Recently, the efforts and initiatives from standard bodies have started to conceptualize the sixth generation mobile networks (6G) [35] and 6G may become an unparalleled transformation to revolutionize the wireless communication systems. Furthermore, intelligent industrial Internet of things (IIoT) in 6G [36] has received considerable attention from both academic and industrial field. In the last decade, the industrial standards and infrastructures have evolved substantially due to the amalgamation of Internet of things (IoT) [37] paradigm with some industrial units and equipment. Sometimes, IIoT is also known as the "Internet of really important stuff, the objects, and machines that powers our life." There may be massive devices connected by the IoT at the end of 2020 [38]. Furthermore, the connected IIoT infrastructures (e.g., actuators, vehicles, and industrial controllers) generate a humongous amount of data that require real-time analysis and evaluations with heterogenous characteristics in terms of size and modes [39]. Specifically, IIoT connects different kinds of industrial assets in industrial environments to enable intelligent operations, such as industrial monitoring, automation, and intelligent control [40]. However, the proliferation of the number of IIoT devices and the ever-increasing computation-intensive applications including augmented reality (AR), real-time online gaming, and ultra-high-definition (UHD) pose great challenges on data processing, architecture rigidity, and resource allocation. To address the aforementioned challenges, it is important to analyze data. Consequently, joint task offloading and resource management have attracted the significant focus from IIoT systems [41, 42].

Generally, mobile cloud computing (MCC) provides a proper paradigm that wireless devices execute the computation offloading in the cloud server [43, 44]. Thus, some wireless devices choose to offload application tasks [45] to the remote cloud server to ameliorate computational velocity, spectrum efficiency, and energy efficiency, which can be utilized widely in the past decade. Nevertheless, due to remote distance between the cloud server and local wireless devices, the long propagation delay, limited channel capacity, and task queuing delay make it hard to process latency-critical and computation-intensive application tasks relying on centralized methods [46].

To address the complex problems, the concept of edge intelligence [47] is conceived for offering powerful computational processing and massive data acquisition at the edge networks. Specifically, the edge server is closer to wireless devices, and hence the offloading scheme for computing tasks can enormously decrease transmission delay and save backhaul bandwidth between cloud servers and wireless devices [48]. At the same time, artificial intelligence (AI) is a promising trend for extracting information from large-scale data and for making efficient resource scheduling strategies in complex environment. By integrating AI into edge networks, the radio networks with service and resource awareness can dynamically adapt to the resource orchestration, which can be viewed as a beneficial remedy for data processing and resource allocation issues [49] in complex IIoT environment. To elaborate a little further, multi-access edge computing (MEC) servers can help alleviate latency, reduce energy consumption, and guarantee the quality of experience (QoE).

6.2.1 System Model

Figure 6.5 shows the network model that consists of multiple IIoT devices (IIoTD) and one BS with the MEC server. In this network model, IIoTD can be defined by $\mathbb{N} = \{1, 2, \ldots N\}$. Besides, the MEC server and the BS are connected with a wired connection (e.g., optical fiber), in which transmission delay between them can be ignored significantly [51]. Each IIoTD has large numbers of application tasks to be processed locally or offloaded to the MEC server with BS. Without

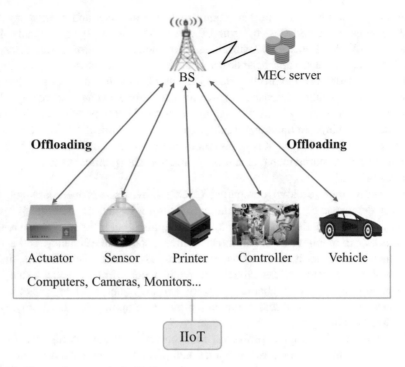

Fig. 6.5 The service scenario for IIoT

loss of generality, assuming that there are Z independent tasks, denoted by $\mathbb{Z} = \{1, 2, \ldots Z\}$, and the computation of each task could not be split for partial offloading or partial computing. That is to say, the task can only be computed in local IIoTD or offloaded to the MEC server but not both. We assume that $l_{n,z}$ is the task size including programming codes and general parameters, and $l_{n,z}$ is the zth task of the nth IIoTD. These parameters are related to features of the task and they can be estimated through task types. Each IIoTD n can choose whether to offload its own computation-intensive task z to the MEC server or not. We define the offloading decision vector A, which can be given by

$$A = [a_{1,1}, a_{1,2}, \ldots a_{n,z}, \ldots a_{N,Z}], \qquad (6.43)$$

where $n \in \{1, 2, \ldots N\}$ and $z \in \{1, 2, \ldots Z\}$ represent the IIoTD and task, respectively. $a_{n,z}$ represents the offloading decision and it belongs to $\{0, 1\}$. In detail, $a_{n,z} = 1$ represents the IIoTD n chooses to offload the task z to the MEC server, and $a_{n,z} = 0$ means that IIoTD n decides to carry out the task z locally. In this way, we can take advantage of parallel computing of IIoTD and MEC servers, which results in a decrease of total delay and energy consumption. We consider $B_{n,z}$ as the optimized wireless channel bandwidth of the zth task of the nth IIoTD. Due to the fact that there only exists one BS, so the interval interference between the BS could be overlooked[52]. Simultaneously, we consider that the assigned channel of application tasks from each agent is orthogonal to each other in the IIoT system model through orthogonal frequency division multiple access (OFDMA). As there is only one BS coverage area and nature characteristics of OFDM, we ignore the co-channel frequency and adjacent channel interference. In terms of Shannon's theory [53], the achievable uplink transmission rate for each task z of the nth IIoTD can be obtained by

$$c_{n,z} = B_{n,z} \log_2 \left(1 + \frac{P_{tran}^n |h_{n,z}|^2}{\sigma^2} \right), \tag{6.44}$$

where P_{tran}^n means the transmission power from the IIoTD, and $h_{n,z}$ represents the channel gain which follows Rayleigh flat fading under the allocated channel bandwidth. σ^2 is noise.

6.2.1.1 Computation Offloading Mode

The MEC server starts to process the task $l_{n,z}$ after it has fully received the IIoTD's task and feeds back information after the entire task z is computed [50]. Because the data size of the feedback message is small in general [54], the feedback energy consumption and delay can be neglected. Subsequently, we formulate transmission time and processing time. Specifically, for the task z of the IIoTD n, the transmission time caused by the uplink channel can be described as

$$T_{tran}^{n,z} = \frac{a_{n,z} l_{n,z}}{c_{n,z}}. \tag{6.45}$$

Similarly, the transmission energy consumption for the task z of the IIoTD n can be denoted by

$$E_{tran}^{n,z} = T_{tran}^{n,z} P_{tran}^n. \tag{6.46}$$

The computation time at the MEC server via the BS can be represented by

$$T_{pro}^{n,z} = \frac{a_{n,z} l_{n,z} e_{n,z}}{F_{server}^{total}}, \tag{6.47}$$

where $e_{n,z}$ represents the number of required CPU cycles, and F_{server}^{total} denotes the computational power. At the same time, we assume that MEC server is pretty powerful and can process all received application tasks concurrently. Furthermore, we model the computational processing energy consumption as the linear function $l_{n,z}$ [55], and it can be written as

$$E_{com}^{n,z} = \beta l_{n,z}, \tag{6.48}$$

where β can be defined as the task weight factor relative to the computational energy consumption from the MEC server and the unit of β is joule per bit. It depends on the application task size and diverse MEC servers. So the total delay can be denoted as

$$T_n^s = \sum_{z=1}^{Z} \left(T_{tran}^{n,z} + T_{pro}^{n,z} \right). \tag{6.49}$$

In addition, the total energy consumption at the computation offloading mode for each IIoTD n is formulated as

$$E_n^s = \sum_{z=1}^{Z} \left(E_{tran}^{n,z} + E_{com}^{n,z} \right). \tag{6.50}$$

6.2.1.2   Local Computing Mode

Next, we formulate the case that each IIoTD decides to execute its task locally. Specifically, the processing energy consumption b_n^l can be represented as

$$b_n^l = k \left(f_n^l \right)^2, \tag{6.51}$$

where f_n^l is the CPU cycle frequency for each IIoTD, and k is a constant interrelated to the hardware performance. So the local processing energy consumption for task z of each IIoTD n can be given by

$$E_{local}^{n,z} = \left(1 - a_{n,z} \right) l_{n,z} b_n^l. \tag{6.52}$$

At the same time, the local processing time can be defined by

$$T_{local}^{n,z} = \frac{l_{n,z} e_{n,z} (1 - a_{n,z})}{f_n^l}. \tag{6.53}$$

Thus, given the task offloading choice $a_{n,z}$, the total local processing delay for each IIoTD n can be denoted as

$$T_n^l = \sum_{z=1}^{Z} T_{local}^{n,z}. \tag{6.54}$$

Meanwhile, the total local processing energy consumption can be depicted as

$$E_n^l = \sum_{z=1}^{Z} E_{local}^{n,z}. \tag{6.55}$$

6.2.1.3   Problem Formulation

In this section, we formulate the joint offloading decision and wireless transmission rate allocation for IIoT system with MEC server as a multi-objective optimization problem. To minimize the total delay and energy consumption, the total cost function $V(L, A, C)$ can be defined as the weighted sum of the task delay and energy consumption, where $L = [l_{1,1}, l_{1,2}, \ldots l_{n,z}, \ldots l_{N,Z}]$ and $C = [c_{1,1}, c_{1,2}, \ldots c_{n,z}, \ldots c_{N,Z}]$. Hence, the total cost function can be denoted as

$$V(L, A, C) = \left\{ \lambda \sum_{n=1}^{N} (T_n^l + T_n^s) + \sum_{n=1}^{N} (E_n^l + E_n^s) \right\}, \tag{6.56}$$

where λ represents the weight on delay relative to total energy consumption. The unit of λ is joule per second and we can adjust λ to attach different importance to delay and energy consumption for various application tasks. In short, the optimization objective can be expressed as

$$\min_{\{L,A,C\}} V(L, A, C), \tag{6.57}$$

$$s.t. \quad a_{n,z} \in \{0, 1\}, \tag{6.58}$$

$$\sum_{n=1}^{N} \sum_{z=1}^{Z} c_{n,z} \le C_{total}, \tag{6.59}$$

$$c_{n,z} > 0. \tag{6.60}$$

Then, we minimize the total cost function $V(L, A, C)$ via choosing the optimal offloading decision vector A and allocating uplink wireless transmission rate vector C under different application tasks L. In fact, L is one state constant vector in the optimization problem and it is diverse in different time slots for network environments. Furthermore, we regard it as the input state vector in the optimization problem. At the same time, decision vector A and transmission rate vector C are considered as two optimized variables. Next, there exist some constraints about minimizing the total cost function $V(L, A, C)$. (16) means offloading decision belongs to 0 or 1, which represents the task is executed locally or the task is offloaded to the MEC server. As the total wireless channel bandwidth is limited for all IIoTD, (17) means that the sum of the achievable transmission rate allocated for each task z must not exceed the maximum C_{total}. Additionally, wireless transmission rate $c_{n,z}$ is closely related to the required channel bandwidth $B_{n,z}$, so optimizing the transmission rate $c_{n,z}$ is equivalent to solving the optimal channel bandwidth $B_{n,z}$. (18) indicates the achievable transmission rate should be positive since each IIoTD is supposed to access the BS with MEC server to guarantee the QoE. In the next section, we show an effective and efficient algorithm based on DRL to resolve this problem. The detailed parameter settings are described in Table 6.2.

Table 6.2 Summary of notations

Notation	Description
L	Application task vector
A	Offloading decision vector
C	Transmission rate vector
N	The number of IIoTD
Z	The number of independent application tasks for each IIoTD
$l_{n,z}$	The task size for zth task of nth IIoTD
$a_{n,z}$	The offloading decision for zth task of nth IIoTD
$B_{n,z}$	The optimized wireless channel bandwidth for zth task of nth IIoTD
P_{tran}^n	Transmission power from the IIoTD
$h_{n,z}$	Channel gain
σ^2	The variance of AWGN channel
$c_{n,z}$	Uplink transmission data rate for zth task of nth IIoTD
F_{server}^{total}	MEC server computational capacity
$e_{n,z}$	The number of cycles required to process each task bit
β	The task weight factor relative to computational energy consumption
k	A constant interrelated to the hardware performance
f_n^l	CPU cycle frequency for each IIoTD
λ	The delay weight factor relative to the total energy consumption
π	The DRL agent policy function
w	Neural network parameter
K	A distance parameter related to the activation function
M	The number of action aggregations
P	Incremental constant in optimal action aggregations

6.2.2 System Optimization

In this section, we firstly introduce a novel DRL-based framework to solve our proposed problems. Then we present the detailed process of how to generate required offloading policy through the DRL framework and give the 2AGT to approximate offloading action $a_{n,z}$. Next, after obtaining the initial offloading action, we transform our proposed initial problem into a convex objective problem and calculate the optimal offloading action by parallel computing in terms of substantial application tasks. Besides, we provide a 3AUS method for setting the number of action aggregations parameter. At the same time, we introduce a DRL network parameter update policy to strengthen the network stability and reduce the over-fitting. The detailed procedures are just shown as follows.

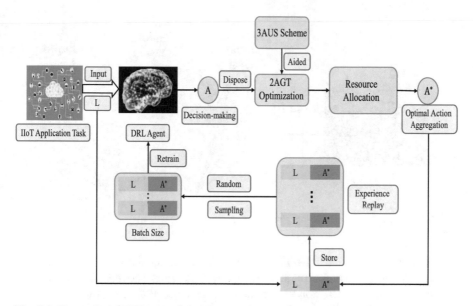

Fig. 6.6 The proposed DRL network structure

6.2.2.1 Deep Reinforcement Learning Framework

As shown in Fig. 6.6, the DRL agent brings the control strategy via interacting with the network environment (e.g., the application tasks) without a precise transition probability and adjusts own behavior depending on the outcomes of actions [56] in order to maximize the discounted reward functions. So DRL means a new exemplification through trial-and-error and delayed incentive mechanism to achieve an optimal behavior policy [57]. While obtaining the application tasks, our DRL network framework contains offloading decision generation, 2AGT optimization, and 3AUS parameters settings. At the same time, the network framework can compute the convex optimization problem in order to select current optimal action aggregation. Each detailed part of the novel DRL agent network framework structure can be illustrated as follows.

6.2.2.2 Offloading Policy Generation

(1) Offloading decision: For the proposed MINPP, our purpose is to generate the offloading action by the DRL agent interacting with the environment (i.e., the application tasks from IIoTD). Specifically, given DRL agent's initial policy function π, we input the application task L to DRL agent, and it can be defined as

$$\pi : \pi_w(A|L) . \tag{6.61}$$

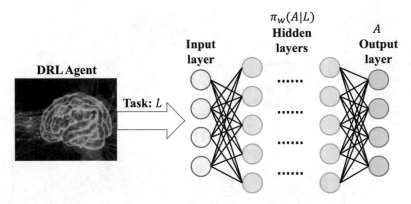

Fig. 6.7 The detailed agent internal structure

The detailed internal network structure of the DRL agent is presented in Fig. 6.7. In this agent network structure, the network structure between different hidden layers employs full connection and we can see that the offloading action depends on the policy function $\pi_w(A|L)$ which is implanted DNN parameter w, i.e., the weights that connect neural neurons between different network layers. In addition, the output layers generate the offloading decision A based on the current policy function $\pi_w(A|L)$. Next, we describe an isotone optimization method in terms of the generated offloading action.

(2) 2AGT optimization: Suppose that we obtain the application task L_s in sth step. The offloading decision A_s can be denoted as

$$A_s = \{0 \le a_{i,j}^s \le 1 | i \in [1, 2, \dots N], j \in [1, 2, \dots Z]\} . \tag{6.62}$$

Then we start to introduce an isotone action method. Inspired by the quantitative technology from signal coding [58], we transform the offloading decision A_s to massive action sets for the sake of obtaining optimal action A_s^*. The number of action aggregations is $1 \le M \le NZ + 1$. A_{s1} can be derived just as follows:

$$A_{s1} = \{a_{ij}^{s1}\} = \begin{cases} 1, & \text{if } a_{ij}^s \ge K, \\ 0, & \text{if } a_{ij}^s < K, \end{cases} \tag{6.63}$$

where K is equal to a discriminant value for offloading decision, and we set it as 0.5 in order to quantify the offloading decision equally. Subsequently, the agent generates other $M - 1$ offloading aggregations with respect to the distance parameter K and reshuffle the originally generated actions, which can be defined by $|a_{11}^s - K| \le |a_{12}^s - K| \dots \le |a_{NZ}^s - K|$. Then, the remaining $M - 1$ offloading aggregations are recalculated according to 2AGT, which can be denoted as two cases.

Case 1: When $a_{ij}^s > a_{m-1m-1}^s$ or $a_{ij}^s = a_{m-1m-1}^s$, $a_{m-1m-1} < K$, the quantitative action value a_{ij}^{sm} can be represented as

$$a_{ij}^{sm} = 1. \tag{6.64}$$

Case 2: When $a_{ij}^s < a_{m-1m-1}^s$ or $a_{ij}^s = a_{m-1m-1}^s$, $a_{m-1m-1} \geq K$, the quantitative action value a_{ij}^{sm} can be represented as

$$a_{ij}^{sm} = 0. \tag{6.65}$$

In terms of above two cases, the total quantitative action aggregations can be calculated as

$$A_{sm} = \{a_{ij}^{sm}\}, \tag{6.66}$$

where $m = 2, 3, 4 \ldots M$, and we can see that there are NZ offloading actions for all application tasks. In addition, we can generate at most $NZ + 1$ action aggregations. Next, by solving the convex optimization problem, the optimal offloading action A_{sm}^* can be denoted as:

$$A_{sm}^* = \underset{A_{sm}}{\arg\min} \, V^* \, (L_s, A_{sm}, C) \, . \tag{6.67}$$

In the next section, we discuss how to adjust the action aggregations parameter.

(3) 3AUS parameter setting: Intuitively, by setting more action aggregations M, a lower total cost function can be calculated followed with higher computational complexity. Instead, setting a proper M may reduce the potential computational complexity without losing the system performance. According to the rolling horizon control (RHC) theory [59], we can update the action aggregations parameter per δ steps. Specifically, when the step s is the integer times of the δ, the DRL agent can choose to renew the aggregations parameter. When $s=1$, RHC parameters are

$$M_s = NZ + 1. \tag{6.68}$$

When $s \bmod \delta = 0$, the update parameter is

$$M_s = \min(\max(m_{s-\delta+1}^*, m_{s-\delta+2}^*, \ldots m_{s-1}^*) + P, NZ + 1) \, , \tag{6.69}$$

where $m_{s-\delta+1}^*$ represents the index of optimal action aggregations. P is a constant in order to allow the number of aggregations to increase during the update period and if it does not reach the update steps δ for other steps, it can be the same as the previous value.

(4) Convex optimization function: According to the 2AGT and 3AUS schemes, we can transfer the initial problem into convex objective [60], as illustrated

in Fig. 6.14. After obtaining the value of offloading action aggregations, the original problem is

$$\min_{\{L,C\}} V(L, C),\tag{6.70}$$

$$s.t. \quad \sum_{n=1}^{N}\sum_{z=1}^{Z} c_{n,z} \leq C_{total},\tag{6.71}$$

$$c_{n,z} > 0.\tag{6.72}$$

Evidently, it is a convex optimization problem and we can solve the Karush–Kuhn–Tucker (KKT) conditions to obtain current optimal cost function.

Proof To demonstrate the optimization problem (6.72) as the convex problem, we need to prove the $V(L,\ C)$, constraint (6.73), and (6.74) as convex function, respectively. Firstly, as L is the time-varying state vector, so the function $V(L,\ C)$ is only related to the optimization variable C. Next, combined with (31) and (32), as $\sum_{n=1}^{N}\sum_{z=1}^{Z} c_{n,z} - C_{total}$ and $-c_{n,z}$ are affine functions, it must be the convex function. Additionally, after obtaining the offloading decision-making, the optimization objective $V(L,\ C)$ can be simply reformulated as

$$V(L,\ C) = \lambda \left(\sum_{n=1}^{N}\sum_{z=1}^{Z}\frac{a_{n,z}l_{n,z}}{c_{n,z}}\right) + \sum_{n=1}^{N}\sum_{z=1}^{Z}\frac{a_{n,z}l_{n,z}P_{tran}^{n}}{c_{n,z}}.\tag{6.73}$$

Lemma 6.3 *For two convex functions $f_1(x)$ and $f_2(x)$, the summation of $f_1(x)$ and $f_2(x)$ is still convex function.*

Proof For any convex function $f(x)$,

$$f(\lambda x_1 + (1-\lambda)x_2) \leq \lambda f(x_1) + (1-\lambda) f(x_2).\tag{6.74}$$

Let $g(x) = f_1(x) + f_2(x)$, where $f_1(x)$ and $f_2(x)$ are two convex functions. Hence,

$$g(\lambda x_1 + (1-\lambda)x_2) = f_1(\lambda x_1 + (1-\lambda)x_2)\tag{6.75}$$

$$+ f_2(\lambda x_1 + (1-\lambda)x_2)\tag{6.76}$$

$$\leq \lambda f_1(x_1) + (1-\lambda)f_1(x_2) + \lambda f_2(x_1) + (1-\lambda)f_2(x_2)\tag{6.77}$$

$$= \lambda(f_1(x_1) + f_2(x_1)) + (1-\lambda)(f_1(x_2) + f_2(x_2))\tag{6.78}$$

$$= \lambda g(x_1) + (1-\lambda)g(x_2),\tag{6.79}$$

which demonstrates that the summation of $f_1(x)$ and $f_2(x)$ is still convex function. As $a_{n,z}$, $l_{n,z}$ and P_{tran}^n are known for $V(L, C)$, $V(L, C)$ can be regarded as the summation of multiple convex functions for optimization variable $c_{n,z}$. Hence, $V(L, C)$ can be proved as a convex function in terms of Lemma 1. Finally, the formulated optimization problem is convex. Then, we choose an action aggregation A_{sm}^* from the optimal total cost function to start to update network parameters. In the next section, we show how to update the network parameters.

6.2.2.3   Network Parameters Update

After the agent obtains the optimal action aggregation A_{sm}^*, the agent can update the network parameters (i.e., the offloading policy $\pi_w(A|L)$). In detail, since the experience of the DRL agent is interrelated, randomly selecting a batch of training samples from replay memory can decrease the interrelation among agent experience and this may help the DRL agent utilize comprehensive experience in order to learn better. So we adopt the experience replay technology [61] to update the network parameters by using the stored data pairs (L_s, A_{sm}^*). Firstly, we keep an empty memory structure. Then the structure supplies new data pairs, and once the memory structure is full, the newly generated data pairs can displace the old. The DRL agent randomly selects several generated data pairs (L_s, A_{sm}^*) in sth step from memory structure to reduce the over-fitting, which can be characterized by total steps S_t. We define the cross-entropy just as follows:

$$O(w_s) = -\frac{1}{|S_t|} \sum_s [(A_{sm}^*)^T \log(\pi_{ws}(A_s|L_s))$$

$$+ (1 - A_{sm}^*)^T \log(1 - (\pi_{ws}(A_s|L_s)))] , \tag{6.80}$$

where $|S_t|$ represents the total number of sampling steps, and the superscript T means the transpose operator. In our simulations, we update our network parameters each ε while collecting enough new data pairs. Meanwhile, the DRL agent only updates from the most recent data pairs, which are produced by a new offloading strategy. The detailed algorithm procedure is described in Algorithm 6.1, where the computational complexity of the proposed algorithm can be derived as $O\left(SL + SMNZ + \frac{S}{K}\right)$.

6.2.3   Experiments and Simulation Results

6.2.3.1   Experimental Settings

In this section, the number of IIoTD can be denoted as $N = 10$ and there are $Z = 5$ tasks to be performed. Simultaneously, the channel bandwidth and

---

Algorithm 6.1 The proposed DRL algorithm

Input:
  Each task L_s;
  Initially the neural network parameter w;
Output:
  The agent outputs A_{sm}^*;
  The cost function $V(L, A, C)$;
 1: **for** $\{1, 2, \dots S\}$ **do**
 2:     Obtain the offloading decision A_s via 2AGT and DRL network structure;
 3:     Choose appropriate M_s in terms of 3AUS;
 4:     **if** $s \bmod \delta = 0$ **then**
 5:        Start to choose the new the number of action aggregations from 3AUS;
 6:     **end if**
 7:     Extend action sets A_s into $\{A_{s1}, A_{s2}, \dots A_{sm}\}$;
 8:     **for** $\{1, 2, \dots |M_s|\}$ **do**
 9:        Calculate $V(L_s, A_s, C)$ for all $\{A_{s1}, A_{s2}, \dots A_{sm}\}$;
10:        Select $A_{sm}^* = \underset{\{A_{s1}, A_{s2}, \dots A_{sm}\}}{\arg\min} \; V(L_s, A_s, C)$;
11:     **end for**
12:     Add the action pairs $\{L_s, A_{sm}^*\}$ into buffer pool;
13:     **if** $s \bmod \varepsilon = 0$ **then**
14:        Stochastically selecting K tuples $\{L_s, A_{sz}^*\}$ and update the DRL agent.
15:     **end if**
16: **end for**

---

transmission power are 100 Mbps and 0.2 W, respectively. The task size $l_{n,z}$ obeys the uniform distribution between (5 MB, 35 MB) and the CPU cycle frequency follows the uniform distribution between (0.6e8 cycle/s, 2.5e8 cycle/s) [62]. Further, $e_{n,z}$ is uniformly distributed between (1000 cycle/bit, 3000 cycle/bit) and F_{server}^{total} is 6.5e9 cycle/s. We set $\sigma^2 = 1 * 10^{-9}$. We utilize the PyCharm community edition as the programming environment for constructing DNN with TensorFlow, and the number of hidden layers is 3 where DNN uses full connection.

6.2.3.2 Convergent Performance Analysis

In this simulation, we input the application tasks L_s into the DRL agent in each step s, where the sample complexity includes 30,000 training samples and 10,000 test samples. After proper time intervals, the DRL agent is retrained again in order to improve its convergent performance. Finally, we obtain a quasi-optimal gain rate and total system cost subject to enumerating actions.

(1) The DRL cost function: As shown in Fig. 6.8a, when the number of training steps increases, the error function of predicted value and optimal value gradu-

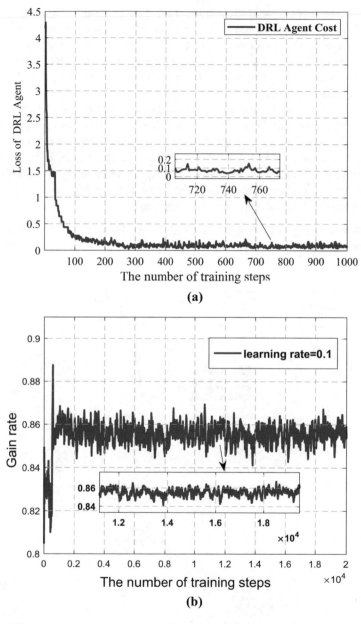

Fig. 6.8 DRL cost function and gain rate. (**a**) The loss function in terms of lr = 0.01. (**b**) The DRL gain rate with lr = 0.1

ally decreases to the minimum value. In fact, when in approximately 250 steps, the cost of the DRL agent is close to 0, which validates that our proposed algorithms have fast convergent speed. At the same time, while receiving different application tasks, the DRL agent has a stronger generalization ability and reduces over-fitting, which demonstrates the effectiveness of buffer pool and stochastically selecting.

(2) Gain Rate: As illustrated in Fig. 6.8b, we can see that the agent will not converge to optimal solutions despite enough training steps. In other words, once more than 250 steps, the agent cannot obtain the optimal offloading action and the gain rate is less than 0.9. It means that the system utility is lower. Further, we have to choose proper network parameters and achieve trade-off between performance and computational complexity. Next, we show the system gain rate with different network parameters settings.

As shown in Fig. 6.9a, we set some different learning rates to illustrate the relationship between gain rate and training steps. For better comparison, the contrast

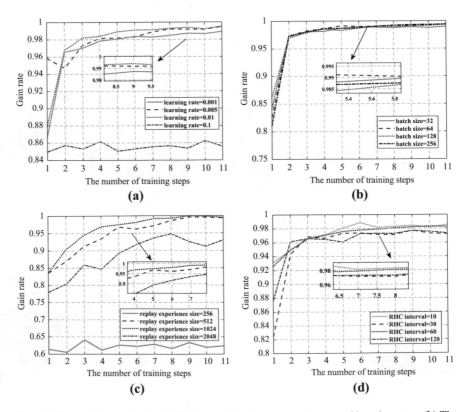

Fig. 6.9 Interrelation analysis. (**a**) The interrelation between gain rate and learning rates, (**b**) The interrelation between gain rate and batch sizes. (**c**) The interrelation between gain rate and replay experience sizes. (**d**) The interrelation between gain rate and RHC intervals

scheme is the optimal solution in terms of extensive search method. Generally, if the learning rate is higher, the convergent speed of the DRL agent can be faster. However, the figure shows that when the learning rate is higher, the gain rate cannot be optimal. If we set a higher learning rate, the DRL agent may obtain a local optimal policy rather than the global optimal. Accordingly, we must select a proper learning rate in terms of a specific network environment.

In Fig. 6.9b, we study the effect of different batch sizes on gain rate. In addition, we can see that a small batch size (e.g., size $= 32$) cannot utilize all data pairs in the memory size, which leads to slow convergent speed. However, if the selected batch size is large enough (e.g., size $= 256$), the agent can frequently use the old data pairs and may reduce the system performance. Hence, we must choose the proper batch size according to the environment states.

Figure 6.9c shows the interrelation between gain rate and different replay experience sizes. At the same time, we set the batch size as 128. We can see that the gain rate is close to 1 when the number of replay experience size is 512 and 1024. Further, as the number of replay experience size is 1024, its convergent speed is faster than others. However, the DRL agent cannot converge to optimal solutions once the replay experience size is 256, since the selected training data pairs are interrelated and lead to a local optimal solution. In addition, when the replay experience size is 2048, the convergent speed of the DRL agent is slow as it cannot fully utilize the collected data pairs to reduce the error loss. Hence, we are supposed to choose the proper replay experience size in terms of different application tasks.

6.2.3.3   3AUS Parameter Interval

In Fig. 6.9d, it shows the interrelation between gain rate and RHC intervals. If the update interval is proper, the aggregation parameter M_s can be renewed frequently, which means the DRL agent decreases its computational complexity with a small aggregation parameter. As the number of RHC interval increases, the gain rate is gradually descendent since the big RHC interval causes the higher computational complexity for the total cost function, which means that the DRL agent must renew the number of action sets with a relatively small RHC interval instead of the big. Hence, we are supposed to choose a proper RHC interval while maintaining system performance.

6.2.3.4   System Performance

In this section, we compare our proposed DRL-based 2AGT method and 3AUS strategy with some benchmarks under a variety of system settings. Simulation results demonstrate the effectiveness of our proposed DRL-based algorithm.

(1) CPU clock speed: As illustrated in Fig. 6.10a, it shows the total system cost of the IIoT model considering the MEC server's CPU clock speed. From

Fig. 6.15c, our proposed method is a quasi-optimal solution compared with extensive search algorithm. Also, there is a small gap between our proposed method and extensive search. At the same time, the full local represents that all application tasks from IIoTD are executed in local devices, while the full offloading indicates that all tasks are offloaded to the MEC server to reduce energy consumption and delay. We can see that our proposed method outperforms the full offloading and local execution. Further, the full local execution is not able to change with the CPU speed clock. This is because the local execution cannot depend on the MEC server resources [63], whereas the full offloading mode decreases the total system cost with the increase of the CPU clock speed. More importantly, we compare our proposed method with two conventional DRL-based algorithms, i.e., conventional deep Q network (C-DQN) and deep deterministic policy gradient (DDPG), which show that our proposed method can further reduce the total system cost in contrast with C-DQN and DDPG. Hence, we can utilize the MEC server's powerful computational resources to help handle the application tasks in terms of our proposed method.

(2) Delay weight factor: In Fig. 6.10b, we plot the total system cost in terms of delay weight factor λ. The delay weight factor reveals the weight on delay relative to total energy consumption in terms of different application tasks from IIoTD. In general, delay weight is more significant than energy consumption for some current IIoT application tasks [64]. Specifically, with the increase of delay weight factor, we compare our proposed method with other offloading strategies including extensive search, full local, full offloading, intelligent C-DQN and DDPG for the total system cost. Under different delay weight factors, we can observe that our proposed method outperforms the full offloading and full local execution schemes in terms of the system cost. Further, our proposed method has lower total system cost than C-DQN and DDPG, which validates the progressiveness and intelligence of our proposed method. Finally, it is also close to extensive search, which means our proposed method can achieve quasi-optimal total system cost.

(3) Task weight factor: In Fig. 6.11a, we show the interrelationship between the total system cost and task weight factor β with different strategies. In detail, the task weight factor suggests the task size is related to the computational energy consumption from the MEC sever. Different task weight factors mean the various application tasks from IIoTD. In general, the larger the task size is, the larger the task weight factor is. From Fig. 6.11a, we can see that our proposed method is superior to the full offloading and full local schemes in terms of the total system cost. Further, our proposed method outperforms C-DQN and DDPG when task weight factor increases, which demonstrates the effectiveness and intelligence of our proposed method compared with C-DQN and DDPG. In addition, when the task weight factor increases, the total system cost is higher for all schemes. Finally, our proposed DRL-based algorithm is close to extensive search for total system cost, which represents that our

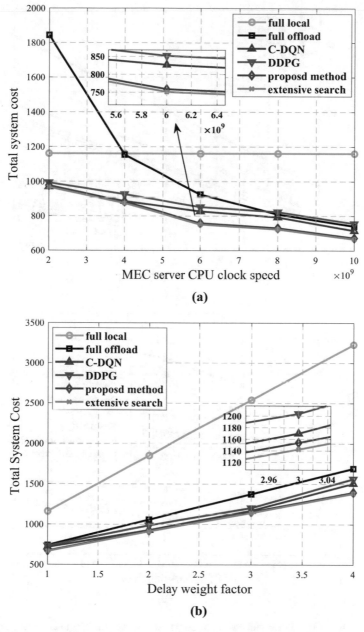

Fig. 6.10 CPU clock speed and delay weight factor analysis. (**a**) The total system cost versus MEC server speed. (**b**) The total system cost versus delay weight factor λ

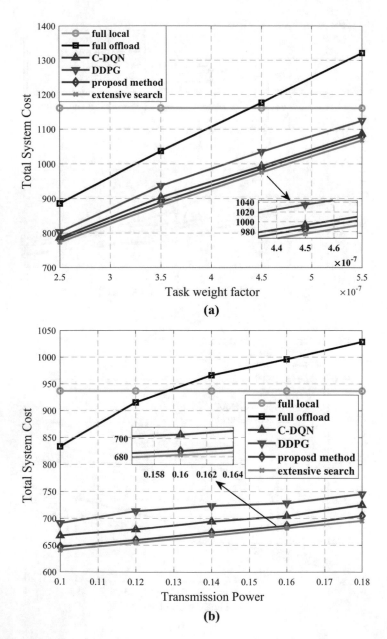

Fig. 6.11 Task weight factor and transmission power analysis. (**a**) The total system cost versus task weight factor of β. (**b**) The total system cost versus transmission power

proposed strategy can achieve a better sub-optimal solution and have superior intelligence.

(4) Transmission power: As shown in Fig. 6.11b, we explore the relationship between different transmission power and total system cost. As DDPG only explores the optimal offloading decision and transmission rate, it has higher computational cost. Additionally, C-DQN can better adapt to the network environments compared with DDPG. However, as we propose 2AGT optimization and 3AUS scheme to help dispose the computation offloading and resource allocation, the total system cost is lower than C-DQN. Additionally, our proposed DRL-based network structure is quasi-optimal compared with extensive search, which further demonstrates the effectiveness and reliability of our proposed method.

6.3 Multi-Agent Driven Resource Allocation for DEN

Since the fifth generation (5G) wireless communication networks cannot meet all requirements for future application scenarios [65], both academic and industrial communities have launched to explore beyond 5G and conceptualize 6G [66]. Compared with foregoing wireless networks, the next-generation networks may undergo unimaginable transformation for promoting wireless communication evolvement from connecting things to connecting intelligence, which can support more rigorous quality of service (QoS) demands, such as extremely high transmission rates (>1 Tera bit/s), ultra low latency (<1 ms), ultra-high reliability ($>99.99999\%$) [67], etc. The burgeoning extensive service metrics derive from the explosive ascent of mobile data and applications from computation-intensive and latency-sensitive tasks, such as mixed reality (XR) [68], industrial control, intelligent health-care, and other cases of Intelligent-of-Everything (IoE) [69, 70].

Deep edge networks (DENs) not only provide computation and decision-making to the edge node but also enable deep integration of wireless communication and computation resources via real-time adaptive collaboration to achieve the prospect of universal intelligence [35]. Moreover, the key capability of DENs is native artificial intelligence (AI), which includes data acquisition, transmission, storage, processing, analysis [72, 73], etc. Additionally, DENs have the ability to push highly distributed intelligence to network edge nodes, which can help to reduce costs, latency, and task risk. Traditional mobile cloud computing [74] structure can process data and tasks from agents through a centralized approach in the cloud server. However, it can cause severe delay and network congestion in terms of massive network traffic and application tasks, which leads to poor quality of experience (QoE). Fortunately, multi-access edge computing (MEC) in DENs can alleviate network traffic pressure and reduce task execution latency by deploying edge cloud server proximal to agents [38, 76].

There have been some existing works about MEC enabling computation offloading and resource scheduling to achieve the network convergence. To illustrate further, Zhang et al. [77] theoretically proposed the structure of mobile center cloud computing to minimize energy consumption in terms of stochastic channels. Zhang et al. [78] proposed an energy-percept task scheduling method to cooperatively process the resource allocation. Besides, Mao et al. [79] illustrated a novel MEC paradigm for the sake of harvested energy and executed an efficient task scheduling strategy. Hong et al. [80] jointly proposed a QoS-aware task scheduling algorithm in robot swarms relying on Stackelberg game theory, which investigated the task scheduling and routing planning for the sake of minimizing the latency while maximizing the energy efficiency. Furthermore, Liu et al. [81] presented a novel mobile vehicle-mounted edge mechanism, which aimed to maximize completed tasks of vehicle-mounted edge with sensitive latency by a gap-adjusted branch and bound algorithm. Lyu et al. [82] established a disturbed Lyapunov optimization model to maximize the performance gain for the sake of balancing the throughput and fairness, which asymptotically optimized the task schedules subject to out-of-date network background [83]. However, some of the above-mentioned methods cannot process complex conditions (e.g., variable channel conditions). Secondly, some of them have high computational complexity for real-time task offloading in the context of complex networks.

6.3.1 System Model

Figure 6.12 shows that DENs are composed of one macro base station (MBS) and multiple collaborative wireless access points (AP), where $\mathbb{M}$ can be represented as $\mathbb{M} = \{1, 2, \ldots\ldots, M\}$. Additionally, we consider $\mathbb{N}$ can be denoted as $\mathbb{N} = \{1, 2, \ldots\ldots, N\}$ in each AP coverage area, where each agent processes one task. We assume that each AP is overlaid by the intelligent MBS. Besides, we ignore the transmission delay cost [84]. Furthermore, the agent receives a task $A_{m,n} = \{L_{m,n}, W_{m,n}, T_{m,n}^d\}$. Here $L_{m,n}$ denotes the task size to be computed, including programming codes and data parameters. $W_{m,n}$ denotes the required CPU cycles and $T_{m,n}^d$ means the maximum latency, which means that execution latency for each task should not exceed the maximum predefined value. The three parameters of each application task can be evaluated by task profiles, so they may be various among different task types.

Currently, with the rising of parallel computing in the computer networks, executing the task in the local agent or the MEC server has attracted huge attention [85–87]. In this paper, assuming that each computation-intensive task is processed by full offloading or local execution. Additionally, we denote that $b_{m,n} \in \{0, 1\}$ is the task scheduling, where $b_{m,n} = 1$ represents the task can be performed at edge cloud, while $b_{m,n} = 0$ means that the task is executed locally on the agent.

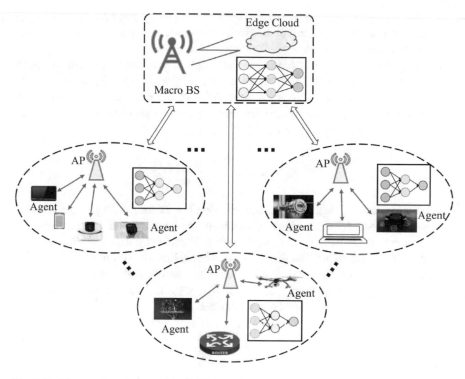

Fig. 6.12 The novel service scenario for DENs

6.3.1.1   Single Edge Network for DENs

(1) Local Execution Mode: We denote that the single edge network is the special scenario of DENs and the number of edge network is $M = 1$, i.e., there is only one edge network area overlaid by one intelligent MBS. Moreover, we formulate f_n^l as the computation capability of each agent. The local computation delay can be represented as

$$t_n^l = \frac{W_n}{f_n^l}. \tag{6.81}$$

Similarly, we can calculate the local execution energy consumption

$$e_n^l = \sum_{w_n=1}^{W_n} \varepsilon \left(f_n^l \right)^2, \tag{6.82}$$

where ε is the efficient switch electric capacity coefficient, which relies on the chip architecture for each agent [88]. For the local execution, we can configure the CPU clock speed utilizing the scalable frequency technology [89] for the sake of reducing the total cost. Furthermore, we assume that the CPU cycle frequency remains unchanged during processing one task.

(2) Computing Offloading Mode: For the single edge network, when the agent chooses to offload the task to edge cloud through the AP, we ignore the transmission cost and delay between AP and edge cloud [90]. Besides, we ignore the channel assignment for each agent for executing the task. Hence, the uplink transmission rate can be written by:

$$c_n = w \log \left(1 + \frac{p_n |h_n|^2}{\sigma^2} \right), \qquad (6.83)$$

where p_n and h_n represent the transmission power for each agent and channel gain between AP and agent, respectively. σ^2 is the noise power. Also, we assume that the total system bandwidth is B and the number of channel is O in each edge network. So the allocated channel bandwidth is $w = B/O$. For the sake of simplicity, we consider there are massive channels and the channel allocation can be ignored (i.e., $O \gg N$). Next, the transmission delay from each agent to the edge cloud can be denoted as

$$t_n^{e1} = \frac{L_n}{c_n}. \qquad (6.84)$$

Similarly, the execution latency on the edge cloud can be represented as

$$t_n^{e2} = \frac{W_n}{F_{total}}, \qquad (6.85)$$

where F_{total} is running speed of the edge cloud, which processes the computational tasks by parallel computing. Hence, the total latency in the computation offloading mode can be represented as

$$t_n^{edge} = t_n^{e1} + t_n^{e2}. \qquad (6.86)$$

Additionally, the transmission energy consumption from the agent can be calculated as

$$e_n^{edge} = p_n t_n^{e1}. \qquad (6.87)$$

In this case, we ignore the task return delay from the edge cloud due to the fact that the task size is very small after being processed via the edge cloud [54].

(3) Optimization Problem Formulation: As illustrated above, we consider the total overhead including latency and energy consumption for the single edge network and the optimization problem can be formulated as follows:

$$P1: \min_{p,b,f} \left\{ \sum_{n=1}^{N} [\lambda_n^t t_n^l + \lambda_n^e e_n^l](1 - b_n) \right\}$$

$$+ \left\{ \sum_{n=1}^{N} [\lambda_n^t t_n^{edge} + \lambda_n^e e_n^{edge}](b_n) \right\}$$

$$\text{s.t.} \quad \text{C1:} \quad b_n \in \{0, 1\}, \tag{6.88}$$

$$\text{C2:} \quad 0 < f_n^l \le f_n^{max},$$

$$\text{C3:} \quad 0 < p_n \le p_n^{max},$$

$$\text{C4:} \quad (1 - b_n)t_n^l + b_n t_n^{edge} \le T_n^d$$

$$\text{C5:} \quad (1 - b_n)e_n^l + b_n e_n^{edge} \le E_n^{max},$$

where λ_n^t represents the delay weight factor and λ_n^e means the energy weight factor. It can be defined by various agents in terms of different application task demands [92]. More specifically, the constraint C1 represents the offloading decision with executing the task C2 means the local CPU cycle frequency cannot exceed the maximum value of the agent. C3 means the transmission power of each agent is limited. C4 guarantees the task completion delay should be under the maximum tolerant delay. C5 ensures that the total energy consumption executing the task cannot exceed the maximum battery capacity of each agent. Furthermore, h_n is the time-varying channel gain and it can be generated by interacting with network environments.

6.3.1.2   Multiple Edge Networks for DENs

Different from the single edge network, the multiple edge scenes contain many edge network areas overlaid by one MBS with edge cloud. Furthermore, we consider the interference management for each agent.

(1) Local Processing Mode: The delay can be formulated as:

$$t_{m,n}^l = \frac{W_{m,n}}{f_{m,n}^l}. \tag{6.89}$$

Similarly, the local processing energy consumption is

$$e_{m,n}^l = \sum_{w_{m,n}=1}^{W_{m,n}} \varepsilon \left(f_{m,n}^l \right)^2. \tag{6.90}$$

(2) Computing Offloading Mode: In this case, each agent employs the orthogonal frequency division multiple access (OFDMA). Hence, the uplink transmission rate is:

$$c_{m,n,o} = w \log \left(1 + \frac{p_{m,n,o}|h_{m,n,o}|^2}{\sigma^2 + D_{m,n,o}} \right), \tag{6.91}$$

where $p_{m,n,o}$ and $h_{m,n,o}$ denote different power and channel gain, respectively. $D_{m,n,o}$ defines the same frequency interference suffering from other proximate areas [90] and the disturbance is

$$D_{m,n,o} = \sum_{k=1,k\neq m}^{M} \sum_{n=1}^{N} \alpha_{k,n,o} p_{k,n,o} |h_{k,n,o}|^2, \tag{6.92}$$

where $\alpha_{k,n,o} \in \{0, 1\}$ means each channel o can be allocated to n for kth edge network in order to offload the task to the access point, and $\alpha_{k,n,o} = 1$ represents o is assigned to n, if not, $\alpha_{k,n,o} = 0$. At the same time, there is no interference in the same area for different agents because they access the channel through OFDMA. Additionally, $p_{k,n,o}$ and $h_{k,n,o}$ are allocated power and channel state in terms of neighboring networks. Also, the information transmission rate can be expressed as

$$c_{m,n} = \sum_{o=1}^{O} \alpha_{m,n,o} c_{m,n,o}. \tag{6.93}$$

Hence, the delay in the offloading mode is

$$t_{m,n}^{em1} = \frac{L_{m,n}}{c_{m,n}}. \tag{6.94}$$

Similarly, the execution delay in this case is

$$t_{m,n}^{em2} = \frac{W_{m,n}}{F_{total}}. \tag{6.95}$$

Hence, when the task is processed, the total task delay can be written by

$$t_{m,n}^{EDGE} = t_{m,n}^{em1} + t_{m,n}^{em2}. \tag{6.96}$$

Furthermore, the transmission power is defined as

$$p_{m,n} = \sum_{o=1}^{O} \alpha_{m,n,o} p_{m,n,o}. \tag{6.97}$$

Moreover, the consumed energy in the offloading mode is

$$e_{m,n}^{EDGE} = p_{m,n} t_{m,n}^{em1}. \tag{6.98}$$

(3) Optimization Problem Formulation: As mentioned above, we jointly consider the task delay and energy consumption for all agents in the multiple edge networks environment. Additionally, with offloading the task to the access point, we consider the interference management from other edge networks and channel allocation for each agent, and the optimization problem in multiple edge networks is

$$P2: \min_{p,b,f,\alpha} \left\{ \sum_{m=1}^{M} \sum_{n=1}^{N} \left[\lambda_{m,n}^{t} t_{m,n}^{l} + \lambda_{m,n}^{e} e_{m,n}^{l} \right] (1 - b_{m,n}) \right\}$$

$$+ \left\{ \sum_{m=1}^{M} \sum_{n=1}^{N} \left[\lambda_{m,n}^{t} t_{m,n}^{EDGE} + \lambda_{m,n}^{e} e_{m,n}^{EDGE} \right] b_{m,n} \right\}$$

$$\text{s.t.} \quad \text{C1:} \quad b_{m,n} \in \{0, 1\},$$

$$\text{C2:} \quad 0 < f_{m,n}^{l} \leq f_{m,n}^{\max},$$

$$\text{C3:} \quad 0 < p_{m,n} \leq p_{m,n}^{\max}, \tag{6.99}$$

$$\text{C4:} \quad (1 - b_{m,n}) t_{m,n}^{l} + b_{m,n} t_{m,n}^{EDGE} \leq T_{m,n}^{d},$$

$$\text{C5:} \quad (1 - b_{m,n}) e_{m,n}^{l} + b_{m,n} e_{m,n}^{EDGE} \leq E_{m,n}^{\max},$$

$$\text{C6:} \quad \sum_{o=1}^{O} \alpha_{m,n,o} \leq 1,$$

$$\text{C7:} \quad D_{m,n,o} \leq D_{\max}.$$

Different from the single edge scenario, C6 indicates n is allocated to one channel in total. C7 indicates that n has interference from other areas and the value cannot exceed the maximum predetermined threshold $D_{\max}$. Other constraints are similar to the single edge network.

6.3.2 Algorithm Design in Single Edge Network

In this section, we firstly formulate a novel DRL-based algorithm framework in terms of extremum methods to tackle our proposed single edge network problem. Generally, the proposed DRL-based framework contains three parts, including offloading decision, CPU cycle allocation, and transmission power assignment. Furthermore, the complete network structure can be found in Fig. 6.13. Specific solutions are explained as follows.

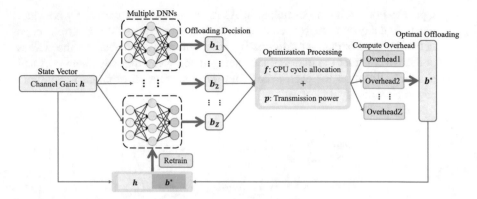

Fig. 6.13 The novel DRL network structure

6.3.2.1 Offloading Decision Generation

As the channel gain is variable in different time slots, we regard the channel gain
as the state vector in the DRL agent and formulate the DRL agent as multiple DNN
structures. After the channel gain is transferred into the DRL agent, numerous DNNs
can generate different offloading decisions according to the channel gain and the
process is illustrated as follows.

$$h \rightarrow b_z, \tag{6.100}$$

where $b_z = \{b_1, b_2, \ldots b_Z\}$ denotes the offloading decision for each agent and $z =
1, 2, \ldots Z$. At the same time, we set the number of DNN as Z and each DNN can
generate the whole offloading decisions for all agents.

6.3.2.2 Optimal Local Execution Overhead

After obtaining the offloading decisions, the original single edge network problem
can be transformed into two subproblems: one is local overhead, and the other is
edge cloud execution overhead. Consequently, the local overhead can be shown as
follows.

$$\min_f [\lambda_n^t t_n^l + \lambda_n^e e_n^l] \tag{6.101}$$

$$s.t. \quad 0 < f_n^l \le f_n^{\max}, \tag{6.102}$$

$$t_n^l \le T_n^d, \tag{6.103}$$

$$e_n^l \le E_n^{\max}. \tag{6.104}$$

We assume that $F_1(f_n^l) = \lambda_n^l t_n^l + \lambda_n^e e_n^l = \lambda_n^l \frac{W_n}{f_n^l} + \lambda_n^e \varepsilon (f_n^l)^2 W_n$, where the local execution overhead for each agent is only related to local CPU cycle frequency. Additionally, the extremum point can be denoted as follows

$$f_n^{l*} = \sqrt[3]{\frac{\lambda_n^t}{2\lambda_n^e \varepsilon}}. \tag{6.105}$$

At the same time, the function $F_1(f_n^l)$ monotonously increases when $f_n^l > f_n^{l*}$; otherwise, the function can decrease monotonously with the increase of f_n^l. Thus, the value f_n^{l*} is the extremum point. Next, we start to observe the constrains from (22) to (24). The constraint (23) can be transformed into

$$f_n^l \geq \frac{W_n}{T_n^d}. \tag{6.106}$$

Likewise, the constraint (24) can be denoted as

$$f_n^l \leq \sqrt{\frac{E_n^{\max}}{\varepsilon W_n}}. \tag{6.107}$$

Combining (22), (26) with (27), the constrains for the local overhead problem can be transferred as

$$f' = \max\left\{0, \frac{W_n}{T_n^d}\right\}. \tag{6.108}$$

$$f'' = \min\left\{\sqrt{\frac{E_n^{\max}}{\varepsilon W_n}}, f_n^{\max}\right\}. \tag{6.109}$$

Consequently, according to the obtained extremum and CPU cycle range, the optimal local execution overhead about energy consumption and latency should be calculated as

$$F_1^*(f_n^l) = \begin{cases} F_1^*(f'), & \text{if } f_n^{l*} \leq f', \\ F_1^*(f_n^{l*}), & \text{if } f' < f_n^{l*} \leq f'', \\ F_1^*(f''), & \text{if } f_n^{l*} > f''. \end{cases} \tag{6.110}$$

6.3.2.3 Optimal Edge Cloud Execution Overhead

Once the agent decides to offload the task to the edge cloud, the execution overhead in the remote edge cloud can be denoted as

$$\min_{p} [\lambda_n^t t_n^{edge} + \lambda_n^e e_n^{edge}] \qquad (6.111)$$

$$s.t. \quad 0 < p_n \le p_n^{\max}, \qquad (6.112)$$

$$t_n^{edge} \le T_n^d, \qquad (6.113)$$

$$e_n^{edge} \le E_n^{\max}. \qquad (6.114)$$

Similar to the local execution overhead, we define the function $F_2(p_n) = \lambda_n^t t_n^{edge} + \lambda_n^e e_n^{edge}$, where the remote execution overhead can be calculated as follows

$$F_2(p_n) = \frac{L_n(\lambda_n^t + \lambda_n^e p_n)}{w \log_2(1 + \frac{p_n |h_n|^2}{\sigma^2})} + \frac{\lambda_n^t W_n}{F_{total}}. \qquad (6.115)$$

We can observe that the function is only related to the transmission power p_n, and we define

$$f_2(p_n) = \frac{L_n(\lambda_n^t + \lambda_n^e p_n)}{w \log_2(1 + \frac{p_n |h_n|^2}{\sigma^2})}. \qquad (6.116)$$

Proposition 1: The function $f_2(p_n)$ is unimodal [93].

As the function $f_2(p_n)$ is unimodal, there is only one extremum. Furthermore, the constraint (33) can be represented as

$$p_n \ge \left(2^{\frac{L_n}{w(T_n^d - \frac{W_n}{F_{total}})}} - 1 \right) \frac{\sigma^2}{|h_n|^2} = p_1. \qquad (6.117)$$

Next, the constraint (34) can be simplified as

$$\frac{p_n}{\log_2(1 + \frac{p_n |h_n|^2}{\sigma^2})} \le \frac{E_n^{\max} w}{L_n}. \qquad (6.118)$$

Denoting that the left part of constraint (38) is subject to proposition 1, the feasible solution for constraint (38) is $p_n \in [p_l, p_h]$, where p_l and p_h are the lower bound and upper bound for the constraint (38). Similarly, combining (32), (37) with (38), the feasible region for function $f_2(p_n)$ can be transformed into

$$p' = \max\{p_1, p_l\}, \qquad (6.119)$$

$$p'' = \min\{p_n^{\max}, p_h\}. \qquad (6.120)$$

Next, the minimum remote execution overhead for the agent can be represented as

$$F_2^*(p_n) = \begin{cases} f_2(p') + \frac{\lambda_n^t W_n}{F_{total}}, & \text{if } p_n^* \le p', \\ f_2(p_n^*) + \frac{\lambda_n^t W_n}{F_{total}}, & \text{if } p' < p_n^* \le p'', \\ f_2(p'') + \frac{\lambda_n^t W_n}{F_{total}}, & \text{if } p_n^* \ge p'', \end{cases} \tag{6.121}$$

where p_n^* is the extremum point for the function $f_2(p_n)$ when the derivative is equal to zero.

6.3.2.4 Training Methods

Next, in terms of the obtained offloading decision b, CPU cycle frequency f, and transmission power p, the agent can compute the total overhead. As there are Z DNNs totally, each DNN can compute the overall overhead for all agents. The optimal overhead for each DNN can be calculated as

$$F^*(h, b, f, p) = \sum_{n=1}^{N} F_1^*(f_n^l) + F_2^*(p_n). \tag{6.122}$$

Then, the agent can compare the obtained total overhead for each DNN, and it can select the current optimal offloading decision b^* in terms of different DNNs $\{1, 2, \ldots Z\}$. Next, the agent can store the state–action pairs into experience replay memory and retrain the DNN in order to update the network parameters after certain steps. Hence, the proposed DRL-based algorithm framework for offloading decision, transmission power, and CPU cycle frequency assignment is illustrated in Algorithm 6.2. The computational complexity of Algorithm 3.1 is in the order of $O(T * D^2 * Z^3)$, where D represents the dataset size.

6.3.3 Algorithm Design in Multiple Edge Networks

In the multiple edge scenes, we jointly intend to optimize the task scheduling, channel assignment, and resource allocation in terms of volatile channel gain. However, the problem P2 is a non-convex and NP-hard problem. To efficiently solve the problem, we firstly transform the problem P2 into P2.1 by optimizing the local CPU cycle frequency, and then utilize the proposed MADDPG for the sake of optimizing the task scheduling, channel allocation, and transmission power.

6.3.3.1 DRL Background Knowledge

In term of the conventional reinforcement learning, we consider that there are state space $S = \{s_1, s_2, \ldots s_T\}$ and action space $A = \{a_1, a_2, \ldots a_T\}$. The agent can

---

Algorithm 6.2 Optimal cost of single edge network

Input:
   Task: $A_{m,n} = \{L_{m,n}, W_{m,n}, T_{m,n}^d\}$;
   Channel Gain: h (channel state of each agent).
Output:
   task scheduling b_z^*; local CPU cycle frequency f; transmission power p;
   optimal overhead for all agents.
1: **while** 1, 2, ... , T **do**
2:    Obtain the task scheduling b_z on the basis of the scheme (20), where $\{z = 1, 2, \ldots Z\}$;
3:    if $b_z == 0$:
      Compute the optimal local execution overhead according to (30) and obtain the total overhead for each agent;
4:    else
      Compute the optimal cloud execution overhead by (41) and obtain the total overhead for each agent.
5:    Compute and compare all DNN overhead from 1 to Z and select the optimal offloading decision b^*
6:    Choose the current channel gain h in time slot t and optimal offloading action b^* to retrain the all DNN from 1 to Z;
7:    Finally, when the number of iterations is T, obtain the optimal offloading decision b^*, local CPU cycle f, transmission power p and total overhead $F^*(h, b, p, f)$ for all agents;
8: **end while**

---

observe the state s_t and takes one action a_t, and then it obtains a reward r_t by interacting with the environment. Next, the agent comes to a new state s_{t+1}. In the RL process, the formulation can be represented as

$$a_t = \pi(s_t). \tag{6.123}$$

When the agent interacts with the environment, the aim is to maximize the total reward from $t = 1$ to T. The total reward R_t can be represented as

$$R_t = \sum_{k=0}^{T} \gamma^k r_{t+k+1}, \tag{6.124}$$

where $\gamma \in [0, 1]$ is a discount factor. Furthermore, combined DNN with RL, DQN is a novel structure to approximately represent Q-function, and it can be represented by

$$Q(s_t, a_t) = \mathbb{E}_\pi[R_t|s_t, a_t] \tag{6.125}$$

which is considered as a Q-function in the RL domain. Then, the loss function of DNN can be calculated as

$$L(w) = \mathbb{E}[y_t - Q(s_t, a_t)|w], \qquad (6.126)$$

where w is the network parameter from DNN and y_t can be formulated as

$$y_t = r_t + \gamma Q(s_{t+1}, a_{t+1}|w'), \qquad (6.127)$$

where s_{t+1} and a_{t+1} are the next state and action of the agent, respectively. Next, to make the DNN more stable and have faster convergent speed, the agent adds the experience replay memory and enables the target network with the same structure. The experience replay memory is to retrain the DNN by randomly sampling a batch of transitions. Simultaneously, the target network can reduce the correlations between current Q-value and target Q-value and it is updated at certain steps. However, the current DQN cannot process the continuous action control problems. Fortunately, the popular actor-critic based deep deterministic policy gradient (DDPG) algorithm can solve the continuous action control problem more easily. Specifically, DDPG consists of two actor networks and two critic networks, and actor network can generate the action and critic network obtains the Q-value function to update the actor network parameters. The policy gradient method can be formulated as

$$\nabla_\theta J = \mathbb{E}\left[\nabla_a Q(s, a|w)_{s-s_t, a-\pi(s_t)} \nabla_\theta \pi(s|\theta)_{s=s_t}\right], \qquad (6.128)$$

where the current critic network can be updated by the loss function (46).

6.3.3.2 MADDPG Algorithm Framework

In this part, we propose a MADDPG method for the sake of minimizing the total cost in the multiple edge networks. Firstly, we optimize the local processing speed and transform the original problem P2 into P2.1. Next, we envision a MADDPG structure in order to obtain the sub-optimal offloading decision, channel allocation, and transmission power for each agent. The optimization process is presented as follows.

(1) CPU Cycle Frequency Optimization: Similar to the single edge network, the CPU frequency is calculated as

$$\min_f [\lambda_{m,n}^l t_{m,n}^{EDGE} + \lambda_{m,n}^e e_{m,n}^{EDGE}] \qquad (6.129)$$

$$s.t. \quad 0 < f_{m,n}^l \le f_{m,n}^{\max}, \qquad (6.130)$$

$$t_{m,n}^l \le T_{m,n}^d, \qquad (6.131)$$

$$e_{m,n}^l \leq E_{m,n}^{\max}. \tag{6.132}$$

Combined with constrains (50), (51), (52), the constraint can be transformed into

$$f_{m,n}^{l'} = \max \left\{ \frac{W_{m,n}}{T_{m,n}^d}, 0 \right\}, \tag{6.133}$$

$$f_{m,n}^{l''} = \min \left\{ f_{m,n}^{\max}, \sqrt{\frac{E_{m,n}^{\max}}{\varepsilon W_{m,n}}} \right\}. \tag{6.134}$$

Assuming that $F_{m,n}^l = [\lambda_{m,n}^t t_{m,n}^{EDGE} + \lambda_{m,n}^e e_{m,n}^{EDGE}]$, the point is

$$f^* = \sqrt[3]{\frac{\lambda_{m,n}^t}{2\varepsilon \lambda_{m,n}^e}}. \tag{6.135}$$

The optimal local execution overhead can be formulated as

$$F_{m,n}^{l*}(f_{m,n}^l) = \begin{cases} F_{m,n}^l(f_{m,n}^{l'}), & \text{if } f^* \leq f_{m,n}^{l'}, \\ F_{m,n}^l(f^*), & \text{if } f_{m,n}^{l'} < f^* \leq f_{m,n}^{l''}, \\ F_{m,n}^l(f_{m,n}^{l''}), & \text{if } f^* > f_{m,n}^{l''}. \end{cases} \tag{6.136}$$

Next, the problem P2 is changed into P2.1

$$P2.1: \min_{h,p,b,\alpha} \left\{ \sum_{m=1}^{M} \sum_{n=1}^{N} F_{m,n}^{l*} (1 - b_{m,n}) \right\}$$

$$+ \left\{ \sum_{m=1}^{M} \sum_{n=1}^{N} \left[\lambda_{m,n}^t t_{m,n}^{EDGE} + \lambda_{m,n}^e e_{m,n}^{EDGE} \right] b_{m,n} \right\}$$

$$\text{s.t.} \quad \text{C1:} \quad b_{m,n} \in \{0, 1\},$$

$$\text{C3:} \quad 0 < p_{m,n} \leq p_{m,n}^{\max}, \tag{6.137}$$

$$\text{C4:} \quad (1 - b_{m,n}) t_{m,n}^l + b_{m,n} t_{m,n}^{EDGE} \leq T_{m,n}^d,$$

$$\text{C5:} \quad (1 - b_{m,n}) e_{m,n}^l + b_{m,n} e_{m,n}^{EDGE} \leq E_{m,n}^{\max},$$

$$\text{C6:} \quad \sum_{o=1}^{O} \alpha_{m,n,o} \leq 1,$$

$$\text{C7:} \quad D_{m,n,o} \leq D_{\max}.$$

(2) MADDPG Algorithm Framework:

We propose the MADDPG structure for the sake of solving P2.1. Assuming that there are $M * N$ agents interacting with the environments totally in the multiple edge

---

Algorithm 6.3 Optimal cost in terms of MADDPG architecture

Input:
  Each task $A_{m,n} = \{L_{m,n}, W_{m,n}, T_{m,n}^d\}$.
Output:
  Calculate the total cost of multi-agents.

 1: **while** each agent **do**
 2:   Initialize all four network parameters $\pi^n(\theta)$, $Q^n(w)$, $\pi^n(\theta')$ and $Q^n(w')$, where $\theta' = \theta$, $w' = w$.
 3: **end while**
 4: **while** $1, 2, \ldots, T$ **do**
 5:   **while** $1, 2, \ldots, M$ **do**
 6:     **while** $1, 2, \ldots, N$ **do**
 7:       We can obtain the state $h_{m,n}(t)$ of each agent;
 8:     **end while**
 9:   **end while**
10:   Obtain all states H_t;
11:   **while** $1, 2, \ldots, M$ **do**
12:     **while** $1, 2, \ldots, N$ **do**
13:       Each agent starts to execute action $a_{m,n}(t) = \pi^n(h_{m,n}(t)|\theta) + \varepsilon$.
14:     **end while**
15:   **end while**
16:   Obtain all actions A_t;
17:   **while** $1, 2, \ldots, M$ **do**
18:     **while** $1, 2, \ldots, N$ **do**
19:       We can obtain the reward $r_{m,n}(t)$ and next state $h_{m,n}(t+1)$;
20:     **end while**
21:   **end while**
22:   Obtain all next states H_{t+1};
23:   **while** $1, 2, \ldots, M$ **do**
24:     **while** $1, 2, \ldots, N$ **do**
25:       Each agent can store $\{H(t), A(t), r_{m,n}(t), H(t+1)\}$ into E;
26:       **if** *learning time reaches* **then**
27:         Each agent can collect K samples from E;
28:         $\theta' = \tau\theta + (1 - \tau)\theta'$;
29:         $w' = \tau w + (1 - \tau)w'$;
30:       **end if**
31:     **end while**
32:   **end while**
33: **end while**
34: Calculate the total cost on the basis of energy consumption and latency.

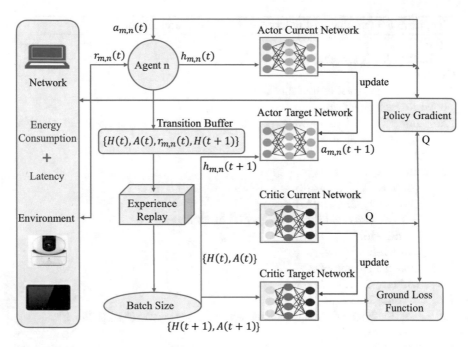

Fig. 6.14 The novel agent network structure

networks, the structure of the agent n is shown in Fig. 6.14. The state $S = \{s_t, t \in (1, 2, \ldots T)\}$ includes the private state $h_{m,n}(t)$ of each agent and other information known by one agent. Additionally, a few actions $A = \{a_t, t \in (1, 2, \ldots T)\}$ consist of private action $a_{m,n}(t) = \{b_{m,n}(t), \alpha_{m,n,o}(t), p_{m,n}(t)\}$ of each agent and some information from other agents. Next, each agent extracts own private information $h_{m,n}(t)$ and then takes own action $a_{m,n}(t)$. At the same time, each agent obtains its reward $r_{m,n}(t)$. Furthermore, the network environment can be transformed into next new state. As shown in Fig. 6.14, each agent has four network structures including actor current network $a_{m,n}(t) = \pi^n(h_{m,n}(t))$, actor target network $a_{m,n}(t + 1) = \pi^{n'}(h_{m,n}(t + 1))$, critic current network $Q^n(H(t), A(t))$, and critic target network $Q^{n'}(H(t + 1), A(t + 1))$. Then, each agent n stores the transition buffer $\{H(t), A(t), r_{m,n}(t), H(t+1)\}$ into the experience replay memory.

Explanation: The proposed MADDPG structure is a centralized-training and distributed execution method. During the training process, each agent takes action $a_{m,n}(t)$ on the basis of private state $h_{m,n}(t)$ in terms of network environment, and the state $H(t)$ and action $A(t)$ are the set of state and action from all agents. In the meantime, all agents can give their own information each other to calculate the reward functions. Additionally, the critic current and target network can receive all states and actions from agents to enable the critic network to train. For the distributed execution period, each agent can obtain its own action $a_{m,n}(t)$ by private state information $h_{m,n}(t)$.

Therefore, we define the private state, action, and reward function of each agent as follows:

(1) $h_{m,n}(t)$ can be represented as

$$h_{m,n}(t) = \left[h_{m,n,1}, h_{m,n,2}, \ldots h_{m,n,O}\right].$$ (6.138)

(2) Action $a_{m,n}(t)$: we define the action of each agent as

$$a_{m,n}(t) = \{b_{m,n}(t), \alpha_{m,n,o}(t), p_{m,n}(t)\}.$$ (6.139)

(3) $r_{m,n}(t)$ is defined as

$$r_{m,n}(t) = -\left\{\sum_{m=1}^{M}\sum_{n=1}^{N} F_{m,n}^{l*}(1 - b_{m,n})\right\} -$$ (6.140)

$$\left\{\sum_{m=1}^{M}\sum_{n=1}^{N} (\lambda_{m,n}^{t} t_{m,n}^{EDGE} + \lambda_{m,n}^{e} e_{m,n}^{EDGE})b_{m,n}\right\}.$$ (6.141)

Then, the absolute state and action are as follows.

(1) H is defined by $H = \{h_{m,n}(t), \forall m, n\}$.
(2) Action A: The absolute action consists of the action from all agents, which is expressed as $A = \{a_{m,n}(t), \forall m, n\}$.

Next, we depict the network structure of the agent n in Fig. 6.14. The agent obtains the state $h_{m,n}(t)$ and chooses the current optimal action $a_{m,n}(t)$. Furthermore, the agent obtains the Q-value from critic current network $Q^n(w)$. Additionally, its target action and target Q-function can be calculated by $\pi^n(\theta')$ and $Q^n(w')$, respectively. The transition buffer is

$$E = \{H(t), A(t), r_{m,n}(t), H(t+1)\},$$ (6.142)

which can be stored into experience replay memory. At the same time, we calculate the temporal difference (TD) error as

$$\triangle_n = r_{m,n}(t) +$$ (6.143)

$$\gamma Q^{n'}(H(t+1), A(t+1)|w') - Q^n(H(t), A(t)|w).$$ (6.144)

Furthermore, the loss function in Fig. 6.14 is

$$G(L(w)) = \mathbb{E}\left[(\triangle_n)^2\right].$$ (6.145)

Then, the policy gradient can be denoted as

Table 6.3 Summary of
simulation parameters

Notation	Description
F_{total}	5e9 cycle/s
$W_{m,n}$	[0.2e9 cycle, 1e9 cycle]
$L_{m,n}$	[500 KB, 1200 KB]
$T_{m,n}^d$	[0.5s, 5s]
M	3
N	2
$f_{m,n}^l$	[0.3e9 cycle/s, 1e9 cycle/s]
$p_{m,n}^{max}$	0.2 W
w	0.5 MHz
O	10

$$\nabla_\theta J = \mathbb{E}\left[\nabla_\theta \pi^n(h_{m,n}(t)|\theta)\nabla_{a_{m,n}(t)} Q(H(t), A(t)|w)\right]. \qquad (6.146)$$

As shown in Algorithm 2, the computational complexity is in the order of $O\left(MN(1 + T + (1 + \frac{1}{K})TD' + TD'')\right)$, where D' and D'' denote the dataset size of actor network and critic network, respectively.

6.3.4 *Experiments and Simulation Results*

In order to demonstrate the utility of the proposed schemes in single edge and multiple edge scenes, we test and validate our analysis by numerous results. Relying on [93], we denote the simulation parameters. Firstly, the coverage area of each edge network is 120 m in radius. Simultaneously, the server speed is set as 6e9 cycle/s and the required CPU cycles are randomly generated from 0.3e9 to 1e9. Additionally, the task size and tolerant delay are randomly distributed from [600 KB, 1200 KB] and [0.2 s, 5 s], respectively. The CPU cycle frequency of each agent is also randomly distributed at [0.4e9, 1e9]. Besides, the transmission power is subject to 0.2 W. We set that there exist 10 dimensions for channel vectors, i.e., 10 channels in multiple edge networks. Next, we set the channel bandwidth as 0.5 MHz. The detailed simulation parameters are shown in Table 6.3.

6.3.4.1 Single Edge Network Scene

As illustrated in Fig. 6.15a, through adjusting proper hyperparameters of DNN, the DNN total loss function gradually decreases with the increase of the number of training steps when learning_rate is 0.01. Additionally, each interval represents 500 training steps in the x-axis. When the number of training steps is approximately up to 200, the total loss of DNN is close to 0.1 and it remains stable in the next training steps, which validates that our proposed algorithm has faster convergent speed and

Fig. 6.15 The total agent overhead versus analysis. (**a**) The number of training steps. (**b**) Delay aware factor. (**c**) MEC server speed

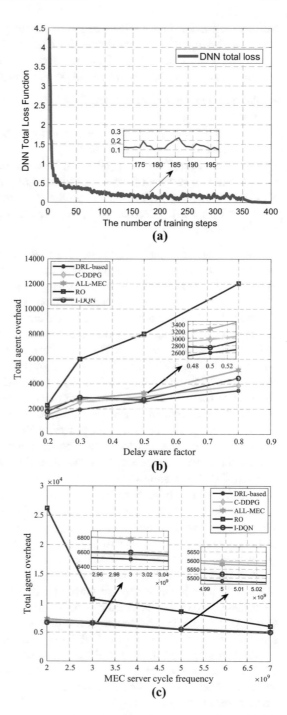

stronger generalization ability. Furthermore, our proposed algorithm can adapt to the time-varying network environments better.

To corroborate the utility of our proposed schemes, we compare it with some baseline algorithms. The details are illustrated as follows.

- All mobile edge computing (ALL-MEC): The tasks from agents are performed in remote edge cloud.
- Random offloading (RO): While executing the tasks, the agent randomly selects the offloading decision, i.e., 0 or 1.
- Intelligent deep Q network (I-DQN): I-DQN represents that the number of DNN is 1, which is different from our proposed DRL-based algorithm.
- Centralized deep deterministic policy gradient (C-DDPG): C-DDPG means that all agents implement the computation offloading via centralized global information.

We compare the performance of five different algorithms in Fig. 6.15b. Delay weight factor and energy weight factor are limited to (0, 1) and their sum is 1. Additionally, if delay weight factor is larger than energy weight factor, it means that the delay of task for each agent is more important than energy consumption. As the delay aware factor increases, it consumes more overhead for all agents, due to the fact that each agent receives more latency-sensitive application tasks. As the delay aware factor increases, DRL-based can undertake the overhead pressure, while I-DQN and ALL-MEC have higher total overhead because they cannot adapt to the time-varying channel conditions better. Additionally, RO has lower efficiency because of randomly choosing different offloading decision. Moreover, C-DDPG can adapt to the dynamic network states, but it is difficult to obtain optimal CPU cycle frequency and transmission power, which causes higher system overhead. Hence, based on the proposed DRL-based algorithm, we can reduce the system overhead and adapt to the various network environments.

Next, Fig. 6.15c shows the impact of the MEC server cycle frequency on the total agent overhead in terms of five different algorithms. Certainly, higher MEC server cycle frequency can reduce the execution latency and total agent overhead. Furthermore, the RO algorithm has higher overhead in terms of different MEC processing speed, while other four schemes can reduce the overhead better. This is because RO cannot find the optimal computation offloading strategy. Additionally, from $3*10e9$ to $5*10e9$ in terms of MEC server speed, we can clearly see that DRL-based algorithm has better performance gain than I-DQN, ALL-MEC, and C-DDPG algorithms since it can help choose better offloading decision-making and allocate CPU cycle frequency and transmission power, which demonstrates the superiority and effectiveness of our proposed DRL-based algorithm.

Furthermore, we validate the effect of Agent_num on the total overhead in terms of five different algorithms in Fig. 6.16a. Firstly, as the number of agents increases, the total overhead can become higher, which is inevitable because more agents bring severe energy consumption and latency. Furthermore, we set the number of agents as 5, 10, 20, and 30 in the single edge network with adequate channels, and the proposed DRL-based algorithm shows better performance than other four

Fig. 6.16 Performance
versus the number of agents.
(**a**) The total agent overhead.
(**b**) The average delay. (**c**)
The energy efficiency

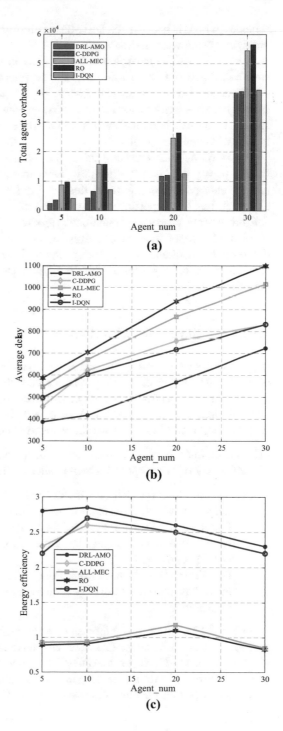

schemes, which is because the DRL-based algorithm can help more agents adapt to the dynamic network environments.

Next, we continue to explore average delay in terms of different agent_num. As shown in Fig. 6.16b, it has lower average delay compared with four other schemes, which means that each agent can acquire lower execution delay. Simultaneously, for certain latency-sensitive application tasks, we can utilize our proposed DRL-based scheme to allocate computational resources for each agent, which can guarantee the quality of experience (QoE).

Figure 6.16c shows the impact of the number of agents on energy efficiency. In this paper, the energy efficiency refers to the ratio of total bits of offloaded tasks to energy consumption. As illustrated in Fig. 6.16c, the proposed DRL-based algorithm has better performance gains, which is because DRL-based framework can orchestrate proper computational resources for each agent. However, as the number of agents increases, energy efficiency is gradually decreasing, which demonstrates centralized processing scheme is not suitable for large-scale agents scenario. Hence, it is vital to choose appropriate number of agents in the single edge network.

6.3.4.2   Multiple Edge Networks Scene

In the multiple edge networks scene, we consider multiple edge areas consisting of many agents and each edge network can share the same spectrum in order to improve the spectrum utilization. Simultaneously, each edge network has 10 variable channels for each agent. Additionally, in this test scenario, we utilize the proposed MADDPG structure to reduce the total energy consumption and latency.

(1) Performance Analysis: In the multi-agent environments, to demonstrate the efficiency of our proposed scheme, other methods are illustrated as follows.

- Random Strategy (RS): RS means it chooses to offload or execute their task randomly, which is different from proposed MADDPG strategy.
- Equal Power Transmission (EPT): EPT indicates all agents in the multiple edge networks can employ the same transmission power for offloading the task to edge cloud.
- Distributed Deep Deterministic Policy Gradient (D-DDPG): D-DDPG represents that each agent can only acquire the local information instead of global information.

In Fig. 6.17a, we set different task size for each agent and its range is from 300 to 700 KB. Certainly, larger task size can cause more computational overhead for all three schemes. However, MADDPG has lower rising speed and fewer computational overhead in four schemes since it can adjust offloading decision-making, transmission power, channel allocation, and CPU cycle frequency to adapt to the

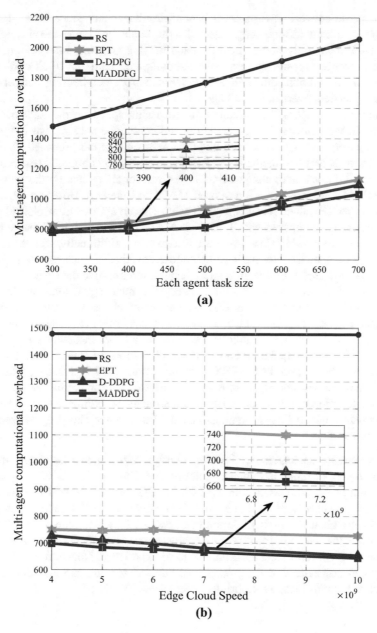

Fig. 6.17 The total computational overhead analysis. (**a**) Each agent task size. (**b**) Edge cloud speed

high-dynamic network environments better. Simultaneously, since RS neglects the interference among different edge networks, this cannot guarantee the channel quality. D-DDPG can only obtain local decision-making and state information, which cannot converge to lower system overhead compared with MADDPG. Moreover, MADDPG can combat the channel interference from neighboring networks and reduce the system overhead for all agents.

Figure 6.17b means the influence of processing speed on the total overhead. First of all, we consider 3 edge networks and each edge network has 2 agents in our simulation parameters. At the same time, we compare our proposed MADDPG scheme with other three methods. As RS means each agent chooses to execute tasks randomly, it causes more computational overhead for all agents. Furthermore, EPT and D-DDPG are close to the proposed MADDPG method. However, MADDPG is superior to EPT since it can help each agent choose optimal transmission power. Simultaneously, compared with D-DPPG, MADDPG can obtain global state and action information, which leads to lower system cost. Additionally, with the increase of edge cloud speed, multi-agent computational overhead is decreasing, so we can utilize the edge cloud service to execute the task and improve the QoE.

In Fig. 6.18a, we test the results between the multi-agent total overhead and energy weight factor. Higher energy weight factor can cause lower multi-agent overhead because delay weight factor has an important role in total system overhead. Meanwhile, although EPT employs the equal transmission power and D-DDPG can adjust the task scheduling and resource allocation schemes in terms of local information, the proposed MADDPG algorithm framework has better performance gains than other baseline algorithms since it utilizes the global information and distributed execution to tailor the orchestration strategy according to various network environments, which further demonstrates the intelligence and progressiveness of our proposed algorithm.

(2) Interference Management: For the multiple edge networks scene, there is too much channel assignment interference from proximal edge networks. Additionally, compared with RS, MADDPG can minimize the total multi-agent overhead and guarantee the QoS better since RS cannot ensure the selected channel quality. Simultaneously, EPT is not able to intelligently allocate the transmission power for each agent and D-DDPG can only profile the computation offloading according to local experience. Hence, from Fig. 6.18b, we can conclude that the proposed MADDPG scheme can face the complex network conditions, which brings lower system overhead.

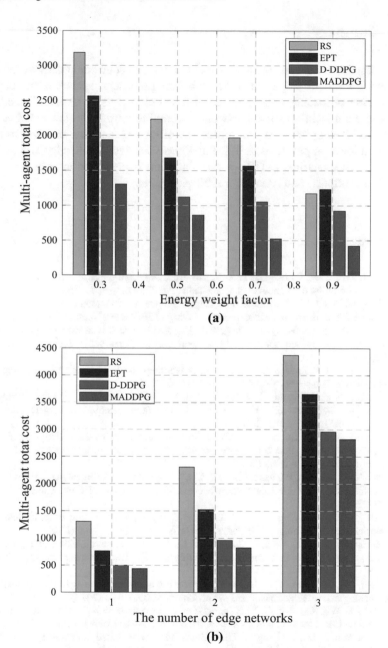

Fig. 6.18 The total multi-agent overhead analysis. (**a**) Energy weight factor. (**b**) The number of edge networks

6.4   Summary

In this chapter, we discuss the resource allocation and edge computation offloading problems in mobile edge networks. We first propose a spectrum sharing and computation offloading scheme for SDN-based ultra dense networks. In addition, we propose a novel DRL-based network structure followed with the 2AGT scheme and 3AUS strategy to jointly optimize the offloading decision and transmission resource allocation problems in IoT. Finally, we consider single edge and multiple edge network scenarios. A DRL-based algorithm is presented to reduce the total energy consumption and latency for multiple agents.

References

1. F. Li, H. Yao, J. Du, C. Jiang, Z. Han, Y. Liu, Auction design for edge computation ofloading in SDN-based ultra dense networks. IEEE Trans. Mobile Comput. **21**, 1580–1595 (2020)
2. Y. Gong, H. Yao, J. Wang, M. Li, S. Guo, Edge intelligence-driven joint offloading and resource allocation for future 6G industrial internet of things. IEEE Trans. Netw. Sci. Eng. (2022)
3. Y. Gong, H. Yao, J. Wang, L. Jiang, F.R. Yu, Multi-agent driven resource allocation and interference management for deep edge networks. IEEE Trans. Vehic. Technol. **71**(2), 2018–2030 (2021)
4. H. Yao, H. Liu, P. Zhang, S. Wu, S. Guo, A learning-based approach to intra-domain QoS routing. IEEE Trans. Veh. Technol. **69**, 6718–6730 (2020)
5. J. Du, C. Jiang, H. Zhang, X. Wang, Y. Ren, M. Debbah, Secure satellite-terrestrial transmission over incumbent terrestrial networks via cooperative beamforming. IEEE J. Sel. Areas Commun. **36**(7), 1367–1382 (2018)
6. C. Qiu, H. Yao, R. Yu, F. Xu, C. Zhao, Deep q-learning aided networking, caching, and computing resources allocation in software-defined satellite-terrestrial networks. IEEE Trans. Veh. Technol. **68**, 5871–5883 (2019)
7. H. Yao, T. Mai, X. Xu, P. Zhang, M. Li, Y. Liu, NetworkAI: an intelligent network architecture for self-learning control strategies in software defined networks. IEEE Int. Things J. **5**, 4319–4327 (2018)
8. H. Yao, T. Mai, J. Wang, Z. Ji, C. Jiang, Y. Qian, Resource trading in blockchain-based industrial internet of things. IEEE Trans. Ind. Informat. **15**, 3602–3609 (2019)
9. Q. Zhang, C. Zhu, L.T. Yang, Z. Chen, Z. Liang, L. Peng, An incremental CFS algorithm for clustering large data in industrial internet of things. IEEE Trans. Ind. Inform. **13**, 1193–1201 (2017)
10. S. Chen, F. Qin, B. Hu, X. Li, Z. Chen, User-centric ultra-dense networks for 5g: challenges, methodologies, and directions. IEEE Wirel. Commun. **23**, 78–85 (2018)
11. F. Zhou, Y. Wu, R.Q. Hu, Q. Yi, Computation rate maximization in UAV-enabled wireless-powered mobile-edge computing systems. IEEE J. Sel. Areas Commun. **36**, 1–15 (2018)
12. F. Li, H. Yao, J. Du, C. Jiang, Y. Qian, Stackelberg game based computation offloading in social and cognitive IIoT. IEEE Trans. Ind. Inform. **16**, 5444–5455 (2019)
13. C. Yang, J. Li, N. Qiang, A. Anpalagan, M. Guizani, Interference-aware energy efficiency maximization in 5g ultra-dense networks. IEEE Trans. Commun. **65**, 728–739 (2017)
14. B. Davie, T. Koponen, J. Pettit, B. Pfaff, M. Casado, N. Gude, A. Padmanabhan, T. Petty, K. Duda, A. Chanda, A database approach to SDN control plane design. Acm Sigcomm. Comput. Commun. Rev. **47**, 15–26 (2017)
15. A. Dixit, F. Hao, S. Mukherjee, T.V. Lakshman, R. Kompella, Towards an elastic distributed SDN controller. Comput. Commun. Rev. **43**, 7–12 (2013)

16. T. Mai, H. Yao, S. Guo, Y. Liu, In-network computing powered mobile edge: Toward high performance industrial IoT. IEEE Netw. **35**(1), 289–295 (2020)
17. G. Mitsis, P.A. Apostolopoulos, E.E. Tsiropoulou, S. Papavassiliou, Intelligent dynamic data offloading in a competitive mobile edge computing market. Future Int. **11**, 118 (2019)
18. L. Duan, J. Huang, B. Shou, Economics of femtocell service provision. IEEE Trans. Mobile Comput. **12**, 2261–2273 (2012)
19. L. Duan, L. Gao, J. Huang, Cooperative spectrum sharing: a contract-based approach. IEEE Trans. Mob. Comput. **13**, 174–187 (2012)
20. Y. Jie, A. Kamal, M. Alnuem, User cooperation solution of multipath streaming application using auction theory, in *IEEE Global Communications Conference*, Washington, DC (2017)
21. J. Du, E. Gelenbe, C. Jiang, H. Zhang, Y. Ren, Contract design for traffic offloading and resource allocation in heterogeneous ultra-dense networks. IEEE J. Sel. Areas Commun. **35**(11), 2457–2467 (2017)
22. B.A.A. Nunes, M. Mendonca, X.N. Nguyen, K. Obraczka, T. Turletti, A survey of software-defined networking: past, present, and future of programmable networks. IEEE Commun. Surv. Tutor. **16**, 1617–1634 (2014)
23. A. Blenk, A. Basta, M. Reisslein, W. Kellerer, Survey on network virtualization hypervisors for software defined networking. IEEE Commun. Surv. Tutor. **18**, 655–685 (2017)
24. H. Yao, S. Ma, J. Wang, P. Zhang, S. Guo, A continuous-decision virtual network embedding scheme relying on reinforcement learning. IEEE Trans. Netw. Service Manag. **17**, 864–875 (2020)
25. M.J. Abdel-Rahman, E.D.A. Mazied, A. Mackenzie, S. Midkiff, M.R. Rizk, M. El-Nainay, On stochastic controller placement in software-defined wireless networks, in *IEEE Wireless Communications and Networking Conference*, San Francisco, CA (2017)
26. S. Zhou, T. Zhao, Z. Niu, S. Zhou, Software-defined hyper-cellular architecture for green and elastic wireless access. IEEE Commun. Maga. **54**, 12–19 (2015)
27. C. Giraldo, F. Gilcastineira, C. Lopezbravo, F.J. Gonzalezcastano, A software-defined mobile network architecture, in *IEEE International Conference on Wireless and Mobile Computing*, Larnaca (2014)
28. R.D.R. Fontes, C.E. Rothenberg, Mininet-WIFI: A platform for hybrid physical-virtual software-defined wireless networking research, in *Proceedings of the 2016 ACM SIGCOMM Conference*, Florianopolis (2016)
29. N. Mckeown, T. Anderson, H. Balakrishnan, G. Parulkar, L. Peterson, J. Rexford, S. Shenker, J. Turner, Openflow:enabling innovation in campus networks. Acm Sigcomm. Comput. Commun. Rev. **38**, 69–74 (2008)
30. M. Jervis, M. Sen, P.L. Stoffa, Network innovation using OpenFlow: a survey. IEEE Commun. Surv. Tutor. **16**, 493–512 (2014)
31. D.B. Rawat, S. Reddy, Recent advances on software defined wireless networking, in *SoutheastCon 2016*, IEEE, Norfolk, VA (2016)
32. C. Singhal, S. De, *Resource Allocation in Next-Generation Broadband Wireless Access Networks* (IGI Global, Pennsylvania, 2011)
33. P. Jehiel, B. Moldovanu, Auctions with downstream interaction among buyers. RAND J. Econ. **31**, 768–791 (2000)
34. K. Bagwell, P.C. Mavroidis, R.W. Staiger, The case for auctioning countermeasures in the WTO, Technical Report, National Bureau of Economic Research (2003)
35. X. You, C.-X. Wang, J. Huang, X. Gao, Z. Zhang, M. Wang, Y. Huang, C. Zhang, Y. Jiang, J. Wang, et al., Towards 6G wireless communication networks: Vision, enabling technologies, and new paradigm shifts. Sci. China Inf. Sci. **64**(1), 1–74 (2021)
36. A. Mukherjee, P. Goswami, M.A. Khan, L. Manman, L. Yang, P. Pillai, Energy-efficient resource allocation strategy in massive IoT for industrial 6G applications. IEEE Int. Things J. **8**(7), 5194–5201 (2020)
37. Y. Gong, J. Wang, T. Nie, Deep reinforcement learning aided computation offloading and resource allocation for IoT, in *2020 IEEE Computing, Communications and IoT Applications (ComComAp)*, Beijing (2020), pp. 01–06

38. Y. Mao, C. You, J. Zhang, K. Huang, K.B. Letaief, A survey on mobile edge computing: The communication perspective. IEEE Commun. Surv. Tutor. **19**(4), 2322–2358 (2017)
39. H. Yao, L. Wang, X. Wang, Z. Lu, Y. Liu, The space-terrestrial integrated network: an overview. IEEE Commun. Mag. **56**(9), 178–185 (2018)
40. H. Yao, T. Mai, J. Wang, Z. Ji, C. Jiang, Y. Qian, Resource trading in Blockchain-based industrial Internet of Things. IEEE Trans. Ind. Inf. **15**(6), 3602–3609 (2019)
41. T. Mai, H. Yao, N. Zhang, L. Xu, M. Guizani, S. Guo, Cloud mining pool aided blockchain-enabled internet of things: an evolutionary game approach. IEEE Trans. Cloud Comput. (2021). https://doi.org/10.1109/TCC.2021.3110965
42. Y. Chen, N. Zhang, Y. Zhang, X. Chen, Dynamic computation offloading in edge computing for internet of things. IEEE Int. Things J. **6**(3), 4242–4251 (2018)
43. K. Kumar, J. Liu, Y.-H. Lu, B. Bhargava, A survey of computation offloading for mobile systems. Mob. Netw. Appl. **18**(1), 129–140 (2013)
44. Z. Hong, W. Chen, H. Huang, S. Guo, Z. Zheng, Multi-hop cooperative computation offloading for industrial IoT–edge–cloud computing environments. IEEE Trans. Parall. Distrib. Syst. **30**(12), 2759–2774 (2019)
45. P. Si, Y. He, H. Yao, R. Yang, Y. Zhang, DAVE: offloading delay-tolerant data traffic to connected vehicle networks. IEEE Trans. Vehic. Technol. **65**(6), 3941–3953 (2016)
46. L. Yang, H. Yao, J. Wang, C. Jiang, A. Benslimane, Y. Liu, Multi-UAV-enabled load-balance mobile-edge computing for IoT networks. IEEE Int. Things J. **7**(8), 6898–6908 (2020)
47. Z. Zhou, X. Chen, E. Li, L. Zeng, K. Luo, J. Zhang, Edge intelligence: paving the last mile of artificial intelligence with edge computing. Proc. IEEE **107**(8), 1738–1762 (2019)
48. T.K. Rodrigues, K. Suto, H. Nishiyama, J. Liu, N. Kato, Machine learning meets computation and communication control in evolving edge and cloud: challenges and future perspective. IEEE Commun. Surv. Tutor. **22**(1), 38–67 (2019)
49. P. Yang, F. Lyu, W. Wu, N. Zhang, L. Yu, X. Shen, Edge coordinated query configuration for low-latency and accurate video analytics. IEEE Trans. Ind. Inf. **16**(7), 4855–4864 (2020)
50. M.-H. Chen, B. Liang, M. Dong, Joint offloading decision and resource allocation for multi-user multi-task mobile cloud. in *IEEE International Conference on Communications. (ICC)*, Kuala Lumpur (2016), pp. 1–6
51. S.R. Bickham, M.A. Marro, J.A. Derick, W.-L. Kuang, X. Feng, Y. Hua, Reduced cladding diameter fibers for high-density optical interconnects. J. Lightwave Technol. **38**(2), 297–302 (2019)
52. H. Widiarti, S.-Y. Pyun, D.-H. Cho, Interference mitigation based on femtocells grouping in low duty operation, in *IEEE Vehicular Technology Conference (VTC)*, Ottawa (2010), pp. 1–5
53. J. Phiri, T.J. Zhao, Using Shannon's information theory and artificial neural networks to implement multimode authentication, in *IEEE International Conference on Communications and Intelligence Information Security (ICCIIS)*, Nanning (2010), pp. 271–274
54. C. You, K. Huang, H. Chae, B.-H. Kim, Energy-efficient resource allocation for mobile-edge computation offloading. IEEE Trans. Wirel. Commun. **16**(3), 1397–1411 (2016)
55. L. Huang, X. Feng, A. Feng, Y. Huang, L.P. Qian, Distributed deep learning-based offloading for mobile edge computing networks. Springer Mob. Netw. Appl. (2018). https://doi.org/10.1007/s11036-018-1177-x
56. Y. Zhan, S. Guo, P. Li, J. Zhang, A deep reinforcement learning based offloading game in edge computing. IEEE Trans. Comput. **69**(6), 883–893 (2020)
57. C. Qiu, F.R. Yu, H. Yao, C. Jiang, F. Xu, C. Zhao, Blockchain-based software-defined industrial internet of things: a dueling deep Q-learning approach. IEEE Int. Things J. **6**(3), 4627–4639 (2018)
58. L. Huang, S. Bi, Y.-J.A. Zhang, Deep reinforcement learning for online computation offloading in wireless powered mobile-edge computing networks. IEEE Trans. Mob. Comput. **19**(11), 2581–2593 (2019)
59. K.J. Åström, *Introduction to Stochastic Control Theory* (Courier Corporation, North Chelmsford, 2012)
60. S. Boyd, S.P. Boyd, L. Vandenberghe, *Convex Optimization* (Cambridge University Press, Cambridge, 2004)

61. B. Luo, Y. Yang, D. Liu, Adaptive Q-learning for data-based optimal output regulation with experience replay. IEEE Trans. Cybern. **48**(12), 3337–3348 (2018)
62. Z. Zhao, R. Zhao, J. Xia, X. Lei, D. Li, C. Yuen, L. Fan, A novel framework of three-hierarchical offloading optimization for MEC in industrial IoT networks. IEEE Trans. Ind. Inf. **16**(8), 5424–5434 (2019)
63. J. Wang, L. Zhao, J. Liu, N. Kato, Smart resource allocation for mobile edge computing: A deep reinforcement learning approach. IEEE Trans. Emergi. Topics Comput. **9**(3), 1529–1541 (2019)
64. J. Wan, S. Tang, Z. Shu, D. Li, S. Wang, M. Imran, A.V. Vasilakos, Software-defined industrial internet of things in the context of industry 4.0. IEEE Sensors J. **16**(20), 7373–7380 (2016)
65. J. Navarro-Ortiz, P. Romero-Diaz, S. Sendra, P. Ameigeiras, J.J. Ramos-Munoz, J.M. Lopez-Soler, A survey on 5G usage scenarios and traffic models. IEEE Commun. Surv. Tutor. **22**(2), 905–929 (2020)
66. J. Wang, C. Jiang, H. Zhang, Y. Ren, K.C. Chen, L. Hanzo, Thirty years of machine learning: the road to pareto-optimal wireless networks. IEEE Commun. Surv. Tutor. **22**(3), 1472–1514 (2020)
67. W. Saad, M. Bennis, M. Chen, A vision of 6G wireless systems: Applications, trends, technologies, and open research problems. IEEE Netw. **34**(3), 134–142 (2019)
68. J. Du, F.R. Yu, G. Lu, J. Wang, J. Jiang, X. Chu, MEC-assisted immersive VR video streaming over terahertz wireless networks: a deep reinforcement learning approach. IEEE Int. Things J. **7**(10), 9517–9529 (2020)
69. M. Giordani, M. Polese, M. Mezzavilla, S. Rangan, M. Zorzi, Toward 6G networks: use cases and technologies. IEEE Commun. Mag. **58**(3), 55–61 (2020)
70. H. Yao, C. Liu, P. Zhang, S. Wu, C. Jiang, S. Yu, Identification of encrypted traffic through attention mechanism based long short term memory. IEEE Trans. Big Data (2019). https://doi.org/10.1109/TBDATA.2019.2940675
71. X. You, C.-X. Wang, J. Huang, X. Gao, Z. Zhang, M. Wang, Y. Huang, C. Zhang, Y. Jiang, J. Wang, et al., Towards 6G wireless communication networks: Vision, enabling technologies, and new paradigm shifts. Sci. China Inf. Sci. **64**(1), 1–74 (2020)
72. Y. Gong, J. Wang, H. Yao, Distributed multi-agent empowered resource allocation in deep edge networks, in *International Wireless Communications and Mobile Computing (IWCMC)*, Harbin (2021), pp. 974–979
73. C. Qiu, X. Wang, H. Yao, J. Du, F.R. Yu, S. Guo, Networking integrated cloud-edge-end in IoT: a blockchain-assisted collective Q-learning approach. IEEE Int. Things J. (2020). https://doi.org/10.1109/JIOT.2020.3007650
74. C. Qiu, H. Yao, C. Jiang, S. Guo, F. Xu, Cloud computing assisted blockchain-enabled internet of things. IEEE Trans. Cloud Comput. (2019). https://doi.org/10.1109/TCC.2019.2930259
75. Y. Mao, C. You, J. Zhang, K. Huang, K.B. Letaief, A survey on mobile edge computing: the communication perspective. IEEE Commun. Surv. Tutor. **19**(4), 2322–2358 (2017)
76. J. Du, F.R. Yu, X. Chu, J. Feng, G. Lu, Computation offloading and resource allocation in vehicular networks based on dual-side cost minimization. IEEE Trans. Vehic. Technol. **68**(2), 1079–1092 (2018)
77. W. Zhang, Y. Wen, K. Guan, D. Kilper, H. Luo, D.O. Wu, Energy-optimal mobile cloud computing under stochastic wireless channel. IEEE Trans. Wirel. Commun. **12**(9), 4569–4581 (2013)
78. J. Zhang, X. Hu, Z. Ning, E.C.-H. Ngai, L. Zhou, J. Wei, J. Cheng, B. Hu, Energy-latency tradeoff for energy-aware offloading in mobile edge computing networks. IEEE Int. Things J. **5**(4), 2633–2645 (2017)
79. Y. Mao, J. Zhang, K.B. Letaief, Dynamic computation offloading for mobile-edge computing with energy harvesting devices. IEEE J. Sel. Areas Commun. **34**(12), 3590–3605 (2016)
80. Z. Hong, H. Huang, S. Guo, W. Chen, Z. Zheng, QoS-aware cooperative computation offloading for robot swarms in cloud robotics. IEEE Trans. Vehic. Technol. **68**(4), 4027–4041 (2019)

81. Y. Liu, Y. Li, Y. Niu, D. Jin, Joint optimization of path planning and resource allocation in mobile edge computing. IEEE Trans. Mob. Comput. **19**(9), 2129–2144 (2019)
82. X. Lyu, W. Ni, H. Tian, R.P. Liu, X. Wang, G.B. Giannakis, A. Paulraj, Optimal schedule of mobile edge computing for internet of things using partial information. IEEE J. Sel. Areas Commun. **35**(11), 2606–2615 (2017)
83. Q. Li, H. Yao, T. Mai, C. Jiang, Y. Zhang, Reinforcement-learning-and belief-learning-based double auction mechanism for edge computing resource allocation. IEEE Int. Things J. **7**(7), 5976–5985 (2019)
84. C. Wang, C. Liang, F.R. Yu, Q. Chen, L. Tang, Computation offloading and resource allocation in wireless cellular networks with mobile edge computing. IEEE Trans. Wirel. Commun. **16**(8), 4924–4938 (2017)
85. J. Wang, C. Jiang, K. Zhang, X. Hou, Y. Ren, Y. Qian, Distributed Q-learning aided heterogeneous network association for energy-efficient IIoT. IEEE Trans. Ind. Inform. **16**(4), 2756–2764 (2019)
86. J. Zhao, Q. Li, Y. Gong, K. Zhang, Computation offloading and resource allocation for cloud assisted mobile edge computing in vehicular networks. IEEE Trans. Vehic. Technol. **68**(8), 7944–7956 (2019)
87. P. Si, Y. He, H. Yao, R. Yang, Y. Zhang, Dave: Offloading delay-tolerant data traffic to connected vehicle networks. IEEE Trans. Vehic. Technol. **65**(6), 3941–3953 (2016)
88. T.D. Burd, R.W. Brodersen, Processor design for portable systems. J. VLSI Signal Process. Syst. Signal Image Video Technol. **13**(2–3), 203–221 (1996)
89. J.M. Rabaey, A.P. Chandrakasan, B. Nikolić, *Digital Integrated Circuits: A Design Perspective* (Pearson Education, Upper Saddle River, 2003)
90. C. Wang, F.R. Yu, C. Liang, Q. Chen, L. Tang, Joint computation offloading and interference management in wireless cellular networks with mobile edge computing. IEEE Trans. Vehic. Technol. **66**(8), 7432–7445 (2017)
91. C. You, K. Huang, H. Chae, B.-H. Kim, Energy-efficient resource allocation for mobile-edge computation offloading. IEEE Trans. Wirel. Commun. **16**(3), 1397–1411 (2016)
92. B.-G. Chun, P. Maniatis, Augmented smartphone applications through clone cloud execution, in *USENIX Workshop on Hot Topics in Operating Systems. (HoTOS)*, Monte Veritłd' (2009), pp. 8–11
93. Y. Wang, M. Sheng, X. Wang, L. Wang, J. Li, Mobile-edge computing: Partial computation offloading using dynamic voltage scaling. IEEE Trans. Commun. **64**(10), 4268–4282 (2016)

Chapter 7
Blockchain-Enabled Intelligent IoT

Abstract The past few years have witnessed an exponential growth of diverse Internet of Things (IoT) devices as well as compelling applications ranging from industrial production, intelligent transport, and warehouse logistics to medical care. Dramatic advances in IoT technology not only bring enormous economic opportunities but also challenges. Recently, with the appearance of blockchain technology, the integration of IoT and blockchain (BCoT) is considered a promising solution to address these issues. Blockchain provides a secure and scalable data management framework for IoT devices. However, the huge computation and energy cost of the consensus process in blockchain prevents it from being directly applied as a generic platform. To overcome this challenge, we first propose a cloud mining pool-aided BCoT architecture. Based on this architecture, we study the mining pool selection problem and analyze the colony behaviors of IoT devices with different pooling strategies. We propose a centralized evolutionary game-based pool selection algorithm for the sake of maximizing the system utility. Secondly, to overcome the power and computation constraints of the IoT devices in the blockchain platform, we introduce the cloud computing service to the blockchain platform for the sake of assisting to offload computational task from the IIoT network itself. Also, we study the resource management and pricing problem between the cloud provider and miners. And a multi-agent reinforcement learning algorithm is conceived for searching the near-optimal policy.

Keywords Blockchain · Cloud mining pool · Evolutionary game · Stackelberg game

The past few years have witnessed an exponential growth of diverse Internet of Things (IoT) devices as well as compelling applications ranging from industrial production, intelligent transport, and warehouse logistics to medical care. Dramatic advances in IoT technology not only bring enormous economic opportunities but also challenges (e.g., privacy and security vulnerabilities). Recently, with the appearance of blockchain technology, the integration of IoT and blockchain (BCoT) is considered a promising solution to address these issues. Blockchain provides a secure and scalable data management framework for IoT devices. However, the huge computation and energy cost of the consensus process in blockchain prevents it from being directly applied as a generic platform. To overcome this challenge, we first propose a cloud mining pool-aided BCoT architecture [1]. The IoT devices can rent the computing resources from the cloud mining pools to offload the mining process. Based on this architecture, we study the mining pool selection problem and analyze the colony behaviors of IoT devices with different pooling strategies. We propose a centralized evolutionary game-based pool selection algorithm for the sake of maximizing the system utility. Secondly, to overcome the power and computation constraints of the IoT devices in the blockchain platform, we introduce the cloud computing service to the blockchain platform for the sake of assisting to offload computational task from the IIoT network itself [2]. Also, we study the resource management and pricing problem between the cloud provider and miners. More explicitly, we model the interaction between the cloud provider and miners as a Stackelberg game, where the leader, i.e., cloud provider, makes the price first, and then miners act as the followers. Moreover, in order to find the Nash equilibrium of the proposed Stackelberg game, a multi-agent reinforcement learning algorithm is conceived for searching the near-optimal policy.

7.1   Cloud Mining Pool-Aided Blockchain-Enabled IoT

In the past decade, the Internet of things (IoT) has attracted a large amount of attention from both academia and industry [3]. The IoT refers to the billions of physical devices that are now connected to and transfer data through the Internet without requiring human-to-human or human-to-computer interaction. These connected IoT devices are slowly entering every aspect of our lives ranging from healthcare to industrial manufacture. According to Gartner's prediction, it is expected more than 25 billion IoT connections in the future year 2025. However, with the large-scale IoT deployments, IoT applications are facing challenges in the aspect of scalability, privacy, and security [4]. The current IoT system adopts a centralized management platform to authenticate, authorize, and connect a massive of heterogeneous IoT devices, which will turn into a bottleneck. Besides, unsecured IoT devices provide an easy target for distributed-denial-of-service (DDoS) attacks, malicious attackers, and data breaches.

In recent years, another breakthrough technology, blockchain, offers significant opportunities to address these challenges [5]. The blockchain is a distributed digital ledger of transactions that is maintained by a community of participants without the intervention of a trusted third party [6]. Within a blockchain community, any new transactions or events must be validated upon the agreement among the majority of the participants through a consensus process (e.g., proof of work (PoW), proof of stake (PoS)) before they are attached to the chain [7]. Such a process creates tamper-resistant records of shared transactions and events among the involved parties. Therefore, no single organization has control over the data generated by IoT devices in blockchain, thereby protecting the privacy of data and enhancing scalability [8]. Moreover, blockchain adds a layer of security in terms of encryption, the removal of a single point of failure, and the ability to quickly identify the weak point in the network [9]. Recently, a large number of applications combining blockchain and the IoT can be seen [4]. For example, Deloitte uses blockchain and IoT technology in supply chain traceability [10].

While blockchain provides a secure and scalable data management framework, there still exist challenges to be addressed before it can serve as a generic platform for IoT. As discussed above, the consensus process (e.g., proof of work (PoW)) in the blockchain is particularly computationally intensive and energy-consuming. The participants, termed miners, have to constantly try to solve a cryptographic puzzle in the form of the hash computation. Considering that the majority of IoT devices are too limited in terms of computing, storage, and energy resources, this computationally intensive process hinders the integration of IoT and blockchain. While some energy-efficient consensus algorithms (e.g., proof of stake (PoS), practical byzantine fault tolerance (PBFT)) are developed, the computation and energy overhead are still inevitable.

To address these challenges, the cloud mining mechanism becomes a viable option. Cloud computing can empower resource-constrained IoT devices with extra sufficient storage and computing resource. In this way, more IoT devices are enabled to participate in the blockchain network, so as to increase the whole system utility. Recently, a large and growing body of literature has investigated cloud mining-based BCoT architecture [9, 11, 12]. These works are mainly focusing on the resource allocation between the devices and cloud servers. For example, in [13], Xiong et al. formulated the resource allocation problem between cloud services and IoT devices as a Stackelberg game and implemented a backward induction algorithm to search the Nash equilibria of this game.

However, with the exponential growth of the number of IoT devices and hash rate, the probability for a single miner to win the mining competition game tends to be slim. Only a few fortunate miners would obtain large rewards and the majority will get no rewards. To seek a steady reward stream, miners are gradually willing to group into several teams, called mining pools. In the mining pool, miners will share the rewards according to their contributed hash power (i.e., computing resource).

Therefore, in this paper, we design a cloud mining pool-aided BCoT architecture, where IoT devices can rent the computing and storage resources from the cloud mining pool. In terms of this architecture, we discuss the mining pool selection problem among IoT devices. Assuming that the IoT devices are rational (i.e., profit-driven), we model the dynamic mining pool selection process as an evolutionary game. To search the evolutionary stable strategy (ESS), we design a centralized evolutionary game-based pool selection algorithm, where a centralized controller is used to synchronize information. Besides, considering the non-cooperative relationship among miners, we propose a distributed reinforcement learning algorithm, termed the "WoLF-PHC" algorithm.

7.1.1 System Model

In this section, we first design a cloud mining pool-aided BCoT architecture. Then, we present the system model and problem formulation of the mining pool selection problem.

7.1.1.1 Cloud Mining Pool-Aided BCoT

As discussed above, the blockchain cannot be directly applied to IoT systems. To fulfill the computing and storage resources required in the consensus process, we adopt the cloud mining paradigm in this paper, where the IoT devices can use the computing power of mining equipment hosted in cloud computing servers without owning or maintaining the equipment. Next, we will detail the mining pool-aided BCoT architecture in the following.

As shown in Fig. 7.1, in our architecture, there exist two types of nodes to participate in the blockchain network, including the cloud services nodes and the IoT devices nodes [4]. The cloud services are responsible for storing the entire blockchain data (e.g., right now the full Bitcoin blockchain data occupies about 200G large) and undertaking computational intensive operations (e.g., consensus process), while IoT devices are only responsible for undertaking some simple operations (e.g., initiating transactions). It is worth mentioning that the IoT devices still need to keep a partial blockchain local for validating the authenticity of transactions.

During the initializing phase, the IoT devices first register as a legitimate entity (i.e., cloud miner) on cloud servers and obtain an identity ID and a public/privacy key. These cloud miners act as the proxy nodes of the IoT devices to offload their mining and storage tasks. Then, these miners will group themselves into several mining pools. In mining process, these pools present themselves to the whole system as single powerful proxy nodes. Combing with more computing resources, the mining pools are able to gain a computation advantage over other individual miners. Note that the miner can choose to redirect its hash power to any other mining pool at any time. In each pool, the cloud provider will place a coordinator in charge of managing the miners. They will work as a task scheduler to guarantee the miners are undertaking different subtasks so that they are not wasting hash power by trying to solve the same sub-cryptographic puzzle. Once successfully mining a block, the coordinator will divide the profit to each miner according to its devoted hash power. We will present the system model in the following.

7.1.1.2 System Model

Consider a set of IoT devices that are interested in participating in the consensus process, which is denoted as $\mathcal{N} = \{1, \ldots, N\}$. We assume that these miners are

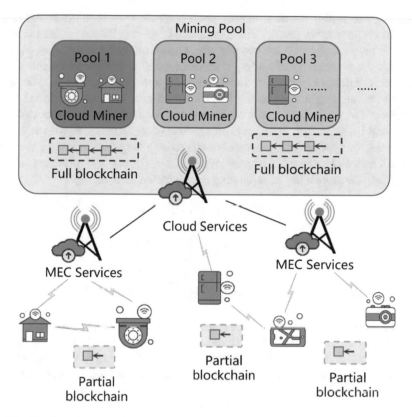

Fig. 7.1 Cloud mining pool-aided BCoT architecture

willing to form $M = \{1, \ldots, M\}$ mining pools, where each mining pool adopts a different pooling strategy with different hash power requirement [14]. Let ω_j represent the hash power required by the pool $j \in M$. According to consensus protocol, the probability of winning the mining game is related to the ratio between local hash power and the total hash power of the entire blockchain network. Therefore, we define a relative hash power α_j of pool j with respect to the entire hash power of all miners, which can be described as:

$$\alpha_j(\boldsymbol{\omega}, x_j, \boldsymbol{x}_{-j}) = \frac{x_j \omega_j}{\sum_{k \in M} x_k \omega_k}, \alpha_j > 0, \tag{7.1}$$

where x_j represents the pool j's population fraction, and the $\boldsymbol{x}_{-j}$ represents the sum of pools' population fraction expect to pool j. Note that α satisfies following condition:

$$\sum_{k \in M} \alpha_k = 1. \tag{7.2}$$

During the mining process, mining pools compete with each other in a race to solve the cryptographic puzzle. The appearance of solving the cryptographic puzzle can be formulated as a Poisson process with a mean random variable $\lambda = \frac{1}{T}$, where T denote the complexity of finding a block (e.g., $T = 600sec$ in Bitcoin). After successfully mining a block, the winner needs to propagate its solution to the entire network for reaching a consensus. Only the first block, which is confirmed by the majority of the participant, could be accepted as a new block. All other candidate blocks will be discarded, called orphaning. According to previous works [15], the propagation time of a block to reach consensus is mainly determined by the set of transaction size Q included in a block. We denote $\tau(Q) = \xi \times Q$ as the propagation time. Then, the probability of orphaning can be formulated as:

$$P^{orphan}(Q) = 1 - e^{-\lambda\tau(Q)}. \tag{7.3}$$

The successful probability of mining pool j to win the mining game can be formulated as:

$$P_j(\alpha_j, Q_j) = \alpha_j \times \left(1 - P_j^{orphan}(Q_j)\right) = \alpha_j \times e^{-\lambda\tau(Q_j)}. \tag{7.4}$$

After successfully mining a block, the winner can obtain a reward, which is composed of a fixed reward $R \geq 0$ and a variable reward ρQ [16]. The variable reward linearly increases with the size of the transaction Q in the block, and the ρ is the linear coefficient. Therefore, the expected reward for pool j can be expressed:

$$u_j(\alpha_j, Q_j) = (R + \rho Q_j)\alpha_j \times e^{-\lambda\tau(Q_j)}. \tag{7.5}$$

And the expected profit of the miner $i \in N$ in pool j can be expressed as:

$$R_i(\alpha_j, Q_j) = \frac{(R + \rho Q_j)}{Nx_j}\alpha_j \times e^{-\lambda\tau(Q_j)}. \tag{7.6}$$

Besides, since the miners rent the computation resource from the cloud servers, the miners have to pay for it [17]. We denote the price of each computing resource unit as p. Thus, the expected reward of the miner $i \in N$ in pool j can be reformulated as:

$$r_i(\alpha_j, Q_j) = \frac{(R + \rho Q_j)}{Nx_j}\alpha_j \times e^{-\lambda\tau(Q_j)} - p\omega_j. \tag{7.7}$$

As shown in Table 7.1, we list the notations of this paper.

Table 7.1   List of main notations

Parameter	Definition
N	Number of miners
M	Number of pools
ω_j	The hash rate required by pool j
Q_j	The set of transactions size included in a block
x_j	Population fraction of pool j
α_j	The relative computing power of pool j with respect to the all system
R	The fixed reward when mining a block
p	The price of each computing and storage resource unit
ρQ	The variable reward when mining a block
$\tau(Q)$	The time needed for a block to propagation
$P_j(\alpha_j, Q_j)$	The successful probability of mining pool j to win the mining game
$u_j(\alpha_j, Q_j)$	The expected reward of pool j
$r_i(\alpha_j, Q_j)$	The expected reward of the device i in pool j

7.1.2   Evolutionary Game Formulation

In this section, we apply the evolutionary game to the mining pool selection problem and present the concepts of replicator dynamics (RD) and evolutionary stable strategy (ESS) [18]. The evolutionary game defines a framework of contests, strategies, and analytic into which colony competition can be modeled. It can capture the strategy adaptation of rational agents according to their fitness. That is, the agent can slowly adjust its strategy (i.e., evolves) based on the environment knowledge. Note that we assume all users are rational (i.e., profit-seeking) [19]. Mathematically, for mining pool selection, the evolutionary game can be formulated as a 4-tuple $\mathcal{G} =< \mathcal{N}, x, \mathcal{M}, R >$, where

- Players: Players are the decision-makers with pre-programmed strategies in the game. In our scenario, each individual miner can be regarded as a player.
- Population: The population $x = [x_1, \ldots, x_M] \in X$ refers to the set of players in a mining pool. The population will present variation among competing players.
- Strategy: The strategy is a set of action $\mathcal{M} = \{1, \ldots, M\}$ that the player can perform. The different strategies will obtain different rewards. The strategy space in our scenario is the all available mining pools.
- Payoff: Payoff r_j reflects the player's expected outcome based on its strategy, where $r_i(\alpha_j, Q_j) = \frac{(R+\rho Q_j)}{Nx_j}\alpha_j \times e^{-\lambda Q_j} - p\omega_j$. Note that the reward is determined not only by the local strategy but also by the other players' strategies.

7.1.2.1 Replicator Dynamics of Pool Selection

We consider an evolutionary game-based pool selection in a blockchain network where a set of players select from a set of the available mining pool. Each mining pool adopts different pooling strategies with different hash rate requirements and the size of transactions. In the evolutionary game, the game is repeated, and each player observes the global average payoff and dynamically adjusts their strategy to obtain a higher expected payoff. To express the evolutionary dynamics in the game, the replicator dynamics function is introduced. The replicator dynamics function is a nonlinear game dynamic used to explain learning as well as evolution in evolutionary game [20]. The core idea of replicator dynamics is that the population will increase (decrease) if fitness is larger (smaller) than the average fitness. In our scenario, the replicator dynamics function of pool j can be described as:

$$\dot{x}_j(t) = \sigma x_j(t)(u_j(\alpha_j(t), Q_j) - \bar{u}(x(t))), \tag{7.8}$$

where $\dot{x}_j(t)$ is the growth rate of the pool j's population, σ is the speed parameter, and $\bar{u}(x)$ is the network average payoff, which can be formulated as:

$$\bar{u}(x(t)) = \sum_{j \in M} u_j(\alpha_j, Q_j)x_j. \tag{7.9}$$

The replicator dynamics functions must satisfy the following condition:

$$\sum_{j \in M} \dot{x}_j(t) = 0. \tag{7.10}$$

From the players' perspective, the miners will slowly adjust their selection strategies, if their payoff is less than the average payoff, otherwise the miners will keep their current strategies.

7.1.2.2 Evolutionary Equilibrium and Stability Analysis

As discussed above, the players constantly adjust their strategies (i.e., evolve) for the sake of a higher expected payoff. Along with the players evolves over time, the whole system will finally converge to the evolutionary stable strategy (ESS). The ESS is phenotypes that can persist in populations and cannot be invaded by any other strategies [21]. We can define that the x^* is an ESS if the following condition is satisfied:

$$\sum_{j \in M} x_j^* u_j((1 - \sigma)x^* + \epsilon x') \geq \sum_{j \in M} x_j' u_j((1 - \sigma)x^* + \sigma x'), \tag{7.11}$$

where x' is the invade state.

According to this definition, in the ESS, none of the players is willing to deviate its selection strategy (i.e., the rate of strategy adaptation is zero). By solving the replicator dynamics functions (i.e., $\dot{x}_j(t) = 0$), a set of fixed points can be obtained. According to [22], these fixed points are stable (i.e., ESS) if all eigenvalues of the Jacobian matrix have negative real parts. Then, the ESS can be defined as a set of stable fixed points, which can be described as follows.

Definition A population state x^* is an ESS, if the condition $(x - x^*)^T R(x^*) = 0$ implies that:

$$(x^* - x)^T R(x) \geq 0, \tag{7.12}$$

where $\forall x \in B - x^*$ is the neighborhood of X. $\qquad\qquad\square$

7.1.2.3 Two Mining Pool Study

To demonstrate the evolutionary stable strategy, in this part, we will present a two mining pools case study. We set the population fraction of two pool as $x_1 = x$, and $x_2 = 1 - x$. Then, we can obtain the Ordinary Differential Equations:

$$\dot{x}_1(t) = x_1 x_2 \left(\frac{\omega_1 k_1 - \omega_2 k_2}{N(x_1 \omega_1 + x_2 \omega_2)} - p(\omega_1 - \omega_2) \right), \tag{7.13}$$

$$\dot{x}_2(t) = x_1 x_2 \left(\frac{\omega_1 k_1 - \omega_2 k_2}{N(x_1 \omega_2 + x_2 \omega_1)} - p(\omega_1 - \omega_2) \right), \tag{7.14}$$

where

$$k_i = (R + \rho Q_i) \times e^{-\lambda \tau(Q_i)}. \tag{7.15}$$

By solving the above formulas, we can obtain the fixed points as $(x*, 1 - x*)$, where $x* = \frac{\omega_1 k_1 - \omega_2 k_2}{pN(\omega_1 - \omega_2)^2} - \frac{\omega_2}{\omega_1 - \omega_2}$. According to the above definition, this fixed point is ESS if all eigenvalues of the Jacobian matrix have negative real parts. For this replicate dynamic system, the Jacobian matrix of the replicator dynamics is

$$J = \begin{pmatrix} \frac{\partial f(x_1)}{x_1} & \frac{\partial f(x_1)}{x_2} \\ \frac{\partial f(x_2)}{x_1} & \frac{\partial f(x_2)}{x_2} \end{pmatrix}.$$

After some tedious mathematical manipulations, the rest point with $x* = \frac{\omega_1 k_1 - \omega_2 k_2}{pN(\omega_1 - \omega_2)^2} - \frac{\omega_2}{\omega_1 - \omega_2}$ is an ESS if the following conditions are satisfied:

$$\begin{cases} \omega_1 k_1 - \omega_2 k_2 < 0 \\ \omega_2 \omega_1 (k_2 - k_1)(\omega_2 - \omega_1) > 0 \end{cases}.$$

7.1.2.4　Delay in Replicator Dynamics

As discussed above, the players can adjust their strategies based on the system's average fitness. However, in actual deployment, the latest fitness information may not be available to all players. They can only rely on historical information to make decisions. Therefore, in this paper, we introduce a certain period delayed τ in our system. The replicator dynamics will be reformulated as:

$$\dot{x}_j(t) = x_j(t - \tau)(u_j(\alpha_j(t - \tau), Q_j) - \bar{u}(x(t - \tau))), \tag{7.16}$$

which is a delay differential equation. To obtain the solution to this equation, the Runge–Kutta method can be applied. Besides, the stability of the delay differential equation has been well studied. In [23], Obando et al. investigated the stability of the replicator dynamics with the effect of a time delay using the Lyapunov method. The theoretical results show that the ESS strategy is stable if the time delay is small enough. The detailed analysis can be founded in [23]. We will evaluate the impact of delay in the experiment section.

7.1.2.5　Evolutionary Game-Based Pool Selection Algorithm

As discussed above, we apply the evolutionary game to the mining pool selection problem. The miners continually adjust their strategies based on the system average fitness for the sake of a higher expected payoff. Along with the players evolves over time, the whole blockchain network can converge to the ESS. Therefore, in this paper, we propose a centralized population evolutionary strategy algorithm. In this approach, a centralized controller is deployed to calculate the average fitness of all players. Then, the average fitness is issued to each player to evaluate the current strategy based on its current payoff (i.e., switch their strategies or keep them) [24]. The centralized pool selection algorithm can be described as follows.

7.1.3　Distributed Reinforcement Learning Approach

Thus far, the paper has argued an ideal model where a centralized controller cloud be placed to guide behaviors of all players. However, in an actual IoT network, considering the non-cooperative relationship among them, the centralized controller may not be available [25]. Each player has to adapt its pool selection decision independently. Therefore, how could the individual devices optimize their strategy in the non-stationary system (i.e., multi-agent system) is a critical challenge. Inspired by the recent success of applying reinforcement learning algorithms in multi-agent system, we introduce a distributed reinforcement learning algorithm, named "WoLF-PHC" [26].

Algorithm 7.1 The population evolution approach for pools selection

---

1: Each pool set ω, Q.
2: All devices randomly choose the pool.
3: **repeat**
4:    Each device compute the payoff from:
5:    $r_i(\alpha_j, Q_j) = \frac{(R+\rho Q_j)}{Nx_j}\alpha_j \times e^{-\lambda Q_j} - p\omega_i$
6:    The payoff information is sent to the controller.
7:    The centralized controller computes average payoff and broadcast it to all devices.
8:    $\bar{r}(x) = \frac{\sum\limits_{i \in N} r_i(\alpha_j, Q_j)}{N}$
9:    **for** $i \in N$ **do**
10:       **if** $r_i < \bar{r}$ **then**
11:          **if** $rand() < (\bar{r} - r_i)/\bar{r}$ **then**
12:             Choose other pool.
13:          **end if**
14:       **end if**
15:    **end for**
16: **until**

7.1.3.1 The Multi-agent System

With the decision-making shift to each miner, these independent miners constitute a multi-agent system, which can be modeled as a decentralized partially observable Markov decision process (Dec-POMDP) [27]. Formally, a Dec-POMDP can be formalized as a 5-tuple $< N, S, O_i, A_i, R >$, where N is the set of agents, S is the global states space, O_i is the local observations space of agent i, A_i is the action space of agent i, and R is the immediate rewards [28]. As shown in Fig. 7.2, at each step, each agent takes an action a_i according to current policy $\pi_i(a_i|o_i)$ and its local observation o_i. Then, the system will generate an immediate reward R, and the state s will transit to a new state s'. Note that compared to the single-agent case where reward only related to its action, the multi-agent case's reward is related to all agents' behaviors.

Specifically, in our scenario, the observation can be described as $O^t = [x_1^{t-1}, \ldots, x_M^{t-1}]$ (i.e., current mining pools' state, which is determined by all agents' action in the last time $t - 1$), the action of each agent can be described as $A^t = [\mathcal{M}]$, where M is the set of the mining pools, and the immediate reward can be formulated as the expected profit r_i. In the following, we will define the three components of the miners:

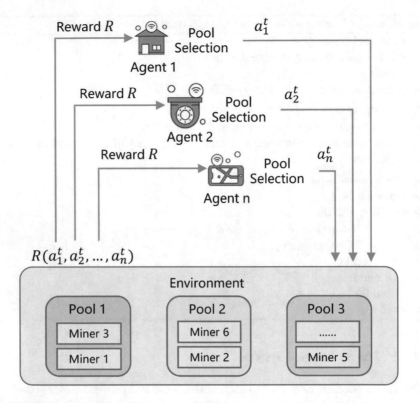

Fig. 7.2 The multi-agent system

Definition The three components of the miners:

- Observation:

$$O^t = \left[x_1^{t-1}, \ldots, x_M^{t-1}\right]$$

- Action:

$$A^t = [\mathcal{M}]$$

- Immediate reward:

$$R_i^t\left(a_1^t, a_2^t, \ldots, a_n^t\right) = \frac{(R + \rho Q_j)}{N x_j}\alpha_j \times e^{-\lambda Q_j} - p\omega_j$$

Note that the reward is only related to the joint action $(a_1^t, a_2^t, \ldots, a_n^t)$.

7.1.3.2 Policy Generation

Learning in a multi-agent system is much more difficult than in a single-agent system. One of the critical challenges is the moving target problem (i.e., non-stationary learning problem), which is caused by the noise signal brought by other agents [29]. Directly applying single-agent reinforcement learning (e.g., Q-learning, Policy gradient) will suffer seriously no-convergence problem [30]. In this paper, we introduce an enhanced policy gradient algorithm, termed WoLF Policy Hill Climbing (WoLF-PHC). It adopts the "wining or learning fast" scheme (i.e., learn slowly while winning or quickly while losing), where a variety of learning rates are used to encourage convergence.

In the WoLF-PHC, the updating rule of the Q value can be described as [31]:

$$Q_i(a_t) \leftarrow (1 - \alpha) Q_i(a_t) + \alpha \left(R_i + \delta \max_{a \in A} Q_i(a_{t+1}) \right), \quad (7.17)$$

where $\delta \in (0, 1]$ is the discount factor, and $\alpha \in (0, 1]$ is the learning rate. The discount factor determines the importance of future rewards and the learning rate determines what extent new knowledge overrides the old knowledge. During the training process, agents continually update their strategies, i.e., $\pi_i(a) :\to Pr(A)$, for the sake of maximizing the cumulative reward by learning from the environment.

To update the $\pi_i(a)$, the WolF-PHC adopts two learning rates θ^{win} and θ^{lose}, where $\theta^{win} > \theta^{lose}$ (i.e., learn slowly while winning or quickly while losing). They are used to update agents' policy depending upon if the agent is winning or losing [32]. To determine the winning or loss of current policy, a baseline is designed. The baseline is the expected reward of the average policy $\overline{\pi}_i(a_t)$, which can be formulated as:

$$\overline{\pi}_i(a_t) \leftarrow \overline{\pi}_i(a_t) + \frac{\pi_i(a_t) - \overline{\pi}_i(a_t)}{N_i(t)}, \ \forall a_t \in A, \quad (7.18)$$

where

$$N_i(t + 1) \leftarrow N_i(t) + 1. \quad (7.19)$$

Then, the θ^{win} is applied to update the policy cautiously in the condition of win, otherwise, θ_m^{lose} is used, i.e.,

$$\theta = \begin{cases} \theta^{win}, & \Pi, \\ \theta^{lose}, & o.w, \end{cases} \quad (7.20)$$

where Π:

$$\sum_{a \in A_i} \pi_i(a_t) Q_i(a_t) > \sum_{a \in A_i} \overline{\pi}_i(a_t) Q_i(a_t). \qquad (7.21)$$

In the learning process, the agents constantly learn and adapt their policy toward maximizing the expected reward, followed by the decrease of the other actions [33]. The update of the pool selection policy of the mining pool can be formulated as:

$$\pi_i(a_t) \leftarrow \pi_i(a_t) + \triangle_{a_t}, \ \forall a \in A, \qquad (7.22)$$

where

$$\triangle_{a_t} = \begin{cases} - \min \left(\pi_i(a_t), \dfrac{\theta_i}{M-1} \right), \Pi', \\[4mm] \sum\limits_{a' \neq a} \min \left(\pi_i(a_t'), \dfrac{\theta_i}{M-1} \right), \ o.w, \end{cases} \qquad (7.23)$$

where

$$\Pi' : a_t \neq \arg\max_{a_t' \in A} Q_i(a_t'), \qquad (7.24)$$

and M is a constant coefficient.

Based on this, the WoLF-PHC algorithm based pool selection policy is described in Algorithm 7.2.

Algorithm 7.2 The WoLF-PHC algorithm for the pool selection

Set $\alpha, \delta, \theta^{win}, \theta^{lose}$
Initialization
repeat
  for $t = 1, 2, 3$ **do**
    Select action a_t according to current policy π_i
    Each miner observes the immediate reward R
    Update $Q_i(a_t)$ by:
    $Q_i(a_t) \leftarrow (1 - \alpha) Q_i(a_t) + \alpha (R_i + \delta \max\limits_{a \in A} Q_i(a_{t+1}))$
    Update $\overline{\pi}_i(a)$ and $\pi_i(a)$ by:
    $\overline{\pi}_i(a_t) \leftarrow \overline{\pi}_i(a_t) + \frac{\pi_i(a_t) - \overline{\pi}_i(a_t)}{N_i(t)}, \ \forall a_t \in A$
    $\pi_i(a_t) \leftarrow \pi_i(a_t) + \triangle_{a_t}, \ \forall a_t \in A,$
  end for
until

7.1.4   *Performance Evaluation*

In this section, we first analyze the colony behaviors of IoT devices in the pool selection problem. Then, we present the experiment results to evaluate the performance of our proposed WoLF-PHC based algorithms.

7.1.4.1   Evolution Analysis

In our experiment, we simulate a blockchain network with 5000 IoT devices (i.e., $N = 5000$). These resources constrained IoT devices are willing to rent the computing resource from the cloud services and evolve to form several mining pools. For the blockchain, we set the fixed reward R as 1000, and the variable reward parameter ρ as 0.01. For the cloud server, we set the price p of computing and storage resources unites as 0.01. The parameters setting can be found in Table 7.2.

We first investigate the dynamic behavior of the players' population. In this case, we deploy two mining pools, where the hash power requirement of two pools is $\omega_1 = 10$ and $\omega_2 = 30$, and the size of transactions size of both two pools is 100. As shown in Fig. 7.3, we plot the phase plane of the replicator in our system. The figure shows that the direction of the adaptation in mining pool selection to the ESS point (i.e., $Population_1 : 0.3$, $Population_2 : 0.7$). For example, when the initial population state is $x = [0.5, 0.5]$, the trajectory of replicator dynamics follows the arrows to reach the ESS.

7.1.4.2   Evolution Analysis with Different Pooling Strategies

Then, we evaluate the evolution behavior of the players population with different pooling strategies. In this case, we design three groups of experiments, where the mining pools' configuration and the ESS point can be found in Table 7.3. As shown

Table 7.2 List of parameters setting

Parameter	Value
Fixed reward R	1000
Number of IoT devices N	5000
Number of agent	30
The price of computation resource unite p	0.01
The variable reward parameter ρ	0.01
The maximum episode numbers	10,000
The learning rate α	0.2
The discount factor β	0.8
The learning rates (win) θ^{win}	0.0025
The learning rates (lose) θ^{lose}	0.01

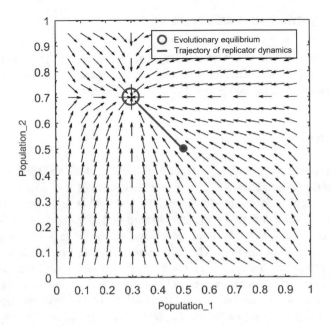

Fig. 7.3 The phase plane of replicator dynamics

Table 7.3 Experiment configuration

Mining pool	Setting	ESS point
Pool 1	$\omega_1 = 10, \omega_2 = 30$	(0.3, 0.7)
Pool 2	$\omega_1 = 10, \omega_2 = 50$	(0.35, 0.65)
Pool 3	$\omega_1 = 30, \omega_2 = 20$	(0.4, 0.6)
Pool 4	$\omega_1 = 30, \omega_2 = 200$	(1, 0)

in Fig. 7.4, we notice that the miner is more willing to join in the pool with less hash power requirement. The increasing hash requirement will reduce the number of players who are willing to joining in. This is mainly caused by the cost of renting the cloud computing resource. The slim profit gained from mining a block cannot meet the exorbitant cost of resource renting. Therefore, to attract more miners to join in, the pool's coordinator should lower the threshold of the hash requirement for each miner.

7.1.4.3   Evolutionary Game-Based Pool Selection Algorithm

In this section, we will evaluate the convergence of the centralized evolutionary game-based pool selection algorithm. The trajectories of players' strategies adaptation over time are shown in Fig. 7.5. We can find that the players with our algorithm can quickly converge to the ESS point. This is mainly because that the centralized

Fig. 7.4 The phase plane of
replicator dynamics with
different pooling strategies.
(**a**) $\omega_1 = 10$, $\omega_2 = 50$. (**b**)
$\omega_1 = 30$, $\omega_2 = 20$. (**c**)
$\omega_1 = 30$, $\omega_2 = 200$

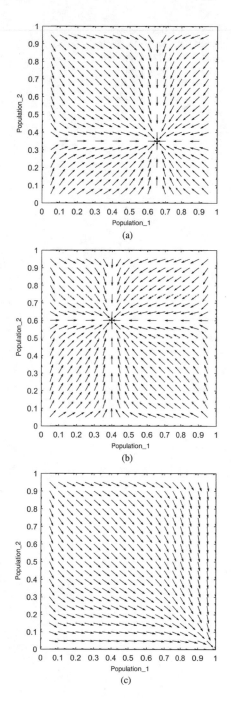

Fig. 7.5 Convergence
analysis of the evolutionary
game-based pool selection
algorithm. (**a**) $\omega_1 = 10$,
$\omega_2 = 50$. (**b**) $\omega_1 = 30$,
$\omega_2 = 20$. (**c**) $\omega_1 = 30$,
$\omega_2 = 200$

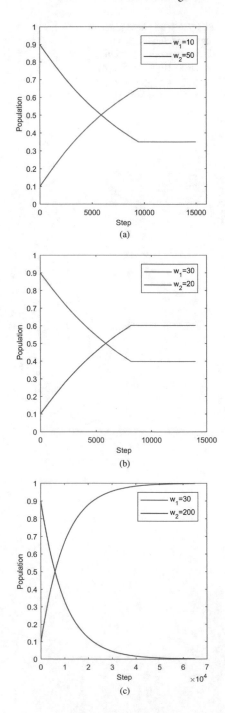

controller can constantly revise the player behavior. The average payoff $\bar{r}$ is issued to each IoT device to evaluate its current strategy. Then, these players can adjust their strategies based on their current payoff (i.e., switch their strategies or keep them).

7.1.4.4   Impact of Delay in Strategy Adaptation

As discussed above, considering the communication latency between miners and the centralized controller, we investigate a certain period of time delay τ in our system. In this section, we evaluate the impact of delay in the process of strategy adaptation. We set four mining pools in our system with the different hash power requirement, where $\omega_1 = 10$, $\omega_2 = 20$, $\omega_3 = 30$, and $\omega_4 = 40$. Also, we set three groups of experiments, where the time delay τ are separately set as 0, 10, and 20. Note that the units of τ are steps in our experiment.

As shown in Fig. 7.6, when the time delay τ is 0, the trajectory of strategy adaptation is relatively smooth. The system can quickly converge to the evolutionary equilibrium. And when the delay is introduced, we notice fluctuating dynamics of strategy adaptation over time toward the ESS. Especially, with the time delay of τ becoming larger, the more fluctuation will be brought. This is because that when outdated knowledge is used by the players, the decisions tend to be inaccurate. But although the trajectory of strategy adaptation is fluctuating, the system can also converge to the near ESS, which means the system still be stable if the time delay is not very large [34].

7.1.5   Wolf-PHC Based Pool Selection

Then, in this section, we evaluate the performance of the WoLF-PHC based algorithm in the mining pool selection problem. We construct a trading environment with 30 IoT devices (i.e., agents) and two mining pools. The blockchain environment setting is consistent with previous experiments. We set the maximum training episode numbers as 10,000, the learning rate α as 0.2, the discount factor β as 0.8, the θ^{win} as 0.0025 and θ^{lose} as 0.01.

7.1.5.1   Convergence Performance of WoLF-PHC

Firstly, we evaluate the convergence of our algorithm. We use Q-Learning as the baseline algorithm, where the learning rate α and the discount factor β are also set as 0.2 and 0.8. 7.17. As shown in Fig. 7.7, the learning process of WoLF-PHC and Q-learning algorithm are demonstrated. We notice that the agents with the Q-learning algorithm exhibit a poor convergence performance. This is caused by the moving target problem in multi-agent system. By contrast, benefiting from the

Fig. 7.6 The impact of delay
in strategy adaptation. (**a**)
Delay = 0. (**b**) Delay = 10.
(**c**) Delay = 20

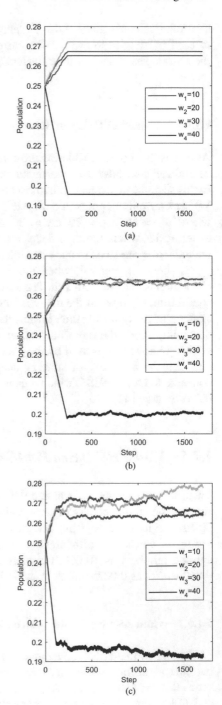

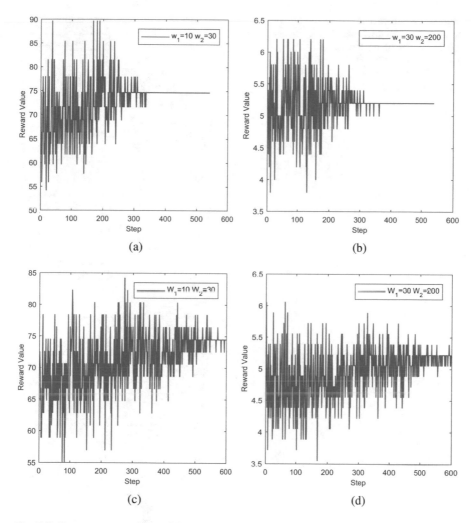

Fig. 7.7 Convergence analysis of Q-learning and WoLF-PHC algorithm. (**a**) WoLF: $\omega_1 = 10$, $\omega_2 = 30$. (**b**) WoLF: $\omega_1 = 30$, $\omega_2 = 200$. (**c**) Q: $\omega_1 = 10$, $\omega_2 = 30$. (**d**) Q: $\omega_1 = 30$, $\omega_2 = 200$

"wining or learning fast" scheme, the WoLF-PHC algorithm presents a much better convergence performance.

Besides, as shown in Fig. 7.8, we present the trajectories of agents' strategies adaptation. In this case, we design three groups of experiments, where the mining pools' configuration can be found in Table 7.3. We notice that while the adaptation process exists fluctuation, after about 400 steps, the system will converge. By comparing to the phase plane of the replicator dynamics, these convergence points are the ESS of the system.

Moreover, we evaluate our proposed algorithm in comparison to some other state-of-the-art reinforcement learning algorithms, including Policy Gradient (PG),

Fig. 7.8 Convergence
analysis of Wolf-PHC based
pool selection algorithm. (**a**)
$\omega_1 = 10$, $\omega_2 = 50$. (**b**)
$\omega_1 = 30$, $\omega_2 = 20$. (**c**)
$\omega_1 = 30$, $\omega_2 = 200$

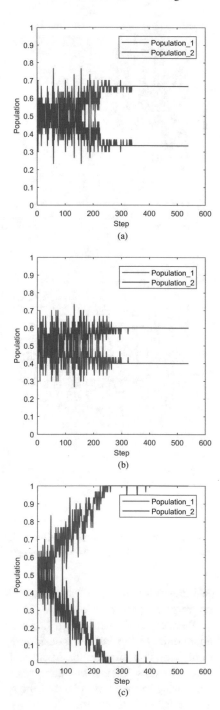

Table 7.4 Convergence performance

Algorithm	Steps	Conv point
WoLF-PHC	320	(0.3, 0.7)
Q-learning	860	(0.33, 0.67)
Policy gradient	1020	(0.3, 0.7)
DQL	2470	(0.27, 0.73)
DDPG	1930	(0.3, 0.7)

Deep Q-learning (DQL), and Deep Deterministic Policy Gradient (DDPG). In this experiment, we set the mining pool's hash power requirement as $\omega_1 = 10$ and $\omega_2 = 30$, and the value of learning rate α and the discount factor β as 0.2 and 0.8. The other algorithm parameters setting of DQL and DDPG can be found in [35]. As shown in Table 7.4, we notice that all algorithms can converge to a small neighborhood of the ESS point $(0.3, 0.7)$. This demonstrates that reinforcement algorithms can adapt to the non-stationary system and converge to the system's ESS point. But different algorithms present different rates of convergence. The WoLF-PHC algorithms exhibit the best performance. It can converge in around 320 steps. In contrast, the convergence of DQL and DDPG are the worst. This is because that they have too many parameters that need to be updated during the learning process. Therefore, we can draw the conclusion that while the complex neural network design enables them to solve complex tasks, simple learning algorithms may be more efficient for simple tasks.

7.1.5.2 The Reward vs. Pooling Strategies

In the following, we evaluate the impact of the pooling strategies to the agent's reward. In this case, we fixed the pool1's hash power requirement as $\omega_1 = 10$, and the set of transactions size as $q = 100$. As shown in Fig. 7.9c, as the hash power requirement of pool2 ω_2 increases, the total reward reduces. This is caused by the cost of renting the cloud computing resource. Because the total reward gained from the blockchain network remains unchanged, renting more computing resources will reduce the total profit of the miners. Besides, we evaluate the impact of the variable reward q on the agent's reward. With the size growing, the more reward will bring to the whole system and therefore improve the agent's profit.

7.1.5.3 The Reward vs. The Number of Miners

Next, we evaluate the impact of the number of miners and the number of pools to the agent's reward. In this experiment, we design three groups of experiments, where the mining pools' configuration can be found in Table 7.5. We set the size of transactions size in a block of all pools as 30. As shown in Fig. 7.9d, as the number of miners increases, the agent's reward reduces, which is caused by the competition

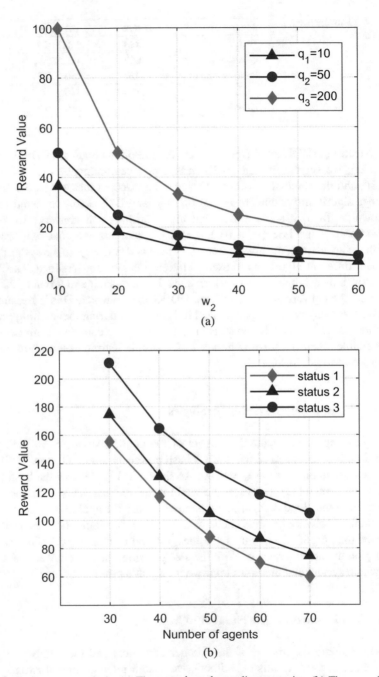

Fig. 7.9 Performance analysis. (**a**) The reward vs. the pooling strategies. (**b**) The reward vs. the number of miners

Table 7.5 Experiment configuration

Mining pool	Setting
Status 1	$\omega_1 = 20, \omega_2 = 30$
Status 2	$\omega_1 = 20, \omega_2 = 30, \omega_2 = 40$
Status 3	$\omega_1 = 20, \omega_2 = 30, \omega_2 = 40, \omega_2 = 50$

among miners. Due to the total gain from the mining block is constant, the single agent's profit will decrease with the number of miners increasing. Besides, we find that the agent's reward will increase with the number of pools growing. This result may be explained by the fact that more pools will offer more opportunities for each agent. More choices will reduce the competition among miners, therefore improving the individual agent's profit.

7.2 Resource Trading in Blockchain-Based IIoT

Recently, Internet of Things has received a large amount of attention from both academics and industries. It is estimated that there may be a total of 20 billion connected IoT devices by the end of 2020. Specifically, industrial IoT as a subset of IoT has shown a significant impact on businesses, safety, and even lives. Compared to other IoT applications, the industrial IoT is focusing on connecting machines and devices in a diverse range of industries, including manufacturing, agriculture, oil and gas, transportation, and health care [36]. However, with the industrial IoT devices numbers and performance requirement continually growing, the traditional centralized IoT architecture poses great challenges, such as device safety, personal privacy, and architecture rigid, especially in the context of the industrial IoT characterized by frequency information exchange and autonomic financial transaction [37]. To address this problem, peer-to-peer architecture was introduced to design trading platform for industrial IoT, where each node can trade their asset (such as surplus energy in microgrids and weather information in the meteorological station) with others directly without third-party organization.

In the past few years, the blockchain has shown its world-changing potential in a range of IoT applications. The blockchain, as an incorruptible digital ledger, is a tamper-proof, distributed database, which is maintained by network nodes without identity authentication in a peer-to-peer (P2P) network [38]. It can not only record financial transactions but also record anything of value using a growing list of the cryptographic hash block. Meanwhile, the concept of decentralized autonomous organizations (DAO) became practical with the development of blockchain, which aims at establishing a fully decentralized and autonomous organization without hierarchical management [39]. A DAO is operated upon the encoding procedure in smart contrasts and is capable of tracking the financial transaction in blockchain [38].

However, the huge computational resources requirement of establishing a DAO platform based on blockchain prevents lightweight industrial IoT devices and smart mobile devices from directly participating. To address this issue, the "cloud mining" mechanism was introduced to provide computation resource (such as CPU, GPU) for the industrial IoT DAO platform, where lightweight devices can offload their computational task to cloud providers [40]. Specifically, relying on the cloud mining, more nodes can participate as the consensus nodes, which is beneficial in terms of improving the robustness of the blockchain. With the cloud computing joining in, a resource management and pricing problem between the resource provider and miners turn up, where the resource provider firstly sets price aiming at maximizing its own reward, while miners purchase the computational resource from the provider for the sake of obtaining the "best buys." Relying on the sequence of two players' actions, in this paper, we model the interaction between the resource provider and miners as a Stackelberg game. In addition, considering the non-cooperative relationship among miners, we propose a multi-agent reinforcement learning to search the Nash equilibrium point. Compared with traditional meticulously designed heuristics approaches, the multi-agent reinforcement learning can converge to the best solution without requiring ideal knowledge about environment system.

7.2.1 Industrial IoT DAO Platform

In this section, we first design a DAO trading platform for industrial IoT based on blockchain network aided with cloud mining. Then, we model the computational resource management and pricing problem between the resource provider and miners as a Stackelberg game.

7.2.1.1 DAO Platform Assisted by Cloud/Fog Computing

The industrial Internet of Things starts up an industrial revolution industry 4.0. Considering the vulnerable and brittle defect in a centralized architecture, to design security and unified distributed platform for billions of devices become reasonable. The blockchain technology is the security guards of the most valuable cryptocurrencies in the world, and its decentralized network and embedded smart contracts are suitable solutions for industrial IoT's security and durability concerns [41].

In blockchain networks, achieving distributed consensus is the core problem. Nakamoto consensus protocol is a decentralized, pseudonymous consensus protocol [42], where a computation-intensive mechanism Proof-of-Work (PoW) is introduced for achieving the consensus from all participants. PoW, also termed as mining, is a process of solving a hash function, which is costly to generate but easy for others to verify [43]. The aforementioned characteristic of PoW mechanism needs to meet certain requirements of updating decentralized shared

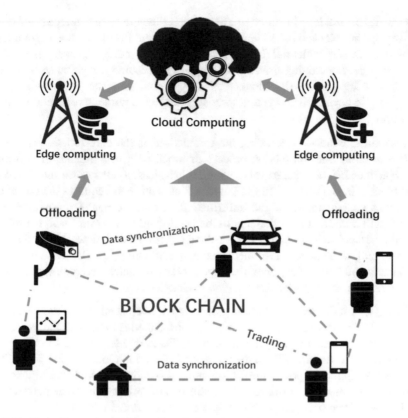

Fig. 7.10 The self-organized DAO trading platform for IoT networks

ledger. Additionally, there is a special type of nodes, namely "miners," which make the decision to devote their computational resources. They may compete with each other to produce a new block and obtain rewards relying on the financial intensive mechanism.

With the aid of PoW, the mobile users in IoT blockchain are incentive to contributing its computation power to maintain the decentralized shared ledger. However, given that the mobile users' power and computational capability are limited, it is difficult for lightweight nodes to directly participate in the PoW process. Hence, as shown in Fig. 7.10, we introduce the cloud mining to the blockchain networks for assisting the nodes to offload their storage and computation task to computation resource provider. Due to the financial intensive mechanism of PoW, the nodes prefer to join in the consensus process. When more nodes change their roles from free riders into consensus nodes, a robust DAO trading platform can be established. Then, we will describe the operation details of this "cloud mining" assisted industrial IoT DAO platform.

System Initialization In order to ensure the authenticity and integrity of digital messages, the blockchain system needs to be initialized by the cryptography technique. In our industrial IoT DAO system, each node registers on a trusted authority agent deploying on the cloud server to become a legitimate entity with an identity ID joining the blockchain system and gets the public/privacy key using elliptic asymmetric cryptography and certificate with curve digital signature algorithm [39].

Transactions Process Recording the transactions is the core function of the IoT trading system. The transaction process of blockchain is a transfer of cryptocurrency(such as Bitcoin, Ether) value which is broadcast to all system and stored into new blocks. In our system, the IoT nodes first send their request (such as energy request and surrounding weather information) to edge computer servers, and edge servers will broadcast their demand to the whole DAO platform. Then, the relevant supplier nodes (such as solar panels or wind generators and meteorological station) will respond to the edge server their inventory and unit price. Based on demand of consumers and stock of supplier, the edge server will match the trading pairs. Then, the consumers and supplier can carry on transactions.

Building Block Based on the PoW process, each node needs to compete with each other to obtain the authority of updating the distributed ledger. With the assisting of cloud mining mechanism, each node can buy the computer service from the cloud computing provider and offload the huge computation to cloud servers. In addition, the more computation power means the more possibility of win financial incentive. Therefore, there exists resource management and pricing problem between IoT devices and cloud provider which we will describe below in detail.

Consensus Process The consensus process is to deal with the agreement of the whole blockchain system onto a truth about their transactions data. Once the solution of the hash puzzle is obtained by a miner, all transactions in the current period are packed into a new block and broadcast to the whole blockchain network. The other nodes in the blockchain apply a hash validation-comparison function and the longest-chain rule to evaluate if they accept this new block. If so, the new block will be linked to the main blockchain in the chronological order, and the node's local view of the blockchain state is updated [44]. In the blockchain, once the consensus process accomplished, details of the transaction (such as ownership, price, and time stamp) are recorded through the distributed ledger guarantees.

7.2.1.2 Problem Formulation

As is described above, a resource management and pricing game between miners and the resource provider turn up. In this paper, we model this relationship as a two stages Stackelberg game which is a strategic game in economics. As shown in Fig. 7.11, the resource provider, termed as the "leader," takes action first and

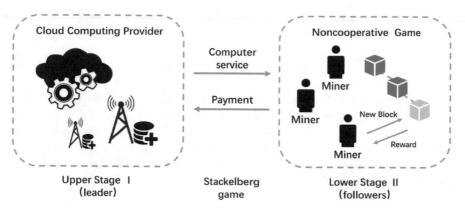

Fig. 7.11 The Stackelberg game model

the miners, termed as the "followers," take actions afterward. Both the leader and followers can constantly adjust their strategies for the sake of earning more profit.

In the upper stage, the resource provider sets the pricing strategy $\{\lambda = [\lambda_i]_{i \in N} : 0 < \lambda_i < \bar{\lambda}\}$ as the unit price of the computational resource, where λ_i is the price for miner i and $\bar{\lambda}$ is the maximum price. In addition, some other overhead c may be costed resulting from the electricity consumption, loss of hardware, as well as operation and maintenance cost. Therefore, the expected reward of the resource provider can be expressed as:

$$R_c = \sum_{i \in N} \lambda_i \mu_i - \sum_{i \in N} c \mu_i, \tag{7.25}$$

where μ_i is the service demand of miner i.

By contrast, in the lower stage, there are a set of nodes denoted as $N = \{1, \ldots, N\}$ which are interested in purchasing computational service and in competing with each other to earn financial incentive by mining blocks. Each miner $i \in N$ decides its computational service demand, which is represented by $\mu_i \in [\underline{\mu}, \bar{\mu}]$. The $\underline{\mu}$ is represented by the minimum computational power to participate in the blockchain for data synchronization, while the maximum computational power of the cloud provider is represented by $\bar{\mu}$. According to Nakamoto consensus protocol, the probability of wining the competition among miners who is the first to solve the PoW-based puzzle is related with the miners' computational power. Therefore, we defined α_i to evaluate the corresponding computational power of miner i among the whole consensus nodes, which can be given by:

$$\alpha(\mu_i) = \frac{\mu_i}{\sum_{j \in N} \mu_j}. \tag{7.26}$$

Furthermore, we define a utility function R_i to evaluate the expected reward by acting μ_i, and we have:

$$R_i = R \times \alpha(\mu_i) - \lambda_i \times \mu_i, \tag{7.27}$$

where R represents to the fixed reward of a successful mining process, while λ_i is the price of the unit computational resource from cloud provider.

As described above, a sequential decision-making process of cloud provider and miners is modeled. The cloud providers first predict the total service demand of miners and set its price to earning more profit. We can formulate the optimization problem of cloud providers as follows:

$$\max_{\lambda} \ R_c(\lambda|\mu)$$

$$s.t. \quad \begin{cases} \lambda \geq 0 \\ \Sigma_{i \in N} \lambda_i \mu_i \geq \Sigma_{i \in N} c\mu_i. \end{cases}$$

Furthermore, observing the price strategies of cloud providers, the miners set its service demand to earn more profit. The optimization problem of miners is denoted as:

$$\max_{\mu_i} \ R_i(\mu_i|\lambda_i)$$

$$s.t. \quad \begin{cases} \mu_i \geq 0 \\ R \times \alpha(\mu_i) \geq \lambda_i \times \mu_i. \end{cases}$$

7.2.1.3   Game Analysis

Hence, we have already formulated the mathematical model of two sides in the Stackelberg game. Both the resource provider and miners are capable of constantly adjusting their strategies to maximize their reward. Specifically, the objective of the Stackelberg game is to find the Nash equilibrium. The Nash equilibrium is the optimal outcome of the game, where no player has an incentive to deviate from its strategy after considering its opponent's choice [45]. In our problem, the Nash equilibrium of the proposed Stackelberg game can be defined as follows.

Definition Let μ^* and λ^* be the optimal unit price of resource provider and service demand of each miner, respectively. Then, the point (μ^*, λ^*) is the Nash equilibrium point if it satisfies:

$$R_i(\mu^*, \lambda^*) \geq R_i(\mu, \lambda^*), \tag{7.28}$$

as well as

$$R_c(\mu^*, \lambda^*) \geq R_i(\mu^*, \lambda). \qquad (7.29)$$

Then, in order to verify the uniqueness and existence of the Nash equilibrium in our Stackelberg game, we take the second order derivatives of the utility function of miners Eq. 7.31 with respect to μ_i and utility function of Eq. 7.25 with respect to λ, which is written as follows:

$$\frac{\partial^2 R_i}{\partial \mu_i{}^2} = -2R \frac{\Sigma_{j \neq i} \mu_j}{(\Sigma_{j \in N} \mu_j)^3} \leq 0, \qquad (7.30)$$

$$\frac{\partial^2 R_c}{\partial \lambda^2} = -\frac{2c}{\lambda^2} \frac{(N-1)R}{N} \leq 0. \qquad (7.31)$$

Therefore, the R_i and R_c are strictly concave. Accordingly, the Nash equilibrium exists in this Stackelberg game.

7.2.2 Multi-agent Reinforcement Learning

In this section, we first introduce the multi-agent reinforcement learning algorithm, termed as "WoLF-PHC." Then, we present how the "WoLF-PHC" can be applied to solve proposed Stackelberg game.

7.2.2.1 The Multi-agent System

The cloud provider and miners constitute a multi-agent system in a Stackelberg game. Considering the non-cooperation relationship among players, each of them can only have incomplete information of the underlying game model. Therefore, it is impractical to implement designed heuristics analysis algorithms, which require accurate environmental information. Inspired by some successful examples of applying machine learning algorithms to game theory, we introduce the reinforcement learning method for the miners and the cloud provider to obtain their own optimal reward. Reinforcement learning algorithms are a kind of model-free algorithms, because the update rule of the policy does not require ideal knowledge about the environment system. In the context of a single agent, the reinforcement learning algorithm can learn a policy mapping from the state to the action by interacting with the underlying environment so as to maximize the cumulative reward. By contrast, for the multi-agent system, compared to the single-agent case which aims for maximizing its own cumulative reward, the multi-agent case aims for maximizing the whole reward of all the agents in the system considering other agents' behaviors [46].

To elaborate, in the learning process, the agent observes the environment states and then takes some actions for getting a reward. Based on such interacting process,

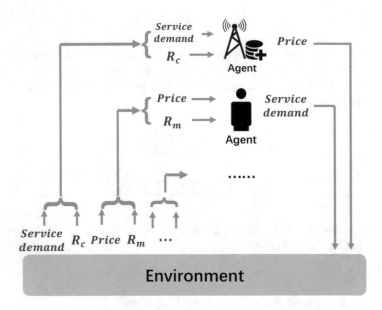

Fig. 7.12 The architecture of multi-agent reinforcement learning algorithm

the agent can update its policy with the purpose of maximizing the cumulative reward as shown in Fig. 7.12. In the following, let us, first of all, define three critical components, i.e., state, action, and immediate reward, of the miners and cloud provider for our proposed reinforcement learning algorithm.

In our paper, let $\mu^t \in A_m$ and $\lambda^t \in A_c$ denote the service demand action of miners and the unit price set by resource provider, respectively, where A_m represents the action space of the miner, while A_c represents the action space of the resource provider. In each time slot, the miners and the resource provider take actions sequentially.

At the beginning of the time slot t, the cloud provider first sets the price λ^t based on the state $s_c^t = [\mu_i^{t-1}]_{i \in N}$ observed from underling game, where μ_i^{t-1} represents the service demand of each miner in time slot $(t - 1)$. The immediate reward is expressed by:

$$R_c = \sum_{i \in N} \lambda_i^t \mu_i^{t-1} - \sum_{i \in N} c\mu_i. \tag{7.32}$$

Similarly, after observing the pricing action of the cloud provider in time slot t, each miner decides its service demand action μ^t based on the state $s_m^t = \lambda^t$, where λ^t represents the pricing action of the cloud provider. Hence, the immediate reward $R_m = \Theta(R) - \lambda^t \times \mu^t$ can be obtained, where the R is the fixed reward of successful mining and $\Theta()$ is to present whether the miner i success in this time stamp. The three critical components of our reinforcement learning algorithm are summarized as follows.

Definition The three components of the miners:

- State:

$$s_m^t = \lambda^t$$

- Action:

$$\mu^t \in A_m$$

- Immediate reward:

$$R_m = \Theta(R) - \lambda^t \times \mu^t$$

Definition The three components of the resource provider:

- State:

$$s_c^t = \left[\mu_i^{t-1}\right]_{i \in N}$$

- Action:

$$\lambda^t \in A_c$$

- Immediate reward:

$$R_c = \sum_{i \in N} \lambda^t \mu_i^t - \sum_{i \in N} c\mu_i^t$$

7.2.2.2 Policy Generation

The key challenges of multi-agent reinforcement learning are the non-stationary learning problem due to the noise signal brought by other agent. Directly applying single-agent reinforcement learning will suffer seriously oscillatory problem and the learning result is hardly to converge [47]. The WoLF-PHC, as the extension of Q-learning, adopts the principle of "win or learn fast" to learn the dynamic target [48], where the variety of learning rates is to effectively encourage convergence of multi-agent in non-stationary environment. Therefore, in this paper, we apply the WoLF-PHC algorithm for the miners and the cloud provider to learn their policy in a multi-agent system.

The updating rule of the Q value is the same as that in the Q-learning algorithm. Let $\alpha_m \in (0, 1]$ denote the learning rate and let $\delta_m \in (0, 1]$ represent the discount factor. The Q-function of the miner with the service demand μ in the state s_m^t can be formulated by $Q_m(s_m^t, \mu^t)$. Hence, the update rule of the Q-value can be given by:

$$
\begin{aligned}
Q_m(s_m^t, \mu^t) &\leftarrow (1 - \alpha_m) Q_m(s_m^t, \mu^t) \\
&+ \alpha_m (R_m + \delta_m V_r(Q_m(s_m^{t+1}))),
\end{aligned}
\tag{7.33}
$$

as well as

$$
V_r(Q_m(s_m^t)) = \max_{\mu \in A_m} Q_m(s_m^t, \mu),
\tag{7.34}
$$

where $V_r(Q_m(s_m^t))$ denotes the maximum Q-value of the miner in the state s_m^t.

In the WoLF-PHC algorithm, miners can update their service demand policy, i.e., $\pi_m :\mapsto Pr(A_m)$, which is a map from the state space to action space, for maximizing the cumulative reward by interacting with the environment and other agents [49]. The $\pi_m(s_m^t, \mu)$ denotes the possibility of choosing action μ in the state s_m^t. The $\pi_m(s_m^t, \mu)$ increases the possibility of the current best action μ^* based on the Q-function, which can be expressed as:

$$
\mu^* = \arg \max_{\mu \in A_m} Q_m(s_m, \mu^t).
\tag{7.35}
$$

Relying on the "win or learn fast" mechanism, a current average policy $\overline{\pi}_m(s_m^t, \mu)$ is introduced as a competitor to judge the "win" or "failure" of the policy $\pi_m(s_m^t, \mu)$. The miner node chooses its learning parameter θ_m from θ_m^{win} and θ_m^{lose}, where $\theta_m^{win} < \theta_m^{lose}$, calculated from the result of the competition of the Q-value of $\pi_m(s_m^t, \mu)$ and $\overline{\pi}_m(s_m^t, \mu)$. In the condition of win, the θ_m^{win} is used to update the policy cautiously. Otherwise, θ_m^{lose} is applied to learn fast from the "failure" condition, i.e.,

$$
\theta_m = \begin{cases} \theta_m^{win}, & \Pi, \\ \theta_m^{lose}, & o.w, \end{cases}
\tag{7.36}
$$

where we have Π:

$$
\sum_{\mu \in A_m} \pi_m(s_m^t, \mu) Q_m(s_m^t, \mu) > \sum_{\mu \in A_m} \overline{\pi}_m(s_m^t, \mu) Q_m(s_m^t, \mu).
\tag{7.37}
$$

In order to calculate the current average policy, $N_m(s_m^t)$ is introduced to record and update the occurrence count of states observed by the agent, which can be calculated by:

$$
N_m(s_m^t) \leftarrow N_m(s_m^t) + 1.
\tag{7.38}
$$

Then, the average service demand of the miner node $\overline{\pi}_m(s_m^t, \mu)$ can be updated by:

$$\overline{\pi}_m(s_m^t, \mu) \leftarrow \overline{\pi}_m(s_m^t, \mu) + \Delta, \ \forall \mu \in A_m, \tag{7.39}$$

where

$$\Delta = \frac{\pi_m(s_m^t, \mu) - \overline{\pi}_m(s_m^t, \mu)}{N_m(s_m^t)}. \tag{7.40}$$

In the learning process, the possibility of the miner choosing a service demand is gradually increased, which can maximize the expected reward, followed by the decrease of the other actions [48]. Therefore, the update of the service demand policy of the miner can be given by:

$$\pi_m(s_m^t, \mu) \leftarrow \pi_m(s_m^t, \mu) + \Delta_{s_m^t, \mu}, \ \forall \mu \in A_m, \tag{7.41}$$

where

$$\Delta_{s_m^t, \mu} = \begin{cases} -\min\left(\pi_m(s_m^t, \mu), \dfrac{\theta_m}{M-1}\right), \Pi', \\[2ex] \sum\limits_{\mu' \neq \mu} \min\left(\pi_m(s_m^t, \mu'), \dfrac{\theta_m}{M-1}\right), o.w. \end{cases} \tag{7.42}$$

where we have Π': $\mu \neq \arg\max\limits_{\mu' \in A_m} Q_m(s_m^t, \mu')$ and where M is a constant coefficient.

As for the complexity of the WoLF-PHC algorithm, during the training process, the agent updates its strategy according to Eq. 7.33. Hence, the complexity of training process of each agent is on the order of $O(S^2 \times A)$, where S represents the number of state space and the A represents the number of action space. As for the complexity of the running process of WoLF-PHC, the most complexity is the Q-table look-up. Therefore, the complexity of the running process of each agent is approximate $O(S)$.

The WoLF-PHC algorithm of pricing policy for the cloud provider is shown in Algorithm 7.3.

7.2.3 Experiments and Simulation Results

In this section, we show the simulation results to evaluate the convergence performance of our proposed algorithm for the multi-agent system. In addition, by performing simulation with different number of miners, we evaluate the miners' and resource provider's policy.

Algorithm 7.3 The WoLF-PHC algorithm for the cloud provider

---

set $\alpha_c, \delta_c, \theta_c^{win}, \theta_c^{lose}$
Initialization
repeat
  for $t = 1, 2, 3$ **do**
    Observe the current state s_c^t
    Select action λ^t at random with the probability policy $\pi_m(s_m^t, \lambda)$
    Observe the next state s_c^{t+1} and immediate reward R_c
    Update $Q_c(s_c^t, \lambda^t)$ and $V_m(s_m^c)$ by:
    $Q_c(s_c^t, \lambda^t) \leftarrow (1 - \alpha_c) Q_c(s_c^t, \lambda^t) + \alpha_c(R_c + \delta_c V_r(Q_c(s_c^{t+1})))$
    Update $\overline{\pi}_c(s_c^t, \lambda)$ and $\pi_c(s_c^t, \lambda)$ by:
    $\overline{\pi}_c(s_c^t, \lambda) \leftarrow \overline{\pi}_c(s_c^t, \lambda) + \Delta, \forall \lambda \in A_c$
    $\pi_c(s_c^t, \lambda) \leftarrow \pi_c(s_c^t, \lambda) + \Delta_{s_c^t, \lambda}, \forall \lambda \in A_c$
  end for
until

7.2.3.1 Convergence Performance of WoLF-PHC

First of all, we evaluate the convergence of the WoLF-PHC algorithm and compare it with the Q-learning algorithm. In our simulations, for simplicity, let the pricing action set and service demand action set of the resource provider and miners be $A_m = (10, 11, \ldots, 100)$ and $A_c = (0, 1, \ldots, 100)$, respectively. The quantitative factor of the cost of the unit resource in cloud provider is $c = 1$, and the reward of the successful mining is $R = 10,000$. For simplicity, we assume a uniform pricing strategy for each miner. In order to ensure that the agent can converge to optimal policy, we set the maximum episode numbers as 5000 for both Q-learning and WoLF-PHC algorithm. In addition, the learning rate is $\alpha = 0.2$, which determines the degree to which the modified Q-value overrides the older one. The discount factor $\beta = 0.8$ quantifies how much importance we focus on future rewards. The learning parameters θ^{win} and θ^{lose} are set as 0.0025 and 0.01, respectively, which can improve the convergence performance [48]. Our experiment environment is developed on Python 3.3.

Firstly, we set only one miner and one cloud provider participating in the blockchain network. We apply the Q-learning and WoLF-PHC algorithm in both side of the resource provider and the miner for learning the optimal policy. The Q-learning learning process is demonstrated in Fig. 7.13 where the learning result cannot converge and shows poor performance. Compared with Q-learning algorithm, the WoLF-PHC algorithm exhibits a good convergence performance due to the automatic adjustment of learning rate as shown in Fig. 7.13. Both the cloud provider and miners converge near to the Nash equilibrium point benefiting from the "wining or learning fast" mechanism of WoLF-PHC.

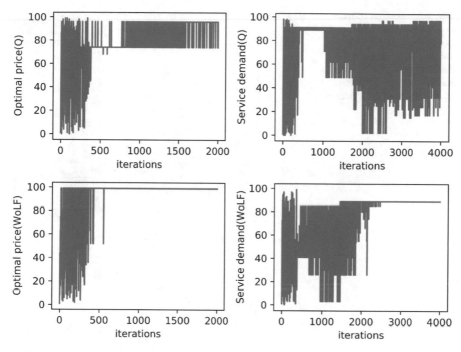

Fig. 7.13 The convergence performance of both Q-learning and WoLF-PHC algorithm

7.2.3.2   The Number of Miners vs. Service Demand

In the following, we evaluate the impact for both cloud provider's price and miners' service demand imposed by the number of miners. It can be obviously seen that the service demand of each miner is decreased with the increase in the number of miners in the blockchain. This is because the competition among miners decreases the possibility of mining a valued block for each miner, which can be explained by Eq. (7.26). Due to the decrease of expected reward, the miners tend to cut down its service demand for reducing financial loss. In addition, Fig. 7.14d shows the impact of the reward. With the increase of the fixed reward from success mining, the service demand of each miner is also increased. This is due to the fact that the increased reward can enhance the miner's expected reward, which stimulates the miner to purchase more computational service to obtain more profits. Therefore, the Nash equilibrium point is increased with the increase of the reward value. As illustrated, we can draw the conclusion that the Nash equilibrium point of service demand is related to the expected reward of each agent, say the service demand increases with the expected reward.

In spite of the service demand of each miner decreasing with the growth of numbers, the total service demand is increasing as shown in Fig. 7.14b. This is due to the fact that the competition among miners distorts the relationship between expected reward and service demand. This competition simulates the miners to have

Fig. 7.14 Service demand
and price analysis. (**a**) Service
demand versus the number of
miners. (**b**) Total service
demand versus the number of
miners. (**c**) Optimal price
versus the number of miners

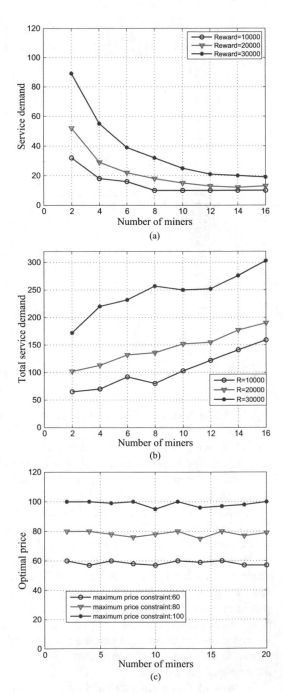

(a)

(b)

(c)

higher service demand to obtain more profit. Therefore, the total service demand is increasing with the competition aggravating.

7.2.3.3 The Number of Miners vs. Price

As for the resource provider, as shown in Fig. 7.14c, the optimal price of the cloud provider is almost invariable with the increase of the number of miners. That is because of the negative correlation between the price and the service demand. With the increase of the number of miners, the total service demand of miners is increased. Therefore, due to the condition of Nash equilibrium, the price needs to slightly increase to meet a new Nash equilibrium point. However, due to the default parameter in our scenario, the optimal price is constrained to the maximum price. Therefore, the optimal price is always close to the maximum price.

7.3 Summary

In this chapter, we discussed the challenges brought by the dramatic advances in IoT technology and gave solutions based on the integration of IoT and blockchain (BCoT). We first propose a cloud mining pool-aided BCoT architecture, where the IoT devices can rent the computing resource from the cloud services to offload their mining tasks. Based on this architecture, we discuss the mining pool selection problem. We proposed a centralized evolutionary game-based pool selection algorithm. A centralized controller is used to guide the behaviors of all players. Besides, considering the non-cooperative relationship among miners, we propose a WoLF-PHC-based pool selection algorithm. We further investigated the resource management and pricing problem in IIoT-based blockchain networks with the aid of cloud mining. We established a self-organized trading platform for presenting how cloud mining can assist the IoT-based blockchain. Moreover, we formulated the interaction between the cloud provider and cloud miners as a Stackelberg game. Then, we invoked a multi-agent reinforcement learning algorithm to achieve the near-optimal policy.

References

1. T. Mai, H. Yao, N. Zhang, L. Xu, M. Guizani, S. Guo, Cloud mining pool aided blockchain-enabled internet of things: an evolutionary game approach. IEEE Trans. Cloud Comput. **11**(1) 692–703 (2023). https://doi.org/10.1109/TCC.2021.3110965
2. H. Yao, T. Mai, J. Wang, Z. Ji, C. Jiang, Y. Qian, Resource trading in blockchain-based industrial internet of things. IEEE Trans. Ind. Inf. **15**(6), 3602–3609 (2019)
3. X. Huang, R. Yu, J. Kang, Z. Xia, Y. Zhang, Software defined networking for energy harvesting internet of things. IEEE Internet Things J. **5**(3), 1389–1399 (2018)

4. H. Dai, Z. Zheng, Y. Zhang, Blockchain for internet of things: a survey. IEEE Int. Things J. **6**(5), 8076–8094 (2019)

5. Z. Zheng, S. Xie, H. Dai, X. Chen, H. Wang, Blockchain challenges and opportunities: a survey. Int. J. Web Grid Serv. **14**(4), 352–375 (2018)

6. Z. Li, J. Kang, R. Yu, D. Ye, Q. Deng, Y. Zhang, Consortium blockchain for secure energy trading in industrial internet of things. IEEE Trans. Ind. Inf. **14**(8), 3690–3700 (2018)

7. I. Bentov, A. Gabizon, A. Mizrahi, Cryptocurrencies without proof of work, in *Financial Cryptography and Data Security: FC 2016 International Workshops, BITCOIN, VOTING, and WAHC*, Christ Church, Barbados, February 26, 2016, Revised Selected Papers 20 (Springer, Berlin, 2016), pp. 142–157

8. C. Qiu, H. Yao, X. Wang, N. Zhang, F.R. Yu, D. Niyato, AI-Chain: blockchain energized edge intelligence for beyond 5G networks. IEEE Netw. **34**(6), 62–69 (2020)

9. C. Qiu, H. Yao, C. Jiang, S. Guo, F. Xu, Cloud computing assisted blockchain-enabled internet of things. IEEE Trans. Cloud Comput. **10**(1), 247–257 (2022). https://doi.org/10.1109/TCC.2019.2930259

10. X. Xu, H. Zhao, H. Yao, S. Wang, A blockchain-enabled energy-efficient data collection system for UAV-assisted IoT. IEEE Internet Things J. **8**(4), 2431–2443 (2020)

11. C. Esposito, A. De Santis, G. Tortora, H. Chang, K.R. Choo, Blockchain: a panacea for healthcare cloud-based data security and privacy?. IEEE Cloud Comput. **5**(1), 31–37 (2018)

12. P. Yang, N. Zhang, Y. Bi, L. Yu, X.S. Shen, Catalyzing cloud-fog interoperation in 5G wireless networks: an SDN approach. IEEE Netw. **31**(5), 14–20 (2017)

13. Z. Xiong, S. Feng, W. Wang, D. Niyato, P. Wang, Z. Han, Cloud/fog computing resource management and pricing for blockchain networks. IEEE Internet Things J. **6**(3), 4585–4600 (2019)

14. X. Liu, W. Wang, D. Niyato, N. Zhao, P. Wang, Evolutionary game for mining pool selection in blockchain networks. IEEE Wireless Commun. Lett. **7**(5), 760–763 (2018)

15. N. Houy, The bitcoin mining game. Available at SSRN 2407834 (2014)

16. A. Kiayias, E. Koutsoupias, M. Kyropoulou, Y. Tselekounis, *Blockchain Mining Games* (2016), pp. 365–382

17. Y. Liu, C. Yang, L. Jiang, S. Xie, Y. Zhang, Intelligent edge computing for IoT-based energy management in smart cities. IEEE Netw. **33**(2), 111–117 (2019)

18. J. Hofbauer, K. Sigmund, Evolutionary game dynamics. Bull. Am. Math. Soc. **40**(4), 479–519 (2011)

19. Z. Liu, N.C. Luong, W. Wang, D. Niyato, P. Wang, Y. Liang, D.I. Kim, A survey on blockchain: a game theoretical perspective. IEEE Access **7**, 47 615–47 643 (2019)

20. C. Taylor, D. Fudenberg, A. Sasaki, M.A. Nowak, Evolutionary game dynamics in finite populations. Bull. Math. Biol. **66**(6), 1621–1644 (2004)

21. D. Friedman, Evolutionary games in economics. Econometrica **59**(3), 637–666 (1991)

22. R. Cressman, C. Ansell, K. Binmore, *Evolutionary Dynamics and Extensive Form Games*, vol. 5 (MIT Press, Cambridge, 2003)

23. F. Mazenc, S. Niculescu, Lyapunov stability analysis for nonlinear delay systems. Syst. Control Lett. **42**(4), 245–251 (2001)

24. T. Mekki, I. Jabri, A. Rachedi, M.B. Jemaa, Vehicular cloud networking: evolutionary game with reinforcement learning based access approach. Int. J. Bio-inspir. Comput. **13**(1), 45–58 (2019)

25. W. He, Y. Liu, H. Yao, T. Mai, N. Zhang, F.R. Yu, Distributed variational Bayes-based in-network security for the internet of things. IEEE Trans. Cloud Comput. **8**(8), 6293–6304 (2021). https://doi.org/10.1109/JIOT.2020.3041656

26. L. Busoniu, R. Babuska, B. De Schutter, Multi-agent reinforcement learning: a survey, in *2006 9th International Conference on Control, Automation, Robotics and Vision* (IEEE, 2006), pp. 1–6

27. M.T.J. Spaan, Partially observable Markov decision processes, in *Reinforcement Learning: State-of-the-Art* (Springer, Berlin, 2012), pp. 387—414

28. X. Yuan, H. Yao, J. Wang, T. Mai, M. Guizani, Artificial intelligence empowered QoS-oriented network association for next-generation mobile networks. IEEE Trans. Cogn. Commun. Netw. **7**(3), 856–870 (2021). https://doi.org/10.1109/TCCN.2021.3065463

29. J. Wang, C. Jiang, H. Zhang, Y. Ren, K.-C. Chen, L. Hanzo, Thirty years of machine learning: the road to pareto-optimal wireless networks. IEEE Commun. Surv. Tutorials **22**(3), 1472–1514 (2020). https://doi.org/10.1109/COMST.2020.2965856

30. P. Hernandezleal, B. Kartal, M.E. Taylor, A survey and critique of multiagent deep reinforcement learning. Auton. Agent. Multi-Agent Syst. **33**(6), 750–797 (2019)

31. J. Wang, C. Jiang, K. Zhang, X. Hou, Y. Ren, Y. Qian, Distributed Q-learning aided heterogeneous network association for energy-efficient IIot. IEEE Trans. Ind. Inf. (2019). https://doi.org/10.1109/TII.2019.2954334

32. D. Bloembergen, K. Tuyls, D. Hennes, M. Kaisers, Evolutionary dynamics of multi-agent learning: a survey. J. Artif. Intell. Res. **53**(1), 659–697 (2015)

33. Y. Zhang, R. Yu, M. Nekovee, Y. Liu, S. Xie, S. Gjessing, Cognitive machine-to-machine communications: visions and potentials for the smart grid. IEEE Netw. **26**(3), 6–13 (2012)

34. D. Niyato, E. Hossain, Dynamics of network selection in heterogeneous wireless networks: an evolutionary game approach. IEEE Trans. Veh. Technol. **58**(4), 2008–2017 (2009)

35. D. Silver, G. Lever, N. Heess, T. Degris, D. Wierstra, M. Riedmiller, Deterministic policy gradient algorithms, in *International Conference on Machine Learning*, PMLR (2014), pp. 387–395

36. Y. Zhang, R. Yu, M. Nekovee, Y. Liu, Cognitive machine-to-machine communications: visions and potentials for the smart grid. Netw. IEEE **26**(3), 6–13 (2012)

37. C. Qiu, X. Wang, H. Yao, J. Du, F.R. Yu, S. Guo, Networking integrated cloud–edge–end in IoT: a blockchain-assisted collective q-learning approach. IEEE Internet Things J. **8**(16), 12 694–12 704 (2020)

38. T.T.A. Dinh, R. Liu, M. Zhang, G. Chen, B.C. Ooi, J. Wang, Untangling blockchain: a data processing view of blockchain systems. IEEE Trans. Knowl. Data Eng. **30**(7), 1366–1385 (2018)

39. J. Kang, R. Yu, X. Huang, S. Maharjan, Y. Zhang, E. Hossain, Enabling localized peer-to-peer electricity trading among plug-in hybrid electric vehicles using consortium blockchains. IEEE Trans. Ind. Inf. **13**(6), 3154–3164 (2017)

40. H. Yang, J. Yuan, H. Yao, Q. Yao, A. Yu, J. Zhang, Blockchain-based hierarchical trust networking for jointcloud. IEEE Internet Things J. **7**(3), 1667–1677 (2019)

41. N. Teslya, I. Ryabchikov, Blockchain-based platform architecture for industrial IoT, in *The 21st Conference of Open Innovations Association (FRUCT)*, Helsinki (2017), pp. 321–329

42. S. Nakamoto, Bitcoin: a peer-to-peer electronic cash system. Working Papers (2008)

43. B. Laurie, R. Clayton, Proof-of-work proves not to work, in *Workshop on Economics and Information, Security* (2004)

44. A. Gervais, G.O. Karame, K. Wüst, V. Glykantzis, H. Ritzdorf, S. Capkun, On the security and performance of proof of work blockchains, in *ACM SIGSAC Conference on Computer and Communications Security*, Vienna (2016), pp. 3–16

45. D. Niyato, E. Hossain, Competitive pricing for spectrum sharing in cognitive radio networks: dynamic game, inefficiency of NASH equilibrium, and collusion. IEEE J. Sel. Areas Commun. **26**(1), 192–202 (2008)

46. L. Busoniu, R. Babuska, B. De Schutter, A comprehensive survey of multiagent reinforcement learning. IEEE Trans. Syst. Man Cybern. **38**(2), 156–172 (2008)

47. Y. Zhang, R. Yu, S. Xie, W. Yao, Home M2M networks: architectures, standards, and QoS improvement. IEEE Commun. Mag. **49**(4), 44–52 (2011)

48. L. Xiao, Y. Li, J. Liu, Y. Zhao, Power control with reinforcement learning in cooperative cognitive radio networks against jamming. J. Supercomput. **71**(9), 3237–3257 (2015)

49. M. Bowling, M. Veloso, Multiagent learning using a variable learning rate. Artif. Intel. **136**(2), 215–250 (2002)

Chapter 8
Conclusions and Future Challenges

8.1 Conclusions

This book mainly discusses the architectures of the Internet of Things as well as the possible techniques and challenges. As shown in Fig. 8.1, with the continuous advancement of global 5G construction and environmental impacts such as smart city guidance, IoT investment continues to increase [1]. However, with the development of the Internet of Things, massive heterogeneous devices and different network protocols continue to increase [2]. How to design a more efficient, flexible, and open network control architecture has become the focus of the current academic and industrial circles. In Chap. 2, as an emerging network mode, the intra-network intelligent driver is a key technology to promote the further development of the Internet of Things, which can effectively improve the manageability, programmability, and reusability of the Internet of Things network.

Then, we discuss the possible machine learning methods for IoT network awareness. With the rapid development of IoT application scenarios such as smart cities, it is very important to strengthen the management of data traffic in smart IoT networks [3]. As a critical part of massive data analysis, traffic awareness plays an important role in ensuring IoT network security and defending traffic attacks. In Chap. 3, we propose an end-to-end IoT traffic classification method relying on a deep learning aided capsule network. Moreover, the attention mechanism was introduced for assisting network traffic classification in the form of the following two models, the attention aided long short term memory (LSTM) as well as the hierarchical attention network (HAN). We also design a machine learning-based in-network DDoS detection framework.

The increasing dynamics and complexity of IoT networks have brought revolutionary changes in its modeling and control, where efficient routing and resource allocation strategies become beneficial. The growth of IoT devices poses great challenges to network service providers [4]. To meet different application requirements, diverse network proprietary hardware have to be implemented in the network. This

© The Author(s), under exclusive license to Springer Nature Switzerland AG 2023
H. Yao, M. Guizani, *Intelligent Internet of Things Networks*, Wireless Networks,
https://doi.org/10.1007/978-3-031-26987-5_8

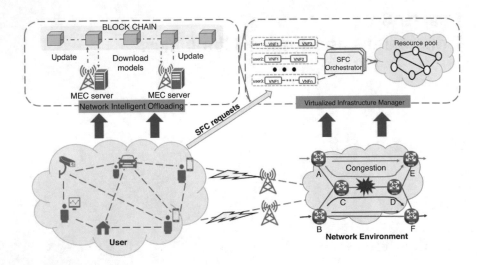

Fig. 8.1 Book organization

rigid paradigm greatly reduces the scalability and flexibility of the network system. In order to achieve effective traffic control, in Chap. 4, we first adopt a centralized training and distributed execution learning paradigm and design a hierarchical social-based DTN architecture. Then, we design a QoS-oriented adaptive routing scheme based on Machine Learning. To satisfy different applications' requirements in the same IoT network, in Chap. 5, we first design a network slicing architecture over the SDN-based long-range wide area network. Then, we propose a Continuous-Decision virtual network embedding scheme relying on Reinforcement Learning.

Based on network control, due to the very limited energy and computing resources of many IoT devices, such resource-constrained IoT devices are not well-equipped to perform complex processing; the computational workload is too heavy for IoT devices. Mobile edge computing (MEC) can provide abundant computing and storage resources to meet the performance requirements of mobile devices (MDs); computational tasks have no need to travel through the core network, allowing IoT data to be processed and results consumed locally with minimal delay [5]. Therefore, in Chap. 6 we propose a second-price auction scheme for ensuring fair bidding for spectrum rent. Moreover, a novel deep reinforcement learning (DRL)-based network structure is proposed to jointly optimize task offloading and resource allocation. In order to alleviate interference in multiple edge scenarios, we also propose a multi-agent aided deep deterministic policy gradient (MADDPG) algorithm to minimize total energy consumption and latency.

However, the issues of privacy and security are vital challenges for computation offloading in the distributed MEC networks [6]. To be specific, the transmitted data may contain private data. This raises the risk of privacy leakage and malicious attacks. Due to the decentralization, anonymity, and trust characteristics of blockchain, the combination of MEC and blockchain has been regarded as a promising technology to solve the security and privacy problems in MEC networks [7]. Specifically, Internet of things applications based on blockchain technologies are usually based on the clouding computing architecture to allocate massive computing resources and offload corresponding tasks; this feature further introduces some intelligent resource allocation algorithms into this field. In Chap. 7, we first discussed the resource allocation method based on the evolutionary game approach and, second, introduced resource management and control mechanism based on the Multi-Agent Reinforcement learning method. Overall, blockchain technologies have promoted further development and evolution of IoT applications. Blockchain-enabled intelligent IoT applications have shown good performance in many fields.

8.2   Future Challenges

8.2.1   IoT Standards and Unified Architecture

NB-IoT and eMTC technologies are oriented towards the 5G mMTC scenario and are the basis for entering the 5G IoT in the future. At present, 3GPP has adopted the subsequent evolution of NB-IoT/ eMTC as the technical standard of 5GmMTC. NB-IoT and eMTC technology have the characteristics of enhanced coverage, large connection, low power consumption, and low cost. With the rapid development of information technology industries such as mobile Internet, cloud computing, and the Internet of Things, information transmission, storage, and processing capabilities have risen rapidly, resulting in an exponential increase in the amount of data [8]. The current technical standards are not enough to meet the development needs of the Internet of Things. It is necessary to formulate relevant standards in the Internet of Things operating system, supporting key applications, and information technologies such as artificial intelligence and 5G. In addition, the traditional three-layer architecture of the Internet of Things has been solidified, and it is necessary to use SDN and network programmable technologies to realize flexible network control, reorganization, and efficient data transmission.

8.2.2   Adopting Emerging Naming and Addressing in IoT

This paper extensively discusses the application of mobile edge computing and blockchain technology in the IoT field and gives some detailed implementation details. However, the application of the above methods in the IoT field mostly relies on traditional IP network architecture, which only provides a "pipeline" function for forwarding without understanding the forwarding content [9]. Moreover, the change of IP address can lead to communication interruption, thus affecting the response time of IoT applications and even causing some security problems. Therefore, future research and challenges will rely on some novel network architectures to explore Emerging Naming and Addressing mechanisms in IoT. For example, there have been some related studies on the combination of Named Data Network(NDN) and IoT applications recently. Those studies decouple the information from the network terminal and sink into the network, finally realizing the deployment of massive IoT applications.

8.2.3   Privacy and Security Issues in IoT

Currently, the Internet of Things has a massive number of nodes and devices. These devices can process data intelligently. The advent of the Internet of Everything era has promoted social and economic development and made information easier to collect and track. IoT data privacy protection has become an urgent problem to be solved. Currently, privacy and security challenges in IoT include data-based privacy threats. IoT devices pose a significant privacy and security risk due to data leakage during data collection, transmission, and processing. This is particularly concerning in the healthcare sector where sensitive patient information is involved, making Health Privacy Security a top priority. The rapid development of information and communication technology has led to an increasing number of IoT devices enabling various e-health scenarios. However, the use of IoT technology in e-health poses risks related to patient identification and the reliability of collected information, which may contradict the Privacy Shield Transparency Principles. Unlike websites, apps, etc., IoT devices and services may not be transparent enough to present their privacy policies to users [10]. The collection of personal information may also lead to leakage without being unified by users.

8.2.4   Quality of Service in IoT

The IoT service is a service applied in the IoT environment. Compared with ordinary services, the biggest difference is that the service faces the IoT environment instead of the traditional Internet. This change brings new challenges to IoT services that are different from traditional services. First, they perceive network heterogeneity caused by different types of devices. Compared with the heterogeneity of the Internet, the Internet of Things is more prominent in heterogeneity. There are various types of sensing devices in the Internet of Things, and the amount of continuously generated sensing data is huge. IoT services are faced with the challenge of massive and heterogeneous sensing data. Moreover, its own limited capabilities and dynamically changing environmental factors lead to the dynamic and changeable links and topology of the Internet of Things, the communication status is very unstable, and the link quality faces severe challenges. IoT services must meet this challenge.

Continuous QoS must be provided in response to network dynamics by dynamically adjusting interactions in the IoT. IoT QoS needs to consider not only the interaction needs of applications but also the needs of resource providers and nodes supporting the communication required by the interacting devices [11]. The support of such multi-dimensional QoS needs to enable a fluent and adaptive framework to support opportunistic interactions in the emerging IoT.

References

1. X. Li, H. Yao, J. Wang, S. Wu, C. Jiang, Y. Qian, Rechargeable multi-UAV aided seamless coverage for QoS-guaranteed IoT networks. IEEE Internet Things J. **6**(6), 10 902–10 914 (2019)
2. S. Wu, H. Yao, C. Jiang, X. Chen, L. Kuang, L. Hanzo, Downlink channel estimation for massive MIMO systems relying on vector approximate message passing. IEEE Trans. Vehi. Technol. **68**(5), 5145–5148 (2019)
3. M. Debashi, P. Vickers, Sonification of network traffic flow for monitoring and situational awareness. PloS One **13**(4), e0195948 (2018)
4. X. Guo, H. Lin, Z. Li, M. Peng, Deep-reinforcement-learning-based QoS-aware secure routing for SDN-IoT. IEEE Internet Things J. **7**(7), 6242–6251 (2019)
5. Z. Zhao, R. Zhao, J. Xia, X. Lei, D. Li, C. Yuen, L. Fan, A novel framework of three-hierarchical offloading optimization for MEC in industrial IoT networks. IEEE Trans. Ind. Inf. **16**(8), 5424–5434 (2019)
6. F. Li, H. Yao, J. Du, C. Jiang, Y. Qian, Stackelberg game-based computation offloading in social and cognitive industrial internet of things. IEEE Trans. Ind. Inf. **16**(8), 5444–5455 (2019)
7. Y. Zuo, S. Jin, S. Zhang, Computation offloading in untrusted MEC-aided mobile blockchain IoT systems. IEEE Trans. Wirel. Commun. **20**(12), 8333–8347 (2021)
8. S.A. Al-Qaseemi, H.A. Almulhim, M.F. Almulhim, S.R. Chaudhry, IoT architecture challenges and issues: lack of standardization, in *2016 Future Technologies Conference (FTC)* (IEEE, Piscataway, 2016), pp. 731–738
9. C. Sobin, A survey on architecture, protocols and challenges in IoT. Wirel. Pers. Commun. **112**(3), 1383–1429 (2020)
10. L. Tawalbeh, F. Muheidat, M. Tawalbeh, M. Quwaider, IoT privacy and security: challenges and solutions. Appl. Sci. **10**(12), 4102 (2020)
11. C.-L. Fok, C. Julien, G.-C. Roman, C. Lu, Challenges of satisfying multiple stakeholders: quality of service in the internet of things, in *Proceedings of the 2nd Workshop on Software Engineering for Sensor Network Applications* (2011), pp. 55–60

Index

Printed in the United States
by Baker & Taylor Publisher Services